FIBRE AND FODDER RESOURCES OF INDIA

FIBRE AND FODDER RESOURCES OF INDIA

S.K. SOOD
BIPIN KAUSHAL

2008

SATISH SERIAL PUBLISHING HOUSE
403, Express Tower, Commercial Complex, Azadpur, Delhi-110033 (India)
Phone : 011-27672469, 27672852, Fax : 91-11-27672046
E-mail :info@satishserial.com, hkjain1975@yahoo.com
Website : www.satishserial.com

Published by :

SATISH SERIAL PUBLISHING HOUSE
403, Express Tower, Commercial Complex, Azadpur,
Delhi-110033 (India)
Phone : 011-27672852, Fax : 91-11-27672046
E-mail : info@satishserial.com, hkjain1975@yahoo.com
Website : www.satishserial.com

First Published 2008

ISBN : 81-89304-52-6

Printed at
SALASAR IMAGING SYSTEMS, DELHI-35

Preface

Indian region is very rich in ethnobotanical heritage as there are about 67.7 million tribal people (which accounts for about 22 per cent of world tribal population) belonging to 573 tribal groups living in different geographic locations with various subsistence patterns. These communities live in biodiversity rich areas and their wisdom acquired over years of quest is vanishing consequent to the assault of modern civilization into the areas, causing much damage both to diversity and to the well being of man. Therefore, for the effective conservation, management and sustainable strategies, the need of the hour is to develop a sound database on their ethnobotanical prudence. This necessitated the preparation of this compendium and it has been the outcome of extensive screening of widely scattered large mass of data in various journals and books relating to different disciplines such as ethnobotany, economic botany, forestry, phytochemistry, plant genetic resources, for which no attempt has been made so far. Layout of textural subject matter encompasses information on General Introduction; Overview of Fibre and Fodder Resources of India; Enumeration of Fibre and Fodder Resources; Fibre and Fodder Resources and Epilogue. To make the reading interesting, 34 photographs, 3 figures and 4 tables were also included in the text. Efforts have also been made to include information on total number of genera and species of fibre and fodder under various divisions of plant kingdom, families of fibre and fodder resources of India, fibre resources of India, predominant genera of fibre resources, fodder resources of India, predominant genera of fodder resources, fibre and fodder resources of India, predominant genera of fibre and fodder resources, index to Botanical names, index to English names, index to Hindi names, index to Sanskrit names as appendices.

The authors owe their deep sense of gratitude and indebtedness to Dr Anil Wilson, Vice-Chancellor, Himachal Pradesh University, Shimla, for his keen interest in this work and kind blessings. Grateful thanks are also due to Prof. Daleep Malhotra, Department of Psychology, Prof. T.N. Lakhanpal, Prof. D.C. Kalia and Prof. V.K. Mattu, Department of Biosciences and fellow botanists for their help and good wishes during the preparation of this compendium. Also, we are delighted to acknowledge Mr Dhiraj Rawat, Mr Suresh, Mr Munish, Ms Sudershna and Mr Bal Krishan for their extensive help in various ways. Last, but not the least, we thank the members of our families, for their moral support and forbearance.

S.K. SOOD
BIPIN KAUSHAL
Department of Biosciences
H.P. University
Shimla 171 005

Contents

1

Introduction

The term 'ethnobotany' was first coined by Harshberger in 1895 to denote the uses of plants by the aboriginals. Subsequently, different workers (Ford, 1978; Jain, 1986; Martin, 1994; Schultes, 1962) greatly enlarged the scope and considered it as an interdisciplinary science, encompassing the entire realm of useful relationship between the plants and tribals. Importantly, the study of the plants in relation to people include both the wild and domesticated plants (Heiser, 1995). However, Alcorn (1984) regarded ethnobotany as the study of contextualized plant use. Wickens (1990) distinguished ethnobotany from economic botany by considering ethnobotany as "the study of useful plants prior to their commercial exploitation and eventual domestication". The importance of ethnobotany stems from the varied economic uses of plants among the primitive human societies, who are considered as the repository of accumulated experience and knowledge of indigenous flora and fauna and have keen practical understanding of the ecological relationships between human societies and their living and nonliving environment. The use of a large number of plants for various purposes by them reveals their great awareness, knowledge and intimate relationship with the plant kingdom (Hajra, 1987), which is generally passed on verbally from one generation to another. Tribal knowledge of plants accumulated due to their long term practices, is important not just for the tribal people themselves but for the wider world (Maheshwari, 1987). According to Mittre (1981), "The utility of all the wild plants for various purposes was not recognized in one day or in one century, it is the result of progressive development of human cultures". Throughout the world these primitive peoples, less affected by modern civilization, constitute the main source of ethnobotanical information. In recent years concern has grown worldwide that the natural genetic variability and traditional uses of plants be safeguarded with an eye to potential future use.

Agenda 21 of the Rio Earth Summit (1992) also emphasized the vital role of indigenous people in environmental management and development.

Man's vital interest in plants, primarily as a source of food, fodder, shelter, medicine and clothing, dates back to very earliest times. Wealth of information on ethnobotany exists in India (Lat. 80^0 4'N to 37^0 06' N; Long. 68^0 07'E to 97^0 25'W) which is considered ethnically, floristically and agriculturally very diverse (Arora, 1990). There are over 550 native or tribal communities under 227 ethnic groups, occupying about 18.74% of the total area of the country (Arora, 1995). Of the estimated 20,000 flowering plants in the country, 9,000 are economically useful (Anon., 1994), and India ranks sixth among 12 megadiversity countries in the world (Khajuria & Chauhan, 2002). Janaki Ammal (1956) lamented that the tribal traditions are fast disappearing due to urbanization, rapid industrialization and changes in sustenance economy. In this regard, Schultes (1960) rightly remarked, "Our challenge is to salvage this precious knowledge before it becomes for ever entombed with the cultures that gave it birth". It is needless to say that ethnobotanical knowledge is important for policy-makers for devising solutions to local and regional problems (Alcorn, 1984).

□□□

2

Overview of Fibre and Fodder Resources of India

One of the earliest and notable contributions from Indian perspective is that of Duthie (1888) who described the fodder grasses of northern India. Later, attempts were made by various workers to study the improvements of fodder and forage in India (Howard, 1923), digestibility coefficient and nutritive value of some tree species (Ram & Ray, 1943), seasonal variation in the composition of some edible tree leaves (Momin & Ray, 1943), and use of *Bauhinia variegata* leaves as cattle feed (Kehar & Goswami, 1950); and nutritive value of some grasses and legumes of Jammu and Kashmir (Chopra *et al.* 1956; Singh *et al.*,1959), 'Kharsoo' and 'Dudhiliya' (Thapliyal, 1962), Himalayan tree fodder 'Bhimal'- *Grewia oppositifolia* (Joshi & Talapatra, 1960), mulberry (*Morus indica*) tree leaves (Jayal & Kehar, 1960), oaks (Purohit 1962), Job's tears- *Coix lacryma-jobi* (Arora, 1971) and *Mallotus phillippnensis* (Bhargava *et al.*, 1977). Other important compilations on the subject by Whyte (1957) and Bor (1960) provided detailed information on the grasslands and fodder resources of India and the grasses of Burma, Ceylon, India and Pakistan (excluding Bambuseae). In subsequent years, contributions from workers like, Dutta *et al.* (1951), Sen (1964), Sawhney & Bedi (1972), and Gupta *et al.* (1975) provided valuable information on trace elements in some Indian grasses, nutritive values of cattle feeds, trace elements of hill feeds and fodders, and *in-vitro* nutrient digestibility of some tree species, respectively. In 1977, Negi published information on the fodder trees in Himachal Pradesh whereas Ranjhan provided valuable data on animal nutrition and feeding practices in India. Besides, various other publications and compilations, viz., fodder trees of India (Singh, 1982), fibre yielding plants of Sikkim (Krishna & Das, 1983), fibre yielding plants in North Eastern region (Islam, 1984), fodder plants of Garhwal hills and their impact on the quantity and quality of milk (Gaur *et al.* 1981), some favourite trees for fuel and fodder (Prakash & Hockins, 1986), fodder trees and shrubs of Central Himalaya (Purohit & Samant, 1995), traditional forage resources of India (Singh & Shankar, 1996), wild fibre yielding plants of Kolli hills, Tamil Nadu (Geetha *et al.*, 1996) and fodder and fuelwood resources of Central Himalaya (Singh *et al.* 1998) have added to our wealth of knowledge on the subject. Not only this, data on some other aspects like fibre quality of *Grewia optiva* (Lata & Verma, 1995), use of *Cannabis sativa* stem for making hempen cloth and cordage (Shah, 1997) and methods describing eco-friendly processing of hand-made paper from *Edgeworthia gardneri* in Eastern Himalaya (Basnet, 1999) has also appeared in literature. Additionally, the entire earlier work on ethnobotany in India has been summarized by Binu *et al.* (1992), and Radhakrishnan *et al.* (2000) and its graphic account given by Jain & Srivastava (2001). So far no systematic attempt has been made to consolidate our data on fibre and fodder resources which is scattered and meagre. The present work is an effort in this direction.

3

Enumeration of Fibre and Fodder Resources

For compiling an exhausting database on fibre and fodder resources of India, the data has been gathered from the scrutiny of various published compendia, books, materia medica, monographs,viz., Dutt (1877) - *Materia Medica of the Hindus, With a Glossary of Indian Plants*; Watt (1889-1896) – *A Dictionary of Economic Products*; Krishnamurti Naidu (1917) – *Commercial Guide to the Forest Economic Products of Mysore*: Chakraberty (1923) – *Comparative Hindu Materia Medica*; Ainsile (1926) – *Materia Medica*; Nadkarni (1927) – *Indian Materia Medica*; Burkill (1935) – *Dictionary of Economic Products of India*; Anonymous (1948-1976) – *The Wealth of India – A Dictionary of Indian Raw Materials and Industrial Products*; Krishnamurthy (1953) – *Horticulture and Economic Plants of the Nilgiris*; Sunderaraj & Balasubramanyam (1959) – *The Useful Plants of India*; Maheshwari & Singh (1965) – *Dictionary of Economic Plants of India*; Atkinson (1982) – *Economic Botany of the Himalayan Region*; Singh *et al.* (1983) – *Dictionary of Economic Plants in India*; Agarwal (1986) – *Economic Plants of India*; Ambasta (1986) – *The Useful Plants of India*; Uphof (1986) – *Dictionary of Economic Plants of South India*; Nayar *et al.* (1989) – *Economic Plants of India*; Jain (1991) – *Dictionary of Indian Folk – Medicine and Ethnobotany*; Saklani & Jain (1994) – *Cross-Cultural Ethnobotany of Northeast India*; Singh (1996) – *Dictionary of Economic Plants in India*; Rama Rao & Henry (1996) – *The Ethnobotany of Eastern Ghats in Andhra Pradesh, India*; Maheshwari (1996) – *Ethnobotany in South Asia*; Varghese (1996) – *Applied Ethnobotany – A Case Study Among the Khasia of Central India*; Singh & Pandey (1998) – *Ethnobotany of Rajasthan*; Kumar (1999) – *The Economic Plants of North East India*; Maheshwari (2000) – *Ethnobotany and Medicinal Plants of Indian Subcontinent*; Varghese & Hembrom (2000) – *Ethnobotanical Lore of the Paharias*; Sood *et al.* (2001) – *Ethnobotany of Cold Desert Tribes of Lahoul – Spiti*

(N. W. Himalaya); Sood & Thakur (2004) – *Ethnobotany of Rewalsar Himalaya*. Besides, more than 1250 references on ethnobotanical work were also screened (Jain *et al.* 1984; Jain, 2002). Not only this, the information on nutritive analysis of over 85 fodder plants has also been presented.

Enumeration presently enlists the use of over 811 plants species for fibre and fodder by tribals and rural populace of India. They are arranged in alphabetical order, furnishing the valid binomial (nomenclature based on Bennet, 1986), synonyms (wherever necessary), family, English, Hindi, Sanskrit and Regional names, distribution, part (s) used, utilization, active constituents. To make the reading interesting, 34 photographs, 3 figures and 4 tables were also included in the text. Appendices include information on: Total number of genera and species of fibre and fodder resources under various divisions of plant kingdom (Appendix I), Families of fibre and fodder resources of India (Appendix II), fibre resources of India (Appendix III), predominant genera of fibre resources (Appendix IV), fodder resources of India (Appendix V), predominant genera of fodder resources (Appendix VI), fibre and fodder resources of India (Appendix VII), predominant genera of fibre and fodder resources (Appendix VIII), Index to Botanical names (Appendix IX), Index to English names (Appendix X), India to Hindi names (Appendix XI), Index to Sanskrit names (Appendix XII).

Abbreviations used in text are: Alc. = Alcoholic; α = Alpha; & = And; Anda. = Andaman; Aq = Aqueous; Ass. = Assam; Beng. = Bengali; β = Beta; Bom. = Bombay; Ca = Calcium; C = Central; Cl = Chlorine; Cu = Copper; E. = Eastern; Eng. = English; *et al.* = *et alia* (and other author); ext. = Extract; etc. = et cetra (so on); Fig.(s) = Figure(s); γ = Gamma; Gar. = Garhwal; g = Gram; Guj. = Gujarat; H.P. = Himachal Pradesh; OH = Hydroxy; i.e. = *id est* (that is); Fe = Iron; I.U. = International Unit; J. & K. = Jammu and Kashmir; Kan. = Kannada; Kash. = Kashmir; Kg = Kilogram; Kon. = Konkan; Kull. = Kullu; Kum. = Kumaon; Ladk. = Ladakh; M.P. = Madhya Pradesh; Mal. = Malalyam; Mar. = Marathi; Mn = Manganese; Mg = Magnesium; m = Meter; μm = Micrometer; mg = Miligram; ml = Mililiter; N = Nitrogen; N.E. = Northeast; N. W. = Northwest; Ori. = Oriya; % = Per cent; P = Phosphorus; +ve = Positive; Ψ = Psi; Pub. = Punjabi; Raj. = Rajasthan; Sans. = Sanskrit; Sec. = Secondary; Sim. = Shimla; S. E. = Southeast; S. W. = Southwest; Sp = Specie; Spp = Species; Syn. = Synonym; Tam. = Tamil; Tel. = Telugu; U. P. = Uttar Pradesh; var. = Variety; Vit. = Vitamin; W. = West / Western; Zn = Zinc.

4

Fibre and Fodder Resources of India

Abelmoschus crinitis Wall.

Syn: *Hibiscus crinitis* (Wall.) G. Don.; *H. cancellantus* Roxb.
Family: Malvaceae.

Vern. Names : Banbhindi, Khobikapah.
Regional Name :

Beng	:	Bankapas, Birkapas.

Distribution	:	Himalaya. N.W. India. Sub-Himalayan Tract from Kashmir Eastwards. W. Bengal upto 1,000 m.
Part Used	:	Whole Plant.
Utilization	:	Herb yields a fibre for ropes and cordage.
References	:	Ambasta (1986), Sood & Thakur (2004), Roy *et al.* (1998).

Abelmoschus esculentus Linn. Moench.

Syn: *Hibiscus esculentus* Linn.
Family: Malvaceae.

Vern. Name: Bhindi.
English, Hindi, Sanskrit and Regional Names:

Eng.	:	Gumbo, Lady's Finger, Okra;
Hindi	:	Bhindi, Bhindi Tori;
Sans.	:	Bhenda, Darivka, Tindisa;
Beng	:	Dheras;

Guj. : Binda;
Kan. : Bendo;
Mal. : Bendai, Venda;
Mar. : Bhendi;
Ori. : Bhendi;
Pub. : Bhinda Tori, Bhindi, Ramturai;
Tam. : Vendai;
Tel. : Benda, Bendi, Penda, Venda.

Distribution : Cultivated Throughout India upto 1,200 m.

Part Used : Stalks.

Utilization : Stalk yields a blast fibre for making ropes.

References : Ambasta (1986), Parrotta (2001), Sood *et al.* (2005).

Abelmoschus ficulneus White & Arn.

Syn: *Hibiscus ficulneus* Linn.
Family: Malvaceae.

English, Hindi and Regional Names:

Eng. : White Wild Musk Mallow;
Hindi : Ranbhindi;
Pub. : Deola, Dula, Kapasiya;
Tam. : Kathivendai;
Tel. : Nella-bende, Parupubenda.

Distribution : W. Bengal, Punjab & Tamil Nadu.

Part Used : Whole Plant.

Utilization : Plant yields a white fibre for twine and light cordage.

References : Ambasta (1986), Roy *et al.* (1998).

Abelmoschus manihot (Linn.) Medic.

Syn: *Hibiscus manihot* Linn.; *H. pungens* Roxb.
Family: Malvaceae.

Regional Names:

Ass. : Usipak;
Guj. : Kantalo-bhendo;

Mar.	:	Jangli-bhendi;
Ori.	:	Ranbhendi.

Distribution : Cultivated & Naturalized in W. Bengal & S. India from Gujarat to Kerala.

Part Used : Stem.

Utilization : Stem yields cotton-like fibre resembling jute in colour and used for making ropes and devices for catching birds.

References : Ambasta (1986), Diwakar & Ansari (2003), Jain (1991), Parrotta (2001), Punjani (2002).

Abelmoschus moschatus Medic.

Syn: *Hibiscus abelmoschus* Linn.
Family: Malvaceae.

English, Hindi, Sanskrit and Regional Names:

Eng.	:	Ambrette Plant, Musk Mallow;
Hindi	:	Muakdana;
Sans.	:	Latakasturikam;
Beng	:	Mushakdana;
Guj.	:	Muskdana;
Kan.	:	Kadukasturi;
Mal.	:	Kattukasthuri;
Mar.	:	Kasturi-bhenda;
Tam.	:	Varttilai-kasthuri, Katukasturi;
Tel.	:	Kasturibendavittu.

Distribution : Hilly Region Throughout Peninsular India as well as in the Himalayan Foot Hills. Cultivated in Maharashtra and U.P.

Part Used : Bark.

Utilization : Bark yields good quality fibre which is used for making a special flat basket for disposal of mud during construction of wells; cordage for the leather hoops of a musical instrument called 'Dholak', 'Saang' and 'Jotri'; and also a pair of straps for tying cart on the neck of the beast of burdens by 'Garasias'.

References : Ambasta (1986), Dhiman (2003), Jain (2001), Mittre (1997), Parrotta (2001), Singh & Kumar (2000), Singh & Pandey (1998), Varghese (1996).

Abies pindrow Royle

Syn: *Abies spectabilis* (Don) Spach.; *A. webbiana* Lindl. var. *pindrow* (Royle) Hook. f.; *Picea smithiana* Loud.; *Pinus spectabilis* Lamb.; *P. striata* Ham., *P. webbiana* Wall.

Family: Pinaceae.

Vern. Name: Tosh.

English, Hindi, Sanskrit and Regional Names:

Eng.	:	Himalayan Silver Fir;
Hindi	:	Talispatra;
Sans.	:	Dhaatipatra, Manduparni, Talispatra;
Gar.	:	Morinda;
Kash.	:	Badar, Buder, Tung;
Tam.	:	Taalispatri;
Tel.	:	Taalispatri.

Distribution : Commonly in W. Himalaya from Kashmir to Nepal upto 2,100-3,600 m.

Part Used : Plant.

Utilization : It is valued for paper making.

References : Dastur (1964), Sahni (1998), Sharma (2004).

Abroma angusta Linn.

Syn: *Abroma angulata* Lam.; *A. fastuosum* Gaertn.; *A. wheeleri* Retz.; *Abroma angusta* Linn. f.

Family: Sterculiaceae.

English, Hindi, Sanskrit and Regional Names:

Eng.	:	Devil's Cotton, Perennial Indian Hemp;
Hindi	:	Ulatkambal;
Sans.	:	Pisacakarpasa, Pivari, Yosini;
Beng	:	Ulatkambal;
Kan.	:	Divvahatti;

Mar. : Olatkambal;
Tam. : Sivappututti.

Distribution : Throughout the Hotter Parts of India.

Part Used : Stem.

Utilization : Stem yields fibre suitable for mixing with jute for for twines, sack-cloth, cordage, matting, fishing-lines and hessian. Bark fibre soft, pale and also employed for making ropes.

References : Agarwal (2003), Ambasta (1986), Anonymous (1948), Dastur (1964), Dey *et al.* (1973), Hooker (1872), Jain (1991), Mittre (1997), Prajapati *et al.* (2003).

Abutilon bedfordianum St.-Hil.

Family: Malvaceae.

Part Used : Stem.

Utilization : Yields a fibre for cordage and paper making.

Reference : Ambasta (1986).

Abutilon fruticosum Guill. *et al.*

Syn: *Abutilon albidum* Webb & Berth.; *A. microphyllum* A. Rich.

Family: Malvaceae.

Part Used : Aerial Plant Parts.

Utilization : Used as fodder.

References : Ambasta (1986), Hooker (1872).

Abutilon glaucum Sweet

Syn: *Abutilon muticum* Sweet.

Family: Malvaceae.

Hindi and Regional Names:

Hindi : Kakrai;
Mar. : Karandi, Kasili.

Parts Used : Aerial Plant Parts. Stem.

Utilization : Aerial plant parts used as fodder and stem for cordage.

References : Agarwal (2003), Ambasta (1986).

Abutilon grandifolium (Willd.) Sweet.

Syn: *Abulition molle* (Ortega) Sweet.

Family: Malvaceae.

Part Used : Stem.

Utilization : Stem fibre used for cordage.

Reference : Ambasta (1986).

Abutilon hirtum (Lam.) Sweet

Syn: *Abutilon graveolens* (Roxb. *ex* Hornem.) W. & A. *ex* Wight. non Britt.

Family : Malvaceae.

Vern. Name: Barlihangi.

Hindi and Regional Names:

Hindi : Barakanghi;
Beng : Jampi, Kanghani;
Kan. : Haktuti;
Ori. : Karpuripotro;
Tam. : Vadattutti.

Distribution : Abundant Throughout Hotter Parts of India.

Part Used : Stem.

Utilization : Stem fibre white, lustrous and coarser than jute but suitable for cordage.

References : Agarwal (2003), Ambasta (1986), Blatter & Hallberg (1984), Das & Agarwal (1991), Hooker (1872), Kirtikar & Basu (1935), Oommachan (1977), Roy *et al.* (1998).

Abutilon indicum (Linn.) Sweet

Syn: *Abutilon asiaticum* (Linn.) Sweet.; *A. indicum* G. Don.; *Sida guinensis* Schumach.; *S. rombifolia* Linn.

Family: Malvaceae.

Vern. Name: Kunch.

English, Hindi, Sanskrit and Regional Names:

Eng. : Chinese Bell-Flowers, Country Mallow, Flowering Maples;

Hindi	:	Atibala, Kanghi;
Sans.	:	Atibala, Jayapateri, Kankatika;
Beng	:	Potari;
Guj.	:	Dabali, Kansaki, Khapat;
Mal.	:	Katuvan, Oorpam, Tutti, Uram, Uruvam, Velluram;
Mar.	:	Chakrabhenda, Mudra, Petari;
Ori.	:	Nakochono, Pedipidika;
Pub.	:	Pilibuti, Sanbal;
Tam.	:	Paniyarattutti, Tutti, Velluram;
Tel.	:	Tuttura.

Distribution : Throughout Warmer Parts of India upto 1,200 m.

Part Used : Stem.

Utilization : Bark fibre white, lustrous with an anticlockwise drying twist coarser than jute and used in the preparation of ropes, strings and for mixing with jute. The bark also have potentials for paper making.

References : Agarwal (2003), Ambasta (1986), Anonymous (1948), Blatter & Hallberg (1984), Caius (2003), Datt (1996), Drury (1978), Hooker (1872), Jain (1991), Khare (2004), Oommachan (1977), Parrotta (2001), Prasad *et al.* (2002), Prasad *et al.* (2003), Punjani (2002), Saklani & Jain (1994), Sharma & Rana (2005), Singh & Pandey (1998).

Abutilon persicum (Burm. f.) Merrill.

Syn: *Abutilon polyandrum* Wight & Arn. *ex* Wight. non G. Don., Mast. in part.

Family: Malvaceae.

Hindi and Regional Names:

Hindi	:	Chotabanse, Tepari;
Mal.	:	Thutthi;
Mar.	:	Madam.

Part Used : Stem.

Utilization : Fibre resemble hemp and fit for ropes.

Reference : Ambasta (1986).

Abutilon theophrastii Medic.

Syn: *Abutilon avicennae* Gaertn.

Family: Malvaceae.

English, Sanskrit and Regional Names:

Eng. : American Jute, China Jute, Indian mallow;

Sans. : Jaya;

Beng : Jaya.

Distribution : Found in N. W. India upto 1,500 m.

Part Used : Stem.

Utilization : Fibre from stem is strong, coarse and lustrous and called Chinese, American Jute. These are used for making rugs, ropes and thread for slippers, paper and caulking materials.

References : Ambasta (1986), Anonymous (1948), Dastur (1964), Jain (1991), Mittre (1997), Prajapati *et al.* (2003).

Acacia caesia Wight & Arn.

Syn: *Acacia intsia* Willd.

Family: Fabaceae.

Regional Names:

Mal. : Incha;

Mar. : Chilar;

Ori. : Dantari;

Tam. : Kyriyandu;

Tel. : Korinta.

Distribution : Sub-Himalayan Tract from the Chenab Eastwards, ascending to 4,000 m.

Parts Used : Branches. Leaves.

Utilization : Used as a fodder plant.

References : Ambasta (1986), Anonymous (1948), Hooker (1948).

Acacia jacquemontii Benth.

Family: Fabaceae.

Hindi and Regional Names:

Hindi : Kikar;
Guj. : Ratabauli;
Pub. : Babul, Bamul, Kikar.

Distribution : Dry Plains of Punjab, Rajputana & N. Gujarat to Sri Lanka.

Parts Used : Branches. Leaves.

Utilization : Tender branches and leaves used as fodder.

References : Ambasta (1986), Anonymous (1948), Hooker (1879).

Acacia leucophloea Willd.

Syn: *Acacia alba* Willd.; *Mimosa alba* Rotter.; *M. leucophloea* Roxb.
Family: Fabaceae.

English, Hindi, Sanskrit and Regional Names:

Eng. : Distiller's acacia, White Kikar;
Hindi : Safed kikar;
Sans. : Shvetabarbura;
Beng : Safed Babul;
Guj. : Haribaval;
Mal. : Pattacharaya - maram;
Mar. : Hewar;
Ori. : Gohira, Goira, Guanria;
Tam. : Velvayalam;
Tel. : Tella - tuma.

Distribution : Dry Tropics from Punjab to Peninsula and Southwards to Tamil Nadu.

Parts Used : Bark.Young Pods.

Utilization : Young pods used as fodder for cattle. Bark fibre strong and used for cordage and fishing nets by Damor tribals.

References : Agarwal (2003), Ambasta (1986), Anonymous (1946), Dastur (1964), Deshmukh *et al.* (2000), Diwakar &

Ansari (2003), Dwivedi (1993), Parrotta (2001), Rastogi & Mehrotra (1995b), Singh (1982), Singh & Pandey (1998), Talbot (1976a).

Acacia mollissima Willd.

Family: Fabaceae.

English and Hindi Names:

Eng. : The Black Wattle;
Hindi : Kalikikar.

Distribution : Nilghiri Hills.

Part Used : Bark.

Utilization : Bark yields fibre of good commerce.

Reference : Sood & Thakur (2004).

Acacia nilotica (Linn) *delile* Subsp. *indica* (Benth.) Brenan.

Syn: *Acacia arabica* Willd. var. *indica* Benth.

Family: Fabaceae.

Vern. Name: Babul.

English, Hindi, Sanskrit and Regional Names:

Eng. : Indian Gum Arabic Tree, White Babool;
Hindi : Kikar;
Sans. : Babbula, Babula, Barbara, Kinkirata;
Beng : Babla;
Guj. : Baval, Kaloabaval;
Mal. : Karivelan;
Mar. : Babhul, Babli, Babul;
Ori. : Babulo, Bambuda, Baubra;
Tam. : Karuvelan, Karuvelei;
Tel. : Nallatumma.

Distribution : Throughout India from Punjab to W. Bengal. Southwards Maharashtra ascending to 800 m.

Parts Used : Slender Twigs. Pods.

Utilization : Pods when green used as fooder for cattle. Slender twigs used for rope making.

References : Ambasta (1986), Diwakar & Ansari (2003), Dwivedi (1993), Mittre (1997), Oommachan (1977), Parrotta (2001), Rastogi & Mehrotra (1991, 1995b), Sahni (1998), Singh (1982), Singh & Pandey (1998).

Acacia planifrons Wight & Arn.

Syn: *Acacia campbellii** Arn.; *A. farnesiana* Wall.; *A. roxburghii* W. & A.; *Mimosa eburnea* Roxb.; *M. horrida* Sm.

Family: Fabaceae.

English, Hindi and Regional Names:

Eng. : Umbrella Thorn;
Hindi : Vilayati babul;
Mar. : Kankari;
Tam. : Kodaivelan;
Tel. : Godugu, Thumma.

Distribution : S. India.

Part Used : Pods.

Utilization : Pods used as fodder for cattle and goats.

References : Anonymous (1948), Hooker (1879), Singh (1982), Sood & Thakur (2004).

Acacia Senegal Willd.

Syn: *Acacia rupestris* Stocks.; *A. verek* Guill.

Family: Fabaceae.

Hindi, Sanskrit and Regional Names:

Hindi : Kher, Kumta;
Sans. : Sneta Khadira;
Bom. : Khor, Kumta;
Guj. : Goradio - babul;
Raj. : Kumta.

Distribution : Rocky slopes in Punjab, Rajasthan & Gujarat.

Parts Used : Leaves. Inflorescence. Stem.

Utilization : Leaves and fallen inflorenceses rapidly browsed by camels and goats. Stem bark fibre used for ropes, cordage and fishing -nets.

References : Agarwal (2003), Ambasta (1986), Anonymous (1948), Bennet *et al.* (1991), Dastur (1964), Hooker (1879), Jain (1991), Mittre (1997), Singh (1982).

Acer acuminatum Wall.

Family: Aceraceae.

English Names:

Eng. : Kandaru, Kanzla.

Distribution : In Forest between 2,000 - 3,600 m.

Part Used : Leaves.

Utilization : Leaves lopped for fodder and considered to be a source of good fodder.

Reference : Purohit & Samant (1995).

Acer caesium Wall.

Family: Aceraceae.

Vern. Names: Kamia, Kanjol.

English, Hindi and Regional Names:

Eng. : Himalayan Maple;
Hindi : Kainsal, Tilpattra;
Pub. : Trekhan.

Distribution : Temperate W. Himalaya from Kashmir - Nepal.

Part Used : Leaves.

Utilization : Leaves lopped for fodder during winter for all classes of livestock and considered a very good fodder.

References : Agarwal (2003), Hooker (1872), Purohit & Samant (1995), Sahni (1998).

Acer cappadocicum Gledtch.

Syn: *Acer pictum* Thunb.

Family: Aceraceae.

Distribution : Throughout the Hills between 2,000-3,000 m.

Part Used : Leaves.
Utilization : Leaves lopped for fodder in rainy season.
Reference : Purohit & Samant (1995).

Acer oblongum Wall.

Syn: *Acer buzimpala* Hamilt. *ex* D. Don.; *A. laurifolim* D. Don.
Family: Aceraceae.

English and Hindi Names:

Eng. : Indian Maple;
Hindi : Pangoi.

Distribution : C. & W. Himalaya from Kashmir to Arunachal Pradesh & Manipur, ascending to 2,500 m.
Part Used : Leaves.
Utilization : Leaves fed to all classes of livestock in the hills during winter and summer.
References : Dwivedi (1993), Hooker (1872), Purohit & Samant (1995).

Acer villosum Wall.

Syn : *Acer sterculiaceum* Wall.
Family : Aceraceae.

Hindi Name:

Hindi : Mandra.

Distribution : Moist Forest between 2,000 - 3,000 m: Kashmir to Nepal.
Part Used : Leaves.
Utilization : Leaves lopped for fodder by livestock owners during rainy season.
References : Dwivedi (1993), Hooker (1872), Purohit & Samant (1995).

Adansonia digitata Linn.

Family: Bombacaceae.

English, Hindi, Sanskrit and Regional Names:

Eng. : Baobab, Monkey Bead Tree;

Hindi	:	Kalpa Vriksha;
Sans.	:	Gangerukie;
Beng	:	Gorakh amali;
Mar.	:	Gorakh chinch;
Tam.	:	Ananipuli.

Distribution : Dry Region of N.W. & S. India. Cultivated in some Parts of U.P., Bihar, Bombay & Madras.

Parts Used : Bark. Wood.

Utilization : Bark yields soft fibre which are occasionally used for ropes and sacking. Paper pulp from wood suitable for wrapping as well as writing paper.

References : Agarwal (2003), Ambasta (1986), Anonymous (1948), Chatterjee & Pakrashi (1991), Das & Agarwal (1991), Dastur (1964), Drury (1978), Hooker (1872), Sahni (1998), Watt (1972).

Ademostemma lavenia (L.) Ktze.

Syn : *Ademostemma viscosum* Forst.

Family : Asteraceae.

Part Used : Plant.

Utilization : Plant used for fodder purposes.

Reference : Islam (2002).

Adina cordifolia Benth. & Hook. f.

Syn : *Nauclea cordifolia* Roxb.

Family : Rubiaceae.

English, Hindi, Sanskrit and Regional Names:

Eng.	:	Yellow Wood;
Hindi	:	Haldu;
Sans.	:	Dharaakadaamba, Girikadambali;
Beng	:	Dakom, Petpuria;
Guj.	:	Aldaven, Haladaravo, Haladhwan, Haridra;
Mal.	:	Manja – Kadamba;
Mar.	:	Heddi;
Ori.	:	Kelikodombo, Holondo;

Tam. : Manja -Kadamba;
Tel. : Pasupu - Kadamba.

Distribution : Throughout India in Dry Forest from Kumaon to Sikkim.

Parts Used : Leaves. Wood.

Utilization : Leaves lopped as fodder. Wood pulp good for manufacturing wrapping, writing and printing papers.

References : Agarwal (2003), Ambasta (1986), Anonymous (1948), Dwivedi (1993), Hooker (1882), Kirtikar & Basu (1984), Rastogi & Mahrotra (1998), Singh (1982).

Aechmanthera gossypina (Nees) Nees.

Syn: *Aechmanthera tomentosa* Nees.
Family: Acanthaceae.

Distribution : Throughout the area between 1,000– 2,000 m.

Parts Used : Bark. Leaves.

Utilization : Leaves considered as an excellent fodder for sheep and goats and fed during rainy and winter seasons. Silvery bark fibre good for weaving, ropes and netting purposes.

References : Purohit & Samant (1995), Sood & Thakur (2004).

Aegle marmelos Correa *ex* Roxb.

Syn: *Crateava marmelos* L.; *C. reliogosa* Ainsite.; *Feronia pellucida* Roxb.
Family: Rutaceae.

Vern. Name: Bil.

English, Hindi, Sanskrit and Regional Names:

Eng. : Bael Tree, Bengal Quince, Holy Fruit, Indian Quince;
Hindi : Bel;
Sans. : Atimangaliya, Bilva, Satyadharma, Tripatra;
Ass. : Bil;
Beng : Bil;
Bom. : Bel;
Guj. : Bil, Billy;
Kan. : Bilpatre;

Mal.	:	Karuvalam, Marivalam, Vilam;
Mar.	:	Bel;
Ori.	:	Bilwa, Sriphola;
Tam.	:	Aluvigam, Iyalbudi, Vilam;
Tel.	:	Maredu.

Distribution : Throughout India; Sub-Himalayan Tract in C. & S. India upto 1,000 m.

Part Used : Leaves.

Utilization : Leaves used as fodder.

Active Constituents : Leaves contain oxazole, pyridine moieties, sitosterol, aegeline and lupeol.

References : Chopra *et al.* (1969), Dwivedi (1993), Prajapati *et al.* (2003), Raju (1998), Rastogi & Mehrotra (1991, 1995b), Sahni (1998), Singh (1982), Thakur *et al.* (1989).

Aeschynomene indica Linn.

Syn: *Aeschynomene aspera* Wall.; *A. diffusa* Willd.; *A. kashmiriana* Camb.; *A. pumila* Linn.; *A. roxburghii* Spreng.; *A. viscidula* Willd.; *Hedysarum neli- tali* Roxb.; *Smithia aspera* Roxb.

Family: Fabaceae.

Hindi and Regional Names:

Hindi	:	Kathshala, Laugauni;
Beng	:	Kath shoal;
Mal.	:	Nellithali;
Ori.	:	Lajuari;
Tam.	:	Nellithali;
Tel.	:	Tiga-jiluga.

Distribution : Wet Grasslands of Bengal, Assam, S. India.

Parts Used : Leaves Bark.

Utilization : Leaves lopped for fodder for sheep and goats. Bark fibre from stem extensively used by tribals for making ropes for affixing the bamboo stripes for making the huts and camping shelter in the hilly tracts, cordages and for brushes.

References : Agarwal (2003), Ambasta (1986), Das *et al.* (1996), Hooker (1879), Kapur & Nanda (1996), Raju (1998), Singh & Pandey (1998).

Aesculus indica Colebr. (Pl. 1A)

Syn: *Pavia indica* Wall.
Family: Sapindaceae.

Vern. Names: Kanhor, Khanor.

English, Hindi and Regional Names:

Eng. : Indian Horse Chestnut Tree;
Hindi : Bankhor, Pangar;
Kum. : Kishing, Pangar;
Pub. : Bankhor, Gu, Gun, Hazara, Khanor;
Tam. : Panger;
Tel. : Tiga jiluga.

Distribution : Hill-Tops of Kashmir, Punjab & U.P.: 900- 3,600 m.

Parts Used : Leaves. Fruits.

Utilization : Leaves lopped for fodder for sheep and goats in the lean period of winter. Fruits also relished by animals.

Active Constituents : Quercetin, sitosterol and escin isolated from the leaves.

References : Agarwal (2003), Ambasta (1986), Anonymous (1948), Kirtikar & Basu (1935), Nautiyal *et al.* (2003), Purohit & Samant (1995), Raju (1998), Rastogi & Mehrotra (1991), Roy *et al.* (1998), Singh (2000).

Aesculus igniarius L.

Family: Sapindaceae.

Distribution : Distributed in Temperate Regions of India.

Part Used : Leaves.

Utilization : Leaves given as fodder.

Reference : Sood & Thakur (2004).

Aganosoma caryophyllata (Roxb. *ex* Sims) G. Don.

Syn: *Aganosoma blumei* Wight.; *Chonemorpha dicthotoma* G. Don Gen.; *Echites caryophyllata* Wall.; *E. dichotoma* Roth.; *E. heynei* Spreng.

Family: Apocynaceae.

Distribution : Lower Bengal.

Part Used : Stem.

Utilization : Stem used as fibre.

References : Hooker (1882), Rao & Henry (1996).

Agathis robusta F. M. Bailey.

Family: Araucariaceae.

Part Used : Wood.

Utilization : Wood suitable as viscose rayon for manufacturing of writing, printing and wrapping papers

Reference : Ambasta (1986).

Agave americana Linn.

Syn: *Agave cantula* Roxb.

Family: Agavaceae.

Vern. Name: Ramban.

English, Hindi, Sanskrit and Regional Names:

Eng. : American Aloe, Carata, Century Plant;
Hindi : Banskorea, Bara Kanwar, Kantala;
Sans. : Kalakantala, Kantala;
Beng : Banskeora, Bilatipat;
Guj. : Jangli- Kunvara;
Kan. : Bhuttale, Kalanaru;
Mal. : Eroppakaita, Panam- Kattazha;
Mar. : Janglianarash, Parkand;
Ori. : Birhot Qkumari;
Pub. : Wilayati Kaitalu;
Tam. : Anaikathaalai;
Tel. : Rakashimatalu.

Distribution : Cultivated in the Gardens Throughout India.

Parts Used : Leaves. Roots.

Utilization : Fibre from leaves are strong and excellent for making ropes for drawing water from the wells and knitting the cots, cordages, lashing calicobale, long lines for ships, mats, carpets, sacks, scrubbing brushes and interior decoration.

Reference : Agarwal (2003), Ambasta (1986), Dastur (1964), Mittre (1981, 1997), Sharma & Rana (2005), Singh & Kumar (2002), Singh & Pandey (1998), Sood & Thakur (2004), Watt (1972).

Agave angustifolia Haw.

Syn: *Agave ixtlioides* Hook.; *A. vivipara* Auct.; *A. wightii* Drummond & Prain.

Family: Agavaceae.

English, Sanskrit and Regional Names:

Eng. : Bastard Aloe, Dwarf aloe;

Sans. : Kantala;

Tam. : Chinna errumaikkattali, Seemakathaali;

Tel. : Balurakkasi, Kittanaral, Samata.

Part Used : Leaves.

Utilization : Leaves considered good source of fibre for mats and baskets.

References : Ambasta (1986), Anonymous (1948), Datt (1996), Kirtikar & Basu (1984).

Agave lurida Act.

Family: Agavaceae.

Distribution : Warmer Regions of India.

Part Used : Leaves.

Utilization : Leaves yield a kind of short fibre for stuffing and netting.

Reference : Sood & Thakur (2004).

Agave sisalana Perr.

Syn: *Agave rigida* Mill.var. *sisalana* Engelon.
Family: Agavaceae.

English Name:

Eng. : Indian Agave.

Distribution : Warmer Regions of India (Assam, Bengal, Madras, Bombay, Mysore) & Sri Lanka.

Part Used : Leaves.

Utilization : Fibre from leaves are of good strength and used as a substitute for Manila Hemp for twines, ropes, marine cordage, fishing-nets, etc. Short fibre used for brushes, craft paper, paper boards, cheap twine and upholstery.

References : Agarwal (2003), Ambasta (1986), Anonymous (1948), Dastur (1964), Harborne & Baxter (2001), Jain (1991), Mittre (1997), Sood & Thakur (2004).

Agave vera-cruz Miller.

Family: Agavaceae.

English, Hindi and Regional Names:

Eng. : Blue Elephant Aloe, Railway Aloe;

Hindi : Kuwar Buti;

Mar. : Latia ghaial;

Ori. : Kektad.

Distribution : Naturalized Throughout India (Assam, Bengal, Bihar & S. India).

Part Used : Leaves.

Utilization : Leaf fibre coarser and stronger than cantala fibre and used for making ropes, cordage and mat.

References : Agarwal (2003), Ambasta (1986), Anonymous (1948), Diwakar & Ansari (2003), Jain (1991).

Ageratum conyzoides L.

Syn: *Ageratum cordifolium* Roxb.
Family: Asteraceae.

English, Hindi and Regional Names:

Eng.	:	Goat Weed, White Weed;
Hindi	:	Koobhi;
Beng	:	Dochunty, Uchunti;
Guj.	:	Ajgandha, Gandhari Sedardi;
Mal.	:	Appa, Muryanpacha;
Mar.	:	Ghaneraosadi;
Ori.	:	Boksunga, Poksunga;
Tam.	:	Pum- pillu.

Distribution : Throughout India; ascending the Himalaya to 1,800 m.

Parts Used : Leaves. Twigs.

Utilization : Aerial parts used for fodder purposes.

References : Hooker (1882), Islam (2002), Islam & Hasin (2003), Parrota (2001), Rastogi & Mehrotra (1991).

Agropyron cristatum Gaertn.

Family: Poaceae.

Part Used : Whole Plant.

Utilization : Considered as a valuable arid pasture plant.

Reference : Ambasta (1986).

Agrostis gigantean Roth.

Family: Poaceae.

English Name:

Eng. : Black bent.

Parts Used : Plant Parts.

Utilization : Suitable for permanent wet pasture.

References : Ambasta (1986), Sharma & Rana (2005).

Agrostis canina Linn.

Syn: *Agraulus caninus* Beauv.; *Trichodium caninum* Schrad.
Family: Poaceae.

Distribution : W. Himalaya: Kashmir 1,000 – 2,000 m.

Part Used : Whole Plant.

Utilization : Suitable for permanent wet pasture.

References : Ambasta (1986), Hooker (1879), Sharma & Rana (2005).

Agrostis stolonifera Linn.

Syn: *Agrostis alba* auct., non Linn.; *A. vertiicillata* Hook. f. in part, non Vill.; *A. wightii* Nees *ex* Steud. in part.
Family: Poaceae.

English Name:

Eng. : Creeping bent.

Part used : Plant.

Utilization : Forms a part of the alpine pasture and useful as a fodder grass.

References : Ambasta (1986), Sharma & Rana (2005).

Aidia tetrasperma (Roxb.) Yamazaki.

Syn: *Randia tetrasperma* Hook. f.
Family: Rubiaceae.

Regional Name:

Pub. : Kikra.

Part used : Plant.

Utilization : Used as fodder for goats.

Reference : Ambasta (1986).

Ailanthus excelsa Roxb.

Syn: *Pongelion wightii* van Tiegh.
Family: Simarubaceae.

English, Hindi, Sanskrit and Regional Names:

Eng. : Tree of Heaven;
Hindi : Maharuk;

Sans.	:	Adoasha, Aralu, Madala, Mahanimba;
Beng	:	Maharukha;
Guj.	:	Ardusi;
Kan.	:	Doddamara;
Mal.	:	Mattipongilyam;
Mar.	:	Maharuk;
Ori.	:	Mahala, Mahanim;
Tam.	:	Perumarum;
Tel.	:	Peddamanu.

Distribution : Bihar, M.P., Gujarat. W. Ghats from Mysore to Travancore.

Utilization : Leaves and shoots used as fodder.

Active Constituents : Leaves contain ailantic acid, beta- sitosterol and vitaxin. Green and yellow summer leaves contain 1.18 and 0.3 mg / g carotenoids, respectively.

References : Ambasta (1986), Dey *et al.* (1973), Dwivedi (1993), Khan & Khanum (2005), Khare (2004), Parrotta (2001), Sharma (2003), Singh (1982), Singh & Pandey (1998).

Aisandra butyracea (Roxb.) Baehni.

Family: Sapotaceae.

Hindi and Regional Names:

Hindi	:	Phalwara;
Kum.	:	Bhelel.

Part Used : Leaves.

Utilization : Leaves serve as good fodder to the cattle.

Reference : Raju (1998).

Alangium chinense (Lour.) Harms.

Syn: *Alangium begoniaefolium* (Roxb.) Baill.; *Marlea bigoniifolia* Roxb.

Family: Alangiaceae.

Regional Names:

Ass.	:	Bhelu, Marli;
Beng.	:	Bonipodo, Marlia, Stolpodo;

Kash. : Prot;
Kum. : Garh, Kimu, Tumri;
Pub. : Budonar, Padlu.

Distribution : Throughout N. India, Bengal, Sikkim, Bhutan & Assam upto 2,700 m.

Part Used : Leaves.

Utilization : Leaves lopped for fodder.

Active Constituents : Leaves rich in Beta- amyrin acetate, tricontanol and beta- sitosterol.

References : Ambasta (1986), Raju (1998), Rastogi & Mehrotra (1995a).

Alangium salvifolium (Linn. f.) Wangerin.

Syn: *Alangium begonifolium* Baill.; *A. lamarckii* Thawites.; *Grewia salvifolium* Linn. f.

Family: Alangiaceae.

English, Hindi, Sanskrit and Regional Names:

Eng. : Sage- Leaved Alangium;
Hindi : Akola, Ankole;
Sans. : Ankola, Ankota, Ankotah;
Beng : Akarkanta, Ankhura;
Guj. : Ankole, Ankola, Ankoli, Onkola;
Kan. : Ankola, Ankole, Ankulemara;
Kum. : Kuilu;
Mal. : Ankolam, Arinjil, Azhinni, Kimiri;
Ori. : Ankula, Ankulo, Dolanku, Kimri;
Tam. : Adigolam, Alangi, Alinjil, Karrikolam;
Tel. : Ankalamu, Chettu, Udagu, Urgu.

Distribution : Distributed Throughout the Hills between 1,000-2,000 m.

Part Used : Leaves.

Utilization : Leaves consider as good fodder for cattle.

Active Constituents : Leaves contain ankorine and choline chloride.

References : Purohit & Samant (1995), Rastogi & Mehrotra (1990).

Albizia julibrissin Durazz.

Syn: *Acacia mollis* Wall.; *A. nemo* Willd.; *A. mollis* Boiv.; *Mimosa arborea* Thunb.; *M. kalkora* Roxb.

Family: Fabaceae.

Vern. Names: Kurmur, Lal Sirin.

English, Hindi and Regional Names:

Eng.	:	Pink Siris, Silk Tree;
Hindi	:	Barau, Kurmura, Lal siris;
Beng	:	Kankora;
Ori.	:	Tentuliya;
Pub.	:	Brind, Mathrishi, Surangru;
Tam.	:	Selaivegai;
Tel.	:	Nallasinduga.

Distribution	:	Himalaya. Sikkim.
Part Used	:	Leaves.
Utilization	:	Leaves considered as an excellent fodder.
References	:	Hooker (1879), Roy *et al.* (1998).

Albizia lebbeck Benth.

Syn: *Acacia lebbeck* Willd.; *Acacia sirisa* Ham.; *Albizia latifolia* Boivin.; *Mimosa lebbeck* L.; *M. sirisa* Roxb.; *M. speciosa* Jacq.

Family: Fabaceae.

English, Hindi, Sanskrit and Regional Names:

Eng.	:	Parrot tree, Siris tree, Sizzling tree;
Hindi	:	Garso, Kalsir, Lasrin;
Sans.	:	Bhandi, Kapitaka, Kapitana. Plavaga, Shankiniphala;
Beng	:	Siris, Sirisha;
Guj.	:	Kalisavas, Kalokansakio;
Kan.	:	Baggera;
Mal.	:	Kattuvaka, Nenmai, Vaka;
Mar.	:	Chichola, Kalashiras, Mothasiras, Shirish;
Ori.	:	Kotvaghe, Vaghe;
Pub.	:	Sarin, Shirin;
Tam.	:	Adukkavajai, Kalindi, Kattuvagai, Pandil, Piram;
Tel.	:	Dirisanamu, Girisam, Siriso.

Distribution : Tropical Himalaya, ascending to 1,900 m, C. Provinces & Khasia upto 1,300 m.

Part Used : Leaves.

Utilization : Leaves form an excellent fodder for camels and cattle in Rajasthan.

Active Constituents : Leaves rich in triterpene, caffeic acid, alkaloides and quercetin.

References : Anonymous (1948), Dwivedi (1993), Hooker (1879), Khare (2004), Oommachan (1977), Parrotta (2001), Rastogi & Mehrotra (1990), Shani (1998), Singh (1982, 2002), Talbot (1976b).

Albizia odoratissima Benth.

Syn: *Acacia lomatocarpus* DC.; *A. micrantha* Boix.; *A. odoratissima* Willd.; *Mimosa lebekkoides* Benth.; *M. marginata* Lam.; *M. odoratissima* Linn.

Family: Fabaceae.

Vern. Name: Siris.

English, Hindi, Sanskrit and Regional Names:

Eng.	:	Black Siris, Ceylon Rosewood;
Hindi	:	Kala siris;
Sans.	:	Shirisha, Sirisa, Sirisha;
Guj.	:	Kalo sarasi;
Kan.	:	Bilvara;
Mal.	:	Pulivaga;
Mar.	:	Chikunda;
Ori.	:	Bonosiris, Sirisi, Tinya;
Pub.	:	Karmaru, Karmbru, Lasrin, Polash;
Tam.	:	Kasuvagei;
Tel.	:	Chinduga.

Distribution : Sub-Himalayan Hills upto 1,500 m. Peninsular India to an elevation upto 900 m.

Parts Used : Leaves. Twigs. Stem.

Utilization : Leaves and twigs lopped for fodder. Stem fibre used for making ropes.

References : Agarwal (2003), Ambasta (1986), Dastur (1964), Hooker (1879), Parrotta (2001), Singh (1982).

Albizia procera (Roxb.) Benth.

Syn : *Acacia procera* (Roxb.) Benth .; *Mimosa procera* (Roxb.) Benth.

Family : Fabaceae.

English, Sanskrit and Regional Names:

Eng. : Safed Siris, White Siris;
Sans. : Kinihi, Sirisa;
Guj. : Karanga;
Mal. : Chalavaka, Kotta Vega;
Mar. : Kinhai;
Ori. : Garso, Sarapatri, Siris;
Tam. : Konda vagei, Valagai;
Tel. : Chegara, Telia.

Distribution : Distributed upto 900 m in the Sub- Himalayan Hill Forests from Yamuna Eastwards to W. Bengal and in Satpura Range between Maharashtra Southwards to Kerela & Tamil Nadu.

Part Used : Aerial Plant Parts.

Utilization : Used as fodder.

References : Agarwal (2003), Ambasta (1986), Dwivedi (1993), Parrotta (2001), Singh (1982).

Albizia stipulata Boiv.

Syn: *Acacia stipulata* DC.; *M. chinensis* Osbeck.; *Mimosa stipulata* Rox.

Family: Fabaceae.

Vern. Name: Oyaee.

Hindi and Regional Names:

Hindi : Sirin;
Ass. : Sau, Sow;
Beng : Chakua;
Bom. : Udala;
Kan. : Hotta bage, Kal -boge;
Mal. : Potta vaga;

Mar.	:	Laeli;
Ori.	:	Reyi;
Pub.	:	Kasir, Oi, Sirin;
Tam.	:	Pili vagei.

Distribution : Sub-Himalayan Tract to Moist Regions of Andaman, Bengal & Assam.

Part Used : Leaves. Twigs.

Utilization : Leaves and twigs used as fodder.

References : Agarwal (2003), Ambasta (1986), Anonymous (1948), Hooker (1879), Kirtikar & Basu (1984), Sharma & Rana (2005), Sood & Thakur (2004).

Alhagi pseudalhagi (Bieb) Desv.

Syn : *Alhagi camelorum* Fisch.; *A. maurorum* Desv. *ex* Baker.; *Hedysarum pseudalhagi* Bieb.

Family: Fabaceae.

English, Hindi, Sanskrit and Regional Names:

Eng.	:	Camel Thorn;
Hindi	:	Bharbarra, Jawasa;
Sans.	:	Bhatuashak, Daralambha;
Guj.	:	Javasa, Jawaso;
Kan.	:	Billiduruva, Durlava;
Mal.	:	Kothithuva, Valiya;
Mar.	:	Jawaso;
Tam.	:	Kanchori, Punaikanjuri;
Tel.	:	Dhanvayasamu, Girikarmika.

Distribution : Drier Regions of Gujarat, Punjab, U. P., Rajasthan & Bihar.

Parts Used : Leaves.Twigs.

Utilization : Leaves and twigs used as fodder.

References : Agarwal (2003), Ambasta (1986), Dwivedi (1993), Sood *et al.* (2005).

Allmania nudiflora R. Br.

Family: Amarantaceae.

Distribution : Sikkim.

Part Used : Herb.

Utilization : Herb eaten by cattle.

References : Islam (2002), Roy *et al.* (1998).

Alnus nepalensis Don.

Syn: *Betula leptostachya* Wall.; *Clethropsis nepalensis* Spach.
Family: Betulaceae.

Vern. Name: Kharaen.

English, Hindi and Regional Names:

Eng. : Indian alder;
Hindi : Utis;
Pub. : Kohi.

Distribution : Throughout Himalaya from Chamba to the Mishmi Hills in Upper Assam.

Part Used : Leaves.

Utilization : Leaves fed to all ruminants during winter.

References : Agarwal (2003), Ambasta (1986), Anonymous (1948), Hooker (1890), Purohit & Samant (1995), Sharma (2004).

Alnus nitida Endl.

Syn: *Clethropsis nitida* Spach.
Family: Betulaceae.

Vern. Name: Bhuj.

Hindi and Regional Names:

Hindi : Kunis;
Pub. : Sharal.

Distribution : Kashmir & Himalaya: 1,000 - 3,000 m.

Part Used : Twigs.

Utilization : Twigs yield fibre for clothing among tribals.

References : Agarwal (2003), Hooker (1890).

Aloe barbadensis Mill.

Syn: *Aloe abyssinica* Lam.; *A. chinensis* Baker.; *A. indica* Royle.; *A. perfoliata* L.; *A. vera* Thorn. *ex* Linn.; *A. vulgaris* Lam.
Family: Liliaceae.

English, Hindi, Sanskrit and Regional Names:

Eng.	:	Barbados Tree; Curacao Aloe, Indian Aloe;
Hindi	:	Ghee Kunwar;
Sans.	:	Kanya, Ghrita - Kumari;
Beng	:	Ghrita- kumari;
Guj.	:	Kumarpathu;
Mal.	:	Kumari;
Mar.	:	Korkand, Korphad;
Ori.	:	Ghikuanri, Kumari, Musaboro;
Pub.	:	Elva;
Tam.	:	Chirukattali, Kattalai;
Tel.	:	Chinakata banda.

Distribution : Cultivated or Wild in the Drier Parts of India.

Part Used : Leaves.

Utilization : Leaves yield fibre.

References : Ambasta (1986), Bhattacharjee (1998), Khare (2004), Prajapati *et al.* (2003).

Alopecurus geniculatus Linn.

Family: Poaceae.

English Name:

Eng. : Marsh Fox Tail.

Distribution : Sub- Tropical Himalaya.

Part Used : Herb.

Utilization : Plant considered a good forage grass.

References : Agarwal (2003), Ambasta (1986), Sharma & Rana (2005).

Alpina allughas Rosc. (Pl. 1B)

Family: Zingiberaceae.

Distribution : Grows in S. India

Parts Used : Leaf. Stem.

Utilization : Stem fibre used for mats and mattresses. Leaf-sheath good for making temporary ropes and cordage.

References : Agarwal (2003), Islam (2002).

Alstonia scholaris R. Br.

Syn: *Echites scholaris* Linn.

Family: Apocynaceae.

Vern. Name: Ditta-bark.

English, Hindi, Sanskrit and Regional Names:

Eng. : Dita bark, Devil's Tree;
Hindi : Chatian, Satiun;
Sans. : Deva-Vriksha, Saptaahva, Saptaparna;
Ass. : Chatian, Satiana;
Beng : Chatittin;
Bom. : Satvin;
Kan. : Chatiun;
Mal. : Pala;
Mar. : Satvin;
Ori. : Chhotina, Kumbaro, Soptorposi;
Tam. : Aclilampali, Pala;
Tel. : Aedakularite-chettu, Palgaruda.

Distribution : Throughtout Moist Regions of India & Southwards to Sri Lanka.

Part Used : Bark.

Utilization : Bark fibre used for paper making.

References : Ambasta (1986), Anonymous (1948), Hooker (1882), Jain (1968), Khan & Khanum (2005), Kirtikar & Basu (1984).

Alternanthera philoxeroides (Mart.) Griseb.

Family: Amarantaceae.

Distribution : Throughout Plains of India.

Part Used : Whole Plant.

Utilization : Commonly used for fodder purposes.

Active Constituents : Glycosides a new flavone present in leaves.

Reference : Islam (2002).

Althaea officinalis Linn.

Family: Malvaceae.

English, Hindi and Regional Names:

Eng. : Marsh-mallow;
Hindi : Gulkheria;
Kan. : Seemetuti;
Mar. : Gulkhair;
Tam. : Simaituti.

Distribution : Kashmir. Punjab.

Parts Used : Stem. Roots.

Utilization : Fibre from stems and roots used for manufacturing paper.

References : Agarwal (2003), Ambasta (1986), Prajapati *et al.* (2003).

Alysicarpus longifolius Wight &Arn.

Syn: *Alysicarpus spicatum* Royl.; *A. vagnelis* Wall.; *Hedysarum bracteatum* Herb.Madr.; *H. longifolium* Rottl.

Family: Fabaceae.

Distribution : Scattered Throughout India.

Parts Used : Leaves. Shoots.

Utilization : Considered a good fodder plant in both green and dried state.

Active Constituents : Leaves rich in myricyl alcohol, β-sitosterol and its acetate, rutin, pinitol and chrysoeriol.

References : Ambasta (1986), Dwivedi (1993), Hooker (1879), Rastogi & Mehrotra (1993, 1995a).

Amaranthus viridis Linn.

Syn: *Amaranthus gracilis* Desf.; *A. fasciatus* Roxb.; *A. polystachus* Ham. in Wall.
Family: Amarantaceae.

Vern. Name: Cholayi.

English, Hindi and Regional Names:

Eng.	:	Chinese Spinach;
Hindi	:	Tandulaya;
Beng	:	Bannote;
Guj.	:	Dhinmado, Dimdo;
Kan.	:	Chilikirae suppy, Daglisoppo;
Mal.	:	Bayam munyit;
Mar.	:	Lhanamat;
Tam.	:	Kuppai kerai, Sinna kerrai, Siru keerai;
Tel.	:	Chalaika thotakura.

Distribution	:	Throughout India, all Tropic & Warm Regions.
Part Used	:	Whole Plant.
Utilization	:	Plant used as cattle fodder.
Active Constituents	:	Total fatty acid from leaves contain palmitic acid (77.8%).
References	:	Ambasta (1986), Bennet *et al.* (1991), Khan & Khanum (2005), Hooker (1885), Rastogi & Mehrotra (1995b), Sood &Thakur (2004).

Ammania baccifera L. (Pl. 1C)

Syn: *Ammania indica* Lamk.; *A. delibis* Ait.; *A. vasicatoria* Roxb.; *A. verticillata* Boiss.; *Hapalocarpum vesicatorium & H. indicum* Miq.
Family: Lythraceae.

English, Hindi, Sanskrit and Regional Names:

Eng.	:	Blistring Ammania;
Hindi	:	Dadmari;
Sans.	:	Agnivarga, Kurandika, Vikata;
Beng	:	Banmarach;
Guj.	:	Jalaagiy, Jarargiyo;

Kan. : Kallurive;
Mal. : Kalluravanchi;
Mar. : Bhar - jamabhuli;
Pub. : Dederbuti;
Tam. : Kallarivi.

Distribution : Throughout Tropical India.

Part Used : Leaves.

Utilization : Used for fodder purposes.

Active Constituents : Leaves rich in vitamin-C, glucose, hentriacontane, dotriacontanol and β-sitosterol.

References : Hooker (1879), Islam (2002), Kirtikar & Basu (1984), Parrotta (2001), Prajapati *et al.* (2003), Rastogi & Mehrotra (1995a).

Ammannia multiflora Roxb.

Family: Lythraceae.

Distribution : Throughout Plains of India.

Part Used : Plant.

Utilization : Commonly used for fodder.

Active Constituents : Rhamnetin and its 3-ramnosyl glucoside are the chief constituents of leaves.

References : Hooker (1879), Islam (2002), Rastogi & Mehrotra (1993).

Ampelocissus rugosa (Wall.) Planch.

Syn: *Vitis lanata* Roxb. var. *rugosa* M.Laws, *Vitis rugosa* Wall.

Family: Vitaceae.

Distribution : Common between 1,800- 2,500 m.

Part Used : Plant.

Utilization : It is considered a very nutritive fodder for all classes of livestock and reported to enhance lactation.

References : Ambasta (1986), Purohit & Samant (1995).

Anagallis arvensis Linn.

Family: Primulaceae.

English, Hindi and Regional Names:

Eng.	:	Birds Eye, Birds Tongue, Female Pimpernel;
Hindi	:	Jhonkhmari, Krishna- neel;
Beng	:	Anaras;
Guj.	:	Anagallide, Golophudi;
Pub.	:	Dhabber.

Distribution : Throughout India ascending to 2,500 m in the Himalaya.

Part Used : Leaves.

Utilization : Used as fodder for cattle.

Active Constituents : Leaves rich in α-spinasterol, avenasterol-7-ene, β-isosterol, stigmasterol, palmitic acid, steracic acid, oleic and linoleic acid.

References : Khanna (2002), Kirtikar & Basu (1984), Rastogi & Mehrotra (1995b).

Ananas comosus (Linn.) Merrill.

Syn: *Ananas sativa* Schult. f.

Family: Bromeliaceae.

English, Hindi, Sanskrit and Regional Names:

Eng.	:	Pineapple;
Hindi	:	Ananas;
Sans.	:	Anamnasam, Bhunetra, Paravati;
Beng	:	Anaras;
Guj.	:	Ananas;
Kan.	:	Anasu;
Mal.	:	Kazhudhachakka;
Mar.	:	Ananas;
Tam.	:	Parangithalli;
Tel.	:	Ananash.

Distribution : Cultivated Throughout Coastal Region of Bengal, Orissa, Assam, Himalayan Tarai and along the W. Coast of Kerala.

Part Used : Leaves. Pressed Peels. Cores.

Utilization : Pressed peels and cores are fed to livestock. Leaf fibre good for rope making

Active Constituents : Leaves possess vit. 63 mg/100gm and ergosterol peroxide.

References : Agarwal (2003), Ambasta (1986), Anonymous (1948), Das & Agarwal (1991), Dastur (1964), Drury (1978), Kirtikar & Basu (1984), Rastogi & Mehrotra (1991).

Annona reticulata L.

Family: Annonaceae.

English, Hindi, Sanskrit and Regional Names:

Eng. : Bullock's Heart;
Hindi : Anta, Luvani;
Sans. : Ramphal, Vasanta;
Beng : Nona;
Guj. : Ramphal;
Kan. : Ramphal;
Mal. : Manilanilan, Ramachchita;
Mar. : Ramphal;
Ori. : Neua, Ramopholo;
Tam. : Ramasitapazham;
Tel. : Ramaseetapandu.

Distribution : Naturalized in Bengal & S. India.

Part Used : Bark.

Utilization : Used as fibre.

References : Agarwal (2003), Anonymous (1948), Kiritikar & Basu (1935).

Annona squamosa L.

Syn: *Annona asiatica* L.
Family: Annonaceae.

English, Hindi, Sanskrit and Regional Names:

Eng. : Custard Apple, Sweet Sop;
Hindi : Sharifa;

Sans.	:	Gandagatra, Krishnabija, Sitaphal, Subha;
Beng	:	Ata;
Guj.	:	Anan, Anuram, Anuri, Sitaphal;
Mal.	:	Attichakka, Sirpha;
Mar.	:	Sitaphal;
Ori.	:	Ato, Sitaphalo;
Pub.	:	Sharifa;
Tam.	:	Atta, Sita, Sita-palam.

Distribution : Cultivated Throughtout India; Naturalised especially in the W. Peninsula.

Part Used : Young Twigs.

Utilization : Young twigs yield an inferior quality of fibre which is used for making brushes and cordages by 'Bhil' and 'Saharia' tribes of Kota district.

References : Anonymous (1948), Singh & Pandey (1998), Sood *et al.* (2005).

Anodendron paniculatum A. DC.

Syn: *Echites coriacea* Wall.; *E. paniculata* Roxb.

Family: Apocynaceae.

English, Hindi and Regional Names:

Eng.	:	Creat;
Hindi	:	Kalmegh;
Kan.	:	Maniballi;
Mar.	:	Kavali.

Distribution : Indo-Malayan Regions from Bengal, Mianmar to Kerala.

Part Used : Aerial Plant Parts.

Utilization : Fibre used for cordage and bow strings.

References : Agarwal (2003), Ambasta (1986), Hooker (1882), Jain (1991), Talbot (1976a).

Anogeissus latifolia Wall. *ex* Bedd.

Syn: *Conocarpus latifolia* Roxb. *ex* DC.
Family: Combretaceae.

English, Hindi, Sanskrit and Regional Names:

Eng.	:	Axle-Wood, Button tree, Dindga tree;
Hindi	:	Dhawa, Dhaura;
Sans.	:	Dhava, Dhavah;
Beng	:	Chakwa, Dhaoyo;
Guj.	:	Dabria, Dhendo;
Mal.	:	Malakanniram, Vellanava;
Mar.	:	Dhaura;
Ori.	:	Dhobu, Dohu, Villaynaga;
Tam.	:	Namai, Namme, Vekkali;
Tel.	:	Chirimanu, Yella maddi.

Distribution : Distributed upto 1,200 m; Commonly in the C. & S. India.

Part Used : Leaves.

Utilization : Used as fodder.

References : Agarwal (2003), Ambasta (1986), Chatterjee & Pakrashi (1997), Dwivedi (1993), Prajapati *et al.* (2003), Parrotta (2001), Singh (1982).

Anogeissus pendula Edgew.

Family: Combretaceae.

Hindi Names :

Hindi : Dhokra, Kardhai.

Distribution : Dry Hot Region between Aravalli Hills in Rajasthan & Bundelkhand in U.P.

Part Used : Leaves.

Utilization : Leaves considered as good fodder.

Active Constituents : Leaves contain 7.60% crude proteins, 19.0% crude fibre, 8.10% ash, 0.10% P and 3.50% Ca.

References : Dwivedi (1993), Singh (1982).

Anthocephalus chinensis (Lamk.) A. Rich. *ex* Walp.

Family: Rubiaceae.

Hindi and Regional Names:

Hindi : Kadam;
Ass. : Roghu;
Kan. : Kadwal;
Tam. : Velli kadambu;
Tel. : Kadambu.

Distribution : Sub-Himalayan Tract from Nepal Eastwards; W. Bengal, Assam, Chotta Nagpur & W. Coast from N. Kanara Southwards to Travancore.

Part Used : Whole Plant Parts.

Utilization : Tree lopped for fodder in Assam.

Reference : Singh (1982).

Antiaris toxicaria Lesch.

Syn: *Antiaris dubia* Spanog.; *A. innoxia* Bl.; *A. saccidora* Dalz.; *Lepuranda saccidora* Nimmo in Grah.

Family: Moraceae.

English, Hindi, Sanskrit and Regional Names:

Eng. : Sacking tree, Uphas tree;
Hindi : Jungali Lakuch;
Sans. : Valkala;
Bom. : Chandla, Chandul, Karvat;
Mal. : Nettavil;
Mar. : Karvat;
Ori. : Bhoonromari;
Tam. : Nittavil.

Distribution : Throughout Hotter Parts of India; on the W. Ghats from the Concan Southwards.

Parts Used : Leaves. Bark.

Utilization : Leaves used as fodder for elephants in the Andamans. Bark yields fibre for cordage, matting, clothing and paper making. Inner bark yields a good fibre for making sacks and hence the popular name 'Sack Tree'.

References : Agrawal (2003), Ambasta (1986), Anonymous (1948), Dastur (1964), Drury (1978), Hooker (1890), Jain (1991), Kirtikar & Basu (1984), Mittre (1997), Sahni (1998), Watt (1972).

Apluda mutica Linn.

Syn: *Apulda aristata* Linn.; *A. mutica* Linn.; *A. varia* Hack.
Family: Poaceae.

Hindi and Regional Names:

Hindi : Ponai;
Beng : Goroma;
Ori. : Goroma;
Tam. : Manda pillu, Mungil pillu.

Part Used : Plant Parts.

Utilization : Young grass eaten by buffaloes.

References : Ambasta (1986), Dwivedi (1993), Rao & Henry (1996), Sharma & Rana (2005), Varghese (1996).

Aquilaria agallocha Roxb.

Syn: *Aquilaria malaccensis* Lamk.; *Gyrinopsis assamica* Benth.
Family: Thymelaeaceae.

English, Hindi, Sanskrit and Regional Names:

Eng. : Aloe Wood, Eagle Wood;
Hindi : Agar;
Sans. : Agaru, Krsnaguruh;
Ass. : Sasi;
Kan. : Krishnagaru;
Tam. : Agar agalicandanam, Krsanguru;
Tel. : Krsangaru.

Distribution : Eastern India, Assam, Hills of Meghalaya, Nagaland, Tripura & Manipur.

Part Used : Bark.

Utilization : Bark fibre used for ropes and home made paper in Assam.

References : Agarwal (2003), Ambasta (1986), Anonymous (1948), Hooker (1890), Prajapati *et al.* (2003).

Artemisia sieversiana Ehrh. *ex* Willd.

Family: Asteraceae.

Hindi, Sanskrit and Regional Names:

Hindi : Dona;
Sans. : Agnidamnaka, Bahukantaka, Damni, Damana;
Bom. : Downa;
Guj. : Damro;
Mar. : Davana, Ranadavana.

Distribution : W. Himalaya: Kashmir to Lahul 2,000- 3,000 m.

Part Used : Plant.

Utilization : Fresh plant fed to pashmina goats as it is claimed to increase pashmina wool production.

Active Constituents : Plant contains 15.5 % proteins and 5.12 % fats.

References : Agarwal (2003), Anonymous (1948), Chaurasia *et al.* (2000), Dwivedi (1993), Hooker (1882), Kirtikar & Basu (1984).

Artocarpus artilis (Park.) Fosberg.

Syn: *Artocarpus communis* Forst.; *A. incisa* Linn. f.

Family: Moraceae.

English and Regional Names:

Eng. : Bread fruits;
Mar. : Vilayati phanas;
Tam. : Seema pila;
Tel. : Seema panasa.

Distribution : Widely Cultivated & Naturalized from sea level to Lower Mountains.

Part Used : Bark.

Utilization : Bark fibre used for ropes.

Reference : Ambasta (1986).

Artocarpus chama Buch.-Ham.

Syn: *Artocarpus chaplasha* Roxb.

Family: Moraceae.

English and Regional Names:

Eng. : Chaplash;

Ass. : Sam;

Beng : Chaplash.

Distribution : N. India.

Parts Used : Bark. Leaves.

Utilization : Bark fibre for making cordage, blankets and coarse bedding. Leaves considered as fodder.

References : Rao & Shanpru (1997), Saklani & Jain (1994), Singh (1982).

Artocarpus heterophyllus Lam.

Syn: *Artocarpus integra* (Thunb.); *A. integrifolia* L.

Family: Moraceae.

English, Hindi, Sanskrit and Regional Names:

Eng. : Jack Fruit Tree;

Hindi : Chakki, Kanthal, Kathar;

Sans. : Asyaha, Panasa;

Beng : Kathal, Kanthal;

Guj. : Phanas, Vanas;

Kan. : Alasa, Halasu;

Mal. : Chakka, Kandakiphala, Pila, Pilavu;

Mar. : Phanas;

Ori. : Kantokalo, Panasa, Panoso;

Tam. : Pala, Pila, Pila pilam;

Tel. : Panasa, Verupanasa.

Distribution : Cultivated Throughout the Hotter Parts of India; Occasionally Wild in S. Garo Hills in Meghalaya.

Parts Used : Leaves. Fruits.

Utilization : Leaves and fruits considered as fodder for cattle and elephants.

Active Constituents : Leaves yield cycloartenone, β-sitosterol, acetylcholine, 11-14% proteins, 18-22% crude fibre, 0.52%-2.15% Ca and 0.11-0.33% P.

References : Chatterjee & Pakrashi (1997), Chopra *et al.* (1969), Diwakar & Ansari (2003), Dwivedi (1993), Drury (1978), Parrotta (2001), Prajapati *et al.* (2003), Sahni (1998), Saklani & Jain (1994), Singh (1982), Varghese (1996).

Artocarpus hirsuta Lamk.

Syn: *A. hirsuta* Lam.
Family: Moraceae.

English, Hindi, Sanskrit and Regional Names:

Eng. : Willd Jack;
Hindi : Ramphalas;
Sans. : Lakukah, Panasah;
Kan. : Hebbhalasu;
Mal. : Aini, ayani;
Mar. : Pat -panas;
Tam. : Anjili;
Tel. : Pejata.

Distribution : W. Ghats from N. Kanara to Malabar, Coorg Southwards upto 1,200 m.

Part Used : Bark.

Utilization : Used as fodder.

References : Agarwal (2003), Anonymous (1948), Parrotta (2001).

Artocarpus lacucha Buch.-Ham.

Syn: *Artocarpus lakoocha* Roxb.

Family: Moraceae.

English, Hindi, Sanskrit and Regional Names:

Eng.	:	Monkey jack;
Hindi	:	Barhal;
Sans.	:	Granthiphal, Lacucha, Lakudi, Pitanaasha;
Beng	:	Dephal;
Guj.	:	Lakucha;
Kan.	:	Vatehuli;
Mal.	:	Chimpa, Lakucham;
Mar.	:	Wotomba;
Ori.	:	Jeuto;
Tam.	:	Hagusam, Irappla;
Tel.	:	Kammaregu.

Distribution : Tropical Himalaya. N. E. & S. India, from Kumaon to Assam. Orissa. M .P. Kerala. Tamil Nadu.

Parts Used : Leaf. Bark.

Utilization : Leaves used as fodder and bark a good source of fibre.

References : Khare (2004), Parrotta (2001), Saklani & Jain (1994), Varghese (1996).

Arundinaria callosa Nakai.

Family: Poaceae.

Distribution : Himalaya & N. E. India.

Parts Used : Stem Bark.

Utilization : Stem bark used for ropes.

References : Hooker (1897), Jain (1991), Saklani & Jain (1994).

Arundinaria falcata Nees.

Syn: *A. utilis* Clegh.

Family: Poaceae.

Vern. Names: Banji, Nagral.

English Name:

Eng. : Himalayan bamboo.

Distribution : W. Himalaya: Ravi to Nepal 1,000 – 4,000 m.

Part Used : Leaves.

Utilization : Leaves fed to livestock during winter.

References : Hooker (1897), Purohit & Samant (1995), Raju (1991), Sharma (2004), Sharma & Rana (2005).

Arundinella nepalensis Trin.

Syn: *A. brassiliensis* Hook. f. in part, non Raddi.

Family: Poaceae.

Regional Name:

Bom. : Dundi.

Distribution : Distributed in the W. Himalaya between 1,000-2,000 m.

Utilization : Eaten by cattle in times of scarcity in W. Himalaya.

References : Ambasta (1986), Duthie (1888).

Asclepias curassavica L.

Family: Asclepiadaceae.

Vern. Names: Ghamera, Parvini.

English, Hindi, Sanskrit and Regional Names:

Eng. : Curassavian Swallow wort;
Hindi : Kokatundi;
Sans. : Kakatundi;
Bom. : Kakatundi, Kuraki;
Kan. : Kakatundi;
Mar. : Karki, Kurki;
Pub. : Kaktundi.

Distribution : Throughout India.

Part Used : Whole Plant.

Utilization : Plant exploited for fibre.

References : Hooker (1885), Parrotta (2001), Singh & Kumar (2000).

Asparagus curillus Buch-Ham. *ex* Roxb.

Syn: *Asparagus nepalensis* Baker.

Family: Asparagaceae.

Regional Name:

Tam. : Jhirni.

Distribution : Himalaya; Garhwal to Nepal (600- 1,200 m).

Utilization : Plants browsed by sheep, goats and other ruminants.

References : Purohit & Samant (1995), Roy *et al.* (1998).

Asparagus racemosus Willd.

Family: Asparagaceae.

English, Hindi, Sanskrit and Regional Names:

Eng. : Willd Asparagus;
Hindi : Shatamuli;
Sans. : Satamuli, Satavari, Svetmuli;
Beng : Satamuli;
Guj. : Satawari;
Mal. : Shatavali;
Mar. : Satawarmul;
Ori. : Chotaru, Mohajolo, Satabari, Satamuli, Sotabori;
Pub. : Bozidon, Satawar;
Tam. : Ammai- kodi, Inli- chedi, Nili – chedi;
Tel. : Chalagadu, Pili- tegalu.

Distribution : Throughout Sub-Tropical and Tropical Regions of India upto 1,500 m.

Utilization : Browsed by sheep and goats.

Active Constituents : Flavonoides, rutin and quercetin -3- glucuronide present in leaves.

References : Agarwal (2003), Parrotta (2001), Purohit & Samant (1995), Rastogi & Mehrotra (1990), Thakur *et al.* (1989).

Astilbe rivularis Buch-Ham.

Syn: *Spiraea barbata* Wall.
Family: Saxifragaceae.

Hindi Name:

Hindi : Buriakhati.

Distribution : Temperate Himalaya. Khasia Hills. Meghalaya: 1,200-1,800 m.

Parts Used : Leaves. Aerial Parts.

Utilization : Leaves and aerial parts used as fodder.

Active Constituents : Leaves contain β- amyrin, astilbic acid, quercetin acid, bergenin and β- sitosterol.

References : Hooker (1879), Rastogi & Mehrotra (1995a), Saklani & Jain (1994).

Astragalus chorostachys Lindl.

Family: Fabaceae.

Distribution : Upto 3,000- 3,600 m.

Part Used : Plant.

Utilization : Used as a fodder plant in W. Himalaya.

Reference : Purohit & Samant (1995).

Astragalus grahamianus Royle *ex* Benth.

Family: Fabaceae.

Distribution : Temperate & Alpine regions. Punjab.

Part Used : Roots.

Utilization : Roots especially in winter are dug and used as fodder.

Reference : Sood *et al.* (2001).

Astragalus marschallianus Fisch.

Family: Fabaceae.

Distribution : Temperate & Alpine Region.

Part Used : Roots.

Utilization : Roots used as nutritious fodder for livestock.

Reference : Sood *et al.* (2001).

Astragalus rhizanthus Royle *ex* Benth.

Family: Fabaceae.

Distribution : Temperate & Alpine Regions upto 3,800 m.

Part Used : Roots.

Utilization : Roots used as fodder for 'Churu'.

References : Hooker (1879), Singh *et al.* (1996), Sood *et al.* (2001).

Asystasia gangetica T. Anders.

Syn: *Asystasia coromandeliana* White *ex* Nees.;
A. violacea Dalz. non C. B. Clarke.

Family: Acanthaceae.

Hindi, Sanskrit and Regional Names:

Hindi : Lavanavalli;
Sans. : Lavanavalli
Mal. : Upputhalli;
Tam. : Medday Keerai.

Distribution : Grows Throughout S. India.

Part Used : Leaves.

Utilization : Leaves used as fodder.

Active Constituent : Thiamine.

References : Agarwal (2003), Ambasta (1986), Anonymous (1948).

Atriplex halimus Linn. (Pl. 1 D)

Family: Chenopodiaceae.

English Names:

Eng. : Mediteranean Saltbush, Sea- Orache.

Part Used : Leaves.

Utilization : Plant considered of high fodder value, suitable for calves and milch cow.

Reference : Ambasta (1986).

Atylosia scarabaeoides (L.) Benth.

Syn: *Cartharospermum paucifloram* W. A.; *Dolichos scarabaeoides* L.

Family: Fabaceae.

Hindi and Regional Names:

Hindi : Walkolu;
Ori. : Bankulthia.

Distribution : Upto 1,800 m in the Himalaya & to Lower elevation in the W. Ghats.

Plant Used : Plant.

Utilization : Used as fodder.

References : Agarwal (2003), Ambasta (1986), Dwivedi (1993), Parrotta (2001).

Avena byzantina C. Koch.

Syn: *Avena sterilis* var. *culta* Linn.

Family: Poaceae.

English, Hindi and Regional Names:

Eng. : Indian Oat;
Hindi : Javi;
Guj. : Jau;
Mar. : Jao;
Tel. : Yavalu.

Distribution : Cultivated in Tropical & Sub-Tropical Regions of India, Found in Bengal to the Indus, Meerut, U.P. Hissar, Karnal district.

Parts Used : Seeds. Whole Plant.

Utilization : Used mainly as fodder for horses and cattle. Seed and 'bhusa' considered an important food for cattle.

References : Ambasta (1986), Anonymous (1948), Sharma & Rana (2005).

Avena sativa Linn.

Family: Poaceae.

Vern. Name: Oat.

English, Hindi, Sanskrit and Regional Names:

Eng.	:	The Oat;
Hindi	:	Jai, Javi;
Sans.	:	Ganer, Kulijud;
Pub.	:	Javi.

Distribution	:	Cultivated in N. India from Bengal to The Indus; and in the Himalaya upto 3,700 m.
Part Used	:	Leaves.
Utilization	:	Considered a good fodder for sheep and cattle.
Active Constituents	:	Fats, proteins 87%, 26 flavonoides of apigenin, luteolin and tricin type isolated from the leaves.
References	:	Anonymous (1948), Caius (2003), Das & Agarwal (1991), Hooker (1879), Karnick (1994), Rastogi & Mehrotra (1991), Sharma & Rana (2005).

Avicennia officinalis Linn.

Syn: *Avicennia resinifera* Forst.; *A. tomantosa* Jack.; *Halodendron thouarsii* Roem.

Family: Verbenaceae.

Vern. Name: Bina.

English, Hindi, Sanskrit and Regional Names:

Eng.	:	White Mangrove;
Hindi	:	Bina;
Sans.	:	Tuvara, Tuvaraka;
Beng	:	Bina;
Bom.	:	Tivar;
Guj.	:	Tavariyan;
Mar.	:	Tivar;
Tam.	:	Kandal, Uppatum, Uppukkutti;
Tel.	:	Mada.

Distribution : Salt Marshes. Tidal Forest of India. Andaman Islands.

Part Used : Leaves.

Utilization : Leaves used as fodder.

Active Constituents : Leaves mainly contain amyrin, taraxerol, taraxerone, betulin, betulinic acid and triacontanol.

References : Agarwal (2003), Ambasta (1986), Anonymous (1948), Das *et al.* (1996), Hooker (1885), Kothari & Rao (2000), Rastogi & Mehrotra (1991, 1993).

Azadirachta indica A. Juss.

Syn: *Antelaea indica* (L.) Adelb.; *Melia azadirachta* L.; *M. indica* (A. Juss.) Brandis.

Family: Meliaceae.

English, Hindi, Sanskrit and Regional Names:

Eng. : Indian Lilac, Margosa, Neem Tree;

Hindi : Nim, Nimb;

Sans. : Aristha, Nimba, Nimbaqh;

Beng : Nim

Guj. : Danujada, Limbado, Limbra;

Mal. : Aryaveppu, Veppa, Ve4ppu;

Mar. : Kadukhajur;

Ori. : Kakapholo, Limbo, Nimbo, Nimo;

Tam. : Vembu, Vepa, Veppam;

Tel. : Vepa, Yapa, Yepa.

Distribution : Shivalik Hills in N. India. Dry Forest of A.P. Tamil Nadu, Karnataka upto 700 m.

Parts Used : Bark. Leaves.

Utilization : Bark yields a fibre for making ropes and mats. Leaves lopped for fodder for cattle and goats.

References : Dastur (1964), Dwivedi (1993), Mittre (1981, 1997), Parrotta (2001), Singh (1982).

Azana lampas (Cav.) Alef.

Syn: *Thespesia lampas* (Cav.) Dalz. & Gibes.; *T. macrophylla* Blume.
Family: Malvaceae.

Hindi and Regional Names:

Hindi	:	Bankapas, Jangali bhendi;
Ass.	:	Bonkopas;
Beng	:	Bankapas;
Guj.	:	Jangali paras- pilla;
Mal.	:	Kattuparati, Katuipuvarasu;
Mar.	:	Renbhendi.

Part Used : Young Stem.

Utilization : Young stem yields a fibre for fabrics, cordage and fishing nets.

References : Ambasta (1986), Jain (1991).

Azolla pinnata R. Br. (Pl. 2A)

Syn: *Salvinia imbricata* Roxb.
Family: Azollaceae.

English Name:

Eng. : Mintue Free- Floating.

Distribution : Throughout India ascending to 900 m.

Part Used : Whole Plant.

Utilization : Used as green forage and poultry feed.

Active Constituents : Rich in proteins, minerals, chlorophyll and carotenoids.

References : Ambasta (1986), Chatterjee & Pakrashi (1997).

Bacopa monniri (L.) Wettsein.

Syn: *Bacopa monnieria* (L.) Wettst.; *Herpestis monmera* Linn.
Family: Scrophulariaceae.

Vern. Name: Jal Brahmi.

English, Hindi, Sanskrit and Regional Names:

Eng.	:	Thyme Leaved Gratiola;
Hindi	:	Brahmi;
Sans.	:	Brahmi, Nira Brahmi, Saraswati;
Ass.	:	Brahmi;
Beng	:	Brihim- sak;
Guj.	:	Jalneveri, Kadaviluni, Neer brahmi;
Mal.	:	Barna;
Mar.	:	Birami, Jalnam, Nirbrahmi;
Ori.	:	Brahmi;
Tam.	:	Nirbrahmi, Sambra ne elaye;
Tel.	:	Sambranicettu.

Distribution : Throughout India upto 1,300 m.

Plant Used : Plant.

Utilization : Plant used as fodder.

Active Constituents : Apigenin-7- glucuronide and luteolin-7- glucuronide isolated from leaves.

References : Anonymous (1948), Islam (2002), Parrotta (2001), Prajapati *et al.* (2003), Thakur *et al.* (1989).

Bambusa arundinacea Willd.

Syn: *Arundo bambos* Linn.; *Bambusa neesiana* Arn. *ex* Munro.; *B. orientalis* Nees.; *B. pungens* Blanco.; *B. spinosa* Roxb.

Family: Poaceae.

English, Hindi, Sanskrit and Regional Names:

Eng.	:	Thorney Bamboo;
Hindi	:	Kantabanse;
Sans.	:	Vanash;
Ass.	:	Kotoha;
Beng	:	Ketua;
Guj.	:	Bans;
Mal.	:	Mungil;
Mar.	:	Bans;
Ori.	:	Bedobaunso;
Tam.	:	Mungil;
Tel.	:	Bonga- veduru.

Distribution : Throughout Low Hills of India upto 2,100 m.

Parts Used : Stem. Leaves.Twigs.

Utilization : Stem yields good quality paper and its fibre used in mat making industry. Leaves and twigs used as fodder.

Active Constituents : Tender shoot comprises of proteins, carbohydrates, calcium, phosphorus, iron, Mg, Na, Cl, vit. C, etc.

References : Agarwal (1986, 2003), Ambasta (1986), Anonymous (1948), Dastur (1964), Hooker (1897), Jain (1991), Khan & Khanum (2005), Kirtikar & Basu (1984), Maya *et al.* (2003), Prajapati *et al.* (2003), Saklani & Jain (1994), Upadhyaya & Chauhan (2000), Varghese (1996).

Bambusa nutans Will. *ex* Munro. (Pl. 2B)

Syn: *Bambusa falconeri* Munro.

Family: Poaceae.

Vern. Name: Ddang.

Regional Name:

Beng : Mahlbans.

Distribution : Commonly Cultivated upto 1,700 m (Jammu to Assam & E. Bengal).

Part Used : Leaves.

Utilization : Leaves used as fodder.

References : Hooker (1897), Sharma (2004), Sharma & Rana (2005).

Bambusa pallida Munro.

Syn: *Bambusa critica* Kurtz in Journ.; *Dendrocalamus criticus* Kurtz.

Family: Poaceae.

Distribution : Arborescent in E. & N. India.

Part Used : Leaves.

Utilization : Leaves used as fodder and also for making baskets and mats.

References : Hooker (1897), Jain (1991), Samati (2004).

Barleria cristata Linn.

Family: Acanthaceae.

Hindi Name:

Hindi : Jinti.

Distribution : Throughout India upto 1,600 m.

Part Used : Plant.

Utilization : Used as fodder.

References : Agarwal (2003), Ambasta (1986).

Bauhina malabarica Roxb.

Syn: *Bauhina tomentosa* Wall.; *Pileostigma malabarica* Benth.

Family: Fabaceae.

English, Hindi, Sanskrit and Regional Names:

Eng. : Mountain Ebmoy;
Hindi : Amli;
Sans. : Amlapatrakah, Asmantaka;
Beng : Karmai;
Guj. : Khatimu, Khatombo, Khatomoi;
Kan. : Cheppura;
Mal. : Arampulli, Kokkavali, Minpuli;
Mar. : Amboti, Amli, Korala;
Ori. : Koteli;
Tam. : Malaiyatti, Mandarai, Vattatti;
Tel. : Pulishinta.

Distribution : Sub- Himalayan Tract. Bengal. Assam. S. India.

Parts Used : Leaves. Bark.

Utilization : Leaves used as fodder. Bark gives cheap fibre for general work.

Active Constituents : Plant contains 2.18% nitrogen, 45.6% carbons and 8.83% ash.

References : Agarwal (2003), Ambasta (1986), Anonymous (1948), Dwivedi (1996), Parrotta (2001), Singh (1982).

Bauhina racemosa Linn. (Pl. 2C)

Syn: *B. parviflora* Vahl.; *B. spicata* Koenig.; *B. timoriensis* Decaisne in Nov.; *Pileostigma racemosa* Benth.

Family: Fabaceae.

English, Hindi, Sanskrit and Regional Names:

Eng.	:	Buddhist Bauhinia, Mountain ebnoy;
Hindi	:	Ashata, Kachnal;
Sans.	:	Svetkanchagna;
Beng	:	Banraj;
Guj.	:	Kosundra, Taur;
Mal.	:	Arambuli, Kotapuli, Mandaram;
Mar.	:	Apta;
Ori.	:	Ambhota, Omborodo;
Tam.	:	Arai;
Tel.	:	Ari.

Distribution : Throughout India upto 1,300 m in the Himalaya.

Parts Used : Bark. Leaves.

Utilization : Bark fibre used for making ropes, fishing-nets and tightening the leather hoops of a musical instrument called 'Dholak'. Leaves regarded as good fodder.

References : Agarwal (2003), Ambasta (1986), Anonymous (1948), Hooker (1897), Jain (1991), Khare (2004), Parrotta(2001), Rao & Henry (1996), Sharma & Rana (2005), Singh & Pandey (1998), Singh & Prakash (1996), Talbot (1976b).

Bauhinia retusa Ham.

Syn: *B. emarginata* Wall.; *Phanera retusa* Benth.

Family: Fabaceae.

Hindi and Regional Names:

Hindi	:	Semla;
Pub.	:	Kural;
Tel.	:	Nirpa.

Distribution : Himalaya, U. P., Orissa, E. Peninsula, M.P.

Parts Used : Bark. Leaves.

Utilization : Leaves used as fodder and bark for paper making.

Active Constituents : Leaves contain 7.84% ash, 7.73% Ca, 46.20% carbon and 2.03% nitrogen.

References : Agarwal (2003), Anonymous (1948), Hooker (1897).

Bauhinia variegata Linn.

Syn: *Bauhinia acuminata* Linn.; *B. candida* Roxb.; *B. purpurea* Wall.; *Phanera variegata* Benth.

Family: Fabaceae.

Vern. Name: Karayalae.

English, Hindi, Sanskrit and Regional Names:

Eng. : Buddhist bauhinia;
Hindi : Kachnar;
Sans. : Ashmantaka, Chamrika, Girija, Uddalaka;
Beng : Raktakanchan;
Guj. : Kavindar;
Mal. : Chuvannamandaram;
Mar. : Kachnar, Thaur;
Ori. : Kanchana;
Tam. : Segapumanchori;
Tel. : Devakanchanamo.

Distribution : Chir Forest Throughtout India upto 1,800 m.

Parts Used : Leaves. Bark.

Utilization : Source of fodder for feeding all classes of livestock throughout the winter. Bark yields a fibre for ropes.

Active Constituents : Plant contains crude proteins 15.6%, carbohydrate 72.36%, starch 14.3% and vit. C 146 mg %. Net energy of fodder has been found out to be 2,466.3 and 1,102.5 K cal / Kg.

References : Agarwal (2003), Ambasta (1986), Anonymous (1948), Dash & Mishra (2000), Dwivedi (1993), Hooker (1879), Parrotta (2001), Purohit & Samant & 1995), Sharma (2004), Singh (1982), Sood & Thakur (2004).

Beaumontia grandiflora Wall.

Syn: *Beaumontia longifolia* Lodd.; *Echites grandiflora* Roxb.
Family: Apocynaceae.

Distribution : E. Himalaya from Nepal to Sikkim.

Part Used : Stem.

Utilization : Young stem also a source of fibre.

References : Agarwal (2003), Ambasta (1986), Anonymous (1948), Hooker (1882).

Berberis affinis Don.

Family: Berberidaceae.

Distribution : Distributed between 3,000 - 3,300 m.

Part Used : Leaves.

Utilization : Browsed by sheep and goats.

Reference : Purohit & Samant (1995).

Berberis asiatica Roxb.

Syn: *Berberis aristata* var. *floribunda* (G. Don) Hook. f.; *B. chitra* Lindley.
Family: Berberidaceae.

Vern. Name: Kashmal.

English, Hindi, Sanskrit and Regional Names:

Eng. : Indian Berbery;
Hindi : Chitra, Darhold;
Sans. : Daruharidra, Kantakaki, Kateri;
Beng : Darhaldi;
Gar. : Chitri;
Guj. : Daruharidra;
Kan. : Maradarishana;
Mal. : Maradarisina, Maramanjal;
Mar. : Daruhald;
Ori. : Daruhaldi, Daruharida;
Tam. : Mullukala, Usikkala;
Tel. : Darhold.

Distribution : Throughout the area between 900- 3,300 m in Himalaya & Nilgiris.

Part Used : Leaves.

Utilization : Browsed by cattle, sheep and goats.

References : Agarwal (2003), Anonymous (1948), Hooker (1872), Purohit & Samant (1995), Raju (1998), Rastogi & Mehrotra (1995b), Sharma (2004), Sood *et al.* (2005).

Berberis kumaonensis Schn.

Family: Berberidaceae.

Distribution : Occurs between 3,000 - 4,000 m.

Part Used : Young Twigs.

Utilization : It is browsed by sheep and goats.

Reference : Purohit & Samant (1995).

Berberis lycium Royle

Syn: *Berberis parkeriana* Schn.

Family: Berberidaceae.

English, Hindi, Sanskrit and Regional Names:

Eng. : Raisin Berberry;
Hindi : Chitra, Darhald, Kashmal;
Sans. : Daruharidra;
Mar. : Darhald.

Distribution : Upto 2,600 m in W. Himalaya from Garhwal to Hazara.

Utilization : Used as fodder plant in W. Himalaya.

References : Agarwal (2003), Ambasta (1986), Hooker (1872), Kirtikar & Basu (1935), Sood & Thakur (2004).

Berberis petiolaris Wall.

Family: Berberidaceae.

Distribution : Distributed between 2,500-3,600 m.

Part Used : Leaves.

Utilization : Leaves browsed by sheep and goats.

Reference : Purohit & Samant (1995).

Berchemia floribunda (Wall.) Brongn.

Family: Rhamnaceae.

Regional Name:

Gar. : Amili.

Distribution : Upto 1,800 m.

Part Used : Plant Parts.

Utilization : Considered a fodder plant in W. Himalaya.

Reference : Ambasta (1986).

Berrya ammonilla Roxb.

Syn: *Berrya cordifolia* Roxb.

Family: Tiliaceae.

English and Regional Names:

Eng. : Trincomalle wood;
Tam. : Chavandalai;

Distribution : Malabar. S. Indian Forest. Andamans.

Part Used : Bark.

Utilization : Used as fibre.

References : Agarwal (2003), Ambasta (1986), Anonymous (1948), Hooker (1872).

Betula alnoides Buch. -Ham. *ex* Don.

Syn: *Betula acuminata* Wall.; *B. affinis, Betulaster cylindrostachya, acuminata, affinis, nitida* Spach.

Family: Betulaceae.

English, Hindi, Sanskrit and Regional Names:

Eng. : Indian Birch;
Hindi : Saur;
Sans. : Bhuja, Bhurz;
Ass. : Dingleen;
Beng : Hlosunle;
Guj. : Bhojpatra;
Tel. : Bhujaptra.

Distribution : Himalaya, Khasia Hills. Manipur.

Part Used : Leaves.

Utilization : Leaves used as fodder.

References : Agarwal (2003), Anonymous (1948), Bennet *et al.* (1991), Dwivedi (1993), Hooker (1890), Purohit & Samant (1995).

Betula utilis Don. (Pl. 2D)

Syn: *Betula bhojpatra* Wall.; *B. jackmontii* Spach.

Family: Betulaceae.

Vern. Name: Dhar.

English, Hindi, Sanskrit and Regional Names:

- **Eng.** : Himalayan Silver Birch;
- **Hindi** : Bhojpatra;
- **Sans.** : Bhujpatrak, Bhuurja, Chitratvark;
- **Beng** : Bhurjapatra;
- **Guj.** : Bhojapatra;
- **Mal.** : Bhurjamaram;
- **Tam.** : Bhurjjamaram;
- **Tel.** : Brjapatramu.

Distribution : Occurs upto 4,200 m (W. Himalaya, Bhutan, Kashmir to Sikkim).

Parts Used : Bark. Leaves.

Utilization : Leaves fed to all classes of livestock and horses during May- November. Bark used as paper for writing in ancient times and roofing houses.

References : Agarwal (2003), Ambasta (1986), Anonymous (1948), Bhattacharjee (1998), Chatterjee & Pakrashi (1997), Dhiman (2003), Dastur (1964), Dwivedi (1993), Hooker (1890), Purohit & Samant (1995), Prajapati *et al.* (2003), Sahni (1998).

Bidens bipinnata Linn. non Wall.

Syn: *Bidens pilosa* Linn.var. *bipinnata* Hook. f.
Family: Asteraceae.

English Names:

Eng. : Beggar Lice, Spinach Needles.

Distribution : Throughout India upto 2,000 m.

Utilization : Plant relished by cattle.

Active Constituents : Leaves rich in vitamin-C and polyacetylene glycoside.

References : Ambasta (1986), Bennet *et al.* (1991), Hooker (1882), Rastogi & Mehrotra (1995b).

Bischofia javanica Bl.

Syn: *Andrachene trifoliata* Roxb.; *Bischofia cumingiana* Decne.; *B. oblongifolia* Decne.; B. *roperiana* Decne.; *B. trifoliata* Hook.; *Microelus roperianas* W. A.; *Stylodiscus trifoliatus* Bennet Plant.
Family: Euphorbiaceae

English, Hindi and Regional Names:

Eng. : Bishop Wood;
Hindi : Paniala;
Ass. : Uriam;
Beng : Kainpal;
Mal. : Nira, Thirippu;
Mar. : Boke;
Ori. : Dingiri;
Tam. : Nannal, Thondi;
Tel. : Nalupumushi.

Distribution : Tropical Himalaya. Meghalaya. Nilgiri Hills. W. Ghats.

Part Used : Leaves.

Utilization : Leaves used as fodder.

Active Constituents : Betulinic acid, friedelinol acetate, hentriacontane present in leaves.

References : Bennet *et al.* (1991), Dwivedi (1993), Hooker (1890), Prajapati *et al.* (2003), Rastogi & Mehrotra (1995a), Saklani & Jain (1994), Sood *et al.* (2005).

Bixa orellana L.

Family: Bixaceae.

English, Hindi, Sanskrit and Regional Names:

Eng. : Annato, Lipstick tree;
Hindi : Latkan;
Sans. : Sindhuri;
Beng : Latkan;
Guj. : Sindhuri;
Mal. : Kuppamanjal;
Mar. : Kesari, Sendari, Shendri;
Ori. : Gulbas, Saka;
Tam. : Japhara;
Tel. : Jafra, Japhara;

Distribution : Cultivated Throughout India.

Part Used : Bark.

Utilization : Bark fibre suitable for making ropes.

References : Agarwal (2003), Ambasta (1986), Das & Agarwal (1991), Dastur (1964), Hooker (1872), Parrotta (2001), Raju (1991).

Blepharis linariaefolia Pers.

Syn: *Blepharis sindica* T. Anders.

Family: Acanthaceae.

Hindi and Regional Names:

Hindi : Bhangari
Bom. : Jasad;
Mar. : Jasad.

Distribution : Maharashtra, Rajasthan.

Part Used : Herb.

Utilization : Herb used as fodder.

References : Ambasta (1986), Blatter & Hallberg (1984), Roy *et al.* (1998).

Blepharis persica (Burn.) Kuntze.

Syn: *Blepharis edulis* Pers.
Family: Acanthaceae.

Hindi, Sanskrit and Regional Names:

Hindi : Chaupatia;
Sans. : Shikhi;
Beng : Shushani;
Bom. : Utangan;
Guj. : Khadakatri;
Mar. : Karadu;
Pub. : Utangan.

Parts Used : Herb.

Utilization : The herb forms a fodder for sheep and camels.

Active Constituents : Blephargenin and sitosterol isolated from the leaves.

References : Ambasta (1986), Bennet *et al.* (1991), Rastogi & Mehrotra (1991).

Blyxa aubertii Rich.

Family: Hydrocharitaceae.

Distribution : Different localities of Assam, Tripura & Manipur.

Part Used : Whole Plant.

Utilization : Plant mainly used as fish food and fodder.

Reference : Islam (2002).

Boehmeria glomerulifera Miq.

Syn: *B. malabarica* Wedd.
Family: Urticaceae.

Part Used : Stem.

Utilization : Stem fibre used for making ropes.

Reference : Saklani & Jain (1994).

Boehmeria macrophylla Horn.

Syn: *Boehrmaria platyphylla* D. Don.; *B. penduliflora* Wall.; *Urtica angustifolia* Ham.; *U. macrostachya* Wall.; *U. pendulifolia* Wall.
Family: Urticaceae.

Hindi Name:

Hindi : Sedeng.

Distribution : Throughout Tropical & Sub-Tropical Regions of India; (Kumaon to Mishmi).

Parts Used : Leaves. Stem.

Utilization : Leaves used as fodder. Stem fibre used for making net, ropes and cordage.

References : Agarwal (2003), Ambasta (1986), Dwivedi (1993), Hooker (1890), Rao & Henry (1996), Parker (1973), Purohit & Samant (1995), Rastogi & Mehrotra (1993), Saklani & Jain (1994), Singh & Kumar (2000), Sood & Thakur (2004).

Boehmeria nivea Gaud.

Syn: *Urtica nivea* Linn.
Family: Urticaceae.

English, Hindi and Regional Names:

Eng. : China Grass, Rhea;
Hindi : Kankura;
Ass. : Rhea;
Beng : Kankura;
Ori. : Bryddhhingi;

Distribution : Bengal. Assam.

Part Used : Stem.

Utilization : Stem fibre excellent, strong and durable and used for fabrics, plushes, paper making, sacks, sail, net and cordage.

References : Agarwal (2003), Ambasta (1986), Anonymous (1948), Jain (1991), Dastur (1964), Drury (1978), Harborne &

Baxter (2001), Kirtikar & Basu (1984), Saklani & Jain (1994), Sharma & Rana (2005).

Boehmeria rugolosa Wedd.

Family: Urticaceae.

Distribution : Throughout the Hills between 500-600 m.

Part Used : Leaves.

Utilization : Leaves considered as good fodder and fed to livestock in winter.

Active Constituents : Leaves contain ash 24.96 %.

Reference : Purohit & Samant (1995).

Boehmaria scbrella Gaud.

Family: Urticaceae.

Distribution : Occurs Throughout the area between 500- 1,000 m.

Part Used : Leaves.

Utilization : Leaves fed to all classes of livestock during rainy season.

Reference : Purohit & Samant (1995).

Bombax ceiba Linn.

Syn: *Bombax malabarium* DC.; *Gossampinus malabaria* (DC.) Merril.; *Salmalia malabarica* (DC.) Schott. & Endl.

Family: Bombacaceae.

Vern. Name: Simbal.

English, Hindi, Sanskrit and Regional Names:

Eng. : Silk Cotton Tree, Simal Tree;
Hindi : Semul, Rakatsenbal;
Sans. : Mocha, Rakatpushpa;
Beng : Rokosimul, Simul;
Bom. : Saer, Somr;
Guj. : Saver, Shemol;
Mal. : Mullilavau, Mullilapula;
Mar. : Saur, Samara, Simlo;

Ori.	:	Bouro, Buroh, Simlo;
Tam.	:	Ilavam, Pullai;
Tel.	:	Mundla- buraga chettu.

Distribution : Throughout Hotter Parts of India.

Parts Used : Leaves. Fruit. Seed.

Utilization : Leaves lopped for fodder. Floss used for stuffing life belts, mattreses, cushions, pillows and upholstery. Fibre spun into yarns for manufacture of plushes. Fruits yield silk fibre for soft domestic articles like pillows, etc.

Active Constituents : Leaves contain crude fibre 1.38, carbohydrates 11.95 and ash 1.09%.

Reference1 : Agarwal (2003), Ambasta (1986), Dash & Misra (2000), Datt (1996), Diwakar & Ansari (2003), Dwivedi (1993), Khanna (2002), Kothari & Rao (2000), Oommachan (1977), Parrotta (2001), Prasad *et al.* (2002), Rao & Henry (1996), Saklani & Jain (1994), Sharma (2004), Singh (1982, 2002), Singh & Pandey (1998), Sood & Thakur (2004), Talbot (1976a), Varghese (1996).

Borassus flabellifer Linn.

Syn: *Borassus flabelliformis* Roxb.

Family: Arecaceae.

English, Hindi, Sanskrit and Regional Names:

Eng.	:	Brab Tree, Char Palm;
Hindi	:	Tar;
Sans.	:	Asvadru, Chirayu, Patri, Tal;
Beng	:	Tal;
Guj.	:	Tad;
Mal.	:	Ampana, Eta, Pana, Talam;
Mar.	:	Tad, Tamar;
Ori.	:	Tad, Tal, Talo, Tanlo;
Tam.	:	Karum panai, Nonkuppanai;
Tel.	:	Taddichetu.

Distribution : Coastal Areas of Peninsular India, Bihar & W. Bengal.

Parts Used : Leaves. Wood. Fruit. Petioles.

Utilization : Leaf stalks yield fibre for brushes and brooms. Petiole fibre used for making ropes, fans, mats, ola bags, water baskets, umbrellas, hats etc. In ancient times leaves were used as writing material as the parallel veins of it served as ruled note books.

References : Agarwal (2003), Ambasta (1996), Caius (2003), Chakhraborty & Bhattacharjee (2003), Dastur (1964), Jain (1991), Kirtikar & Basu (1984), Rao & Henry (1996), Saklani & Jani (1994), Sahni (1998), Upadhyaya & Chauhan (2000).

Borreria articularis (Linn. f.) F. N. William.

Syn: *Borreria hispida* (Linn.) K. Schum.;
Spermacoce articularis Linn. f., Roxb. *ex* Fleming in Asiat.
Family: Rubiaceae.

English, Hindi, Sanskrit and Regional Names:

Eng. : Shaggy button weed;
Hindi : Guthari;
Sans. : Madana ghanta;
Beng : Madana - bantakadu;
Guj. : Maduri jedi;
Mar. : Ghanti - chi - bhaji;
Ori. : Solaghanti;
Tam. : Nattai churi, Mutti Chori;
Tel. : Madana grandhi.

Distribution : Shimla Eastwards to Assam & Southwards upto Sri Lanka.

Part Used : Leaves.

Utilization : Used as fodder.

References : Ambasta (1986), Das & Agarwal (1991).

Boswellia serrata Roxb.

Syn: *Boswellia thurifera* Roxb. *ex* Fleming in Asiat.
Family: Burseraceae.

Vern. Name: Salar.

English, Hindi, Sanskrit and Regional Names:

Eng.	:	Indian Olibanum Tree;
Hindi	:	Salai;
Sans.	:	Kanduru, Sallaki;
Beng	:	Salai;
Bom.	:	Salai, Salgu;
Guj.	:	Dhup, Gugali;
Mal.	:	Kantturukam, Mukandam;
Mar.	:	Salai;
Ori.	:	Loban;
Tam.	:	Parangisambrani;
Tel.	:	Parangisambrani.

Distribution	:	W. Himalaya. Bihar to Rajputana, Southwards into the Deccan and Concan.
Part Used	:	Leaves.
Utilization	:	Leaves used as fodder.
Active Constituents	:	Sitosterol isolated from the leaves.
References	:	Anonymous (1948), Dhiman (2003), Hooker (1872), Parrotta (2001), Singh (1982, 2002, 2003).

Bothriochola intermedia (R. Br.) A. Camus.

Syn: *Andropogon intermedius* R .Br. var., *intermedia* R. Br.
Family: Poaceae.

Part Used	:	Plant.
Utilization	:	An excellent fodder grass.
Active Consitutes	:	Leaves possess 4.7% crude proteins, 0.08% P, 33.0% crude fibre and fat 1.7%.
References	:	Dwivedi (1993), Islam (2002), Sharma & Rana (2005).

Bothriochloa pertusa (Willd.) A. Camus.

Syn: *Andropogon pertusus* Willd.
Family: Poacace.

English, Hindi and Regional Names:

Eng. : Pitted Blue Grass;
Hindi : Sandhur;
Mar. : Gohhaya;
Tam. : Chinna- karai pullu;
Tel. : Janu Gaddi.

Utilization : Much valued as a fodder grass for grazing.

Active Constituents : Protein 8.8%.

References : Agarwal (2003), Ambasta (1986), Sharma & Rana (2005).

Brachiaria distachya Staph.

Family: Poaceae.

Part Used : Herb.

Utilization : Commonly used as fodder.

Reference : Islam (2002).

Brachiaria ramosa (Linn.) Staph.

Syn: *Panicum ramosum* Linn.
Family: Poaceae.

Regional Names:

Tam. : Pala pul;
Tel. : Anda Korra.

Distribution : Throughout the Plains of India.

Part Used : Herb.

Utilization : Used as fodder.

References : Das *et al.* (1996), Roy *et al.* (1998), Sharma & Rana (2005).

Brachiaria reptans (Linn.) Gard. & C. E. Hubbard.

Syn: *Urochola reptans* Stapf.
Family: Poaceae.

Hindi and Regional Names:

Hindi : Andakora;
Tam. : Shani Pillu.

Part Used : Plant.

Utilization : Fodder liked by cattle.

References : Agarwal (2003), Ambasta (1986), Sharma & Rana (2005).

Brachiaria paspaloides (Persl.) C. E. Hubbard.

Syn: *Panicum ambiguum* Trin.
Family: Poaceae.

English, Hindi and Regional Names:

Eng. : Crab Grass;
Hindi : Takri, Takria;
Mal. : Chikhari, Karkoljodi.

Part Used : Plant.

Utilization : A good fodder relished by cattle, sheep and goats.

References : Agarwal (2003), Ambasta (1986).

Brassica napus Linn.

Syn: *Brassica compestris* Hook. F. & Thoms. Sub sp., *napus* Hook. f. & T. Anders. in part.
Family: Brassicaceae.

Vern. Name: Banarasi Ryi.

English, Hindi, Sanskrit and Regional Names:

Eng. : Mustard, Rape;
Hindi : Kali sarson , Lutni, Toria;
Sans. : Kattaka, Rajakshavaka, Siddhaartha;
Beng : Sadaeai.

Distribution : Cultivated Throughout India.

Part Used : Leaves.

Utilization : Leaves fed to cattle.

References : Agarwal (2003), Ambasta (1986), Anonymous (1948), Khare (2004).

Brassica oleracea Linn. Convar. *acephala* (DC.) Alef. var. *viridis* Linn.

Syn: *Brassica oleracea* var., *acephala* DC. in part.
Family: Brassicaceae.

English, Hindi, Sanskrit and Regional Names:

Eng. : Borecole, Marrow Cabbage;
Hindi : Karam Sag;
Sans. : Kembukah;
Guj. : Pangoli;
Mal. : Kyabej;
Mar. : Knolkhol;
Tam. : Koshipoo;
Tel. : Kosugadd.

Distribution : Cultivated in Assam, Maharashtra, Gujarat, Kashmir.

Part Used : Leaves.

Utilization : Grown for fodder in some parts of U. P. as it is believed to help in better growth of wool.

Active Constituents : Leaves rich in mineral matter 1.4, iron 1-3 mg, vit. C, 66 mg/100gm, quercetin and isorhamnetin.

References : Agarwal (2003), Ambasta (1986), Anonymous (1948), Prajapati *et al.* (2003), Rastogi & Mehrotra (1991), Singh *et al.* (1996).

Bridelia montana Hook. f. &Th.

Syn: *Cluytia montana* Roxb.
Family: Euphorbiaceae.

Hindi, Sanskrit and Regional Names:

Hindi : Gondni;

Sans.	:	Asana, Pasanabheda;
Guj.	:	Asano;
Mar.	:	Asana;
Ori.	:	Nuanali;
Pub.	:	Vengaimaram;
Tam.	:	Vengaimaram;
Tel.	:	Panchanivani, Pantegi, Pantenga.

Distribution : Punjab to Bhutan: 1,300- 3000 m.

Part Used : Leaves.

Utilization : Leaves considered as good fodder and fed during rainy and winter season.

Active Constituents : Leaves contain β-sitosterol.

References : Parrotta (2001), Purohit & Samant (1995), Rastogi & Mehrotra (1991), Singh & Pandey (1998).

Bridelia retsua Spreng.

Syn: Andrachne doonkyboisca Heyne.; *Bridelia amoena* Wall. *ex* Baill.; *B. spinosa* Roxb.; *Cluytia retsua* Linn.; *C. spinosa* Willd.

Family: Euphorbiaceae.

English, Hindi, Sanskrit and Regional Names:

Eng.	:	Asasna, Ekadivi, Feather Foil;
Hindi	:	Ekdania, Gondui, Khaja;
Sans.	:	Asasna, Ekadivi, Ekavira;
Beng	:	Geio;
Guj.	:	Asana, Monj;
Mal.	:	Mukkayini;
Mar.	:	Asana, Kutki;
Ori.	:	Kasi, Kosi;
Tam.	:	Mullu- vengai;
Tel.	:	Bonthayepi.

Distribution : Throughout India along the Foot Hills of Himalaya 1,500 m.

Part Used : Leaves.

Utilization : Leaves used as fodder.

Active Constituents : Leaves contain 10.14-10.755 ash, 1.30% Ca, 44.20% C and 2.32- 3.01% N.

References : Agarwal (2003), Ambasta (1986), Anonymous (1948), Drury (1978), Dwivedi (1993), Hooker (1890), Khan & Khanum (2005), Parrotta (2001), Singh (1982).

Bromelia pinguin Linn.

Family: Bromeliaceae.

English Names:

Eng. : Pingvin, Wild Pine apple.

Distribution : Introduced in India.

Part Used : Leaves.

Utilization : Isolated leaf fibre commonly known as wild pine apple fibre exploited for mats and carpets.

Reference : Ambasta (1986).

Bromus catharticus Vahl.

Syn: *Bromus unioloides* H. B. & K.

Family: Poaceae.

Distribution : Grows in Darjeeling & Nilgiris.

Part Used : Plant.

Utilization : Used for fodder purposes.

References : Agarwal (2003), Ambasta (1986).

Bromus conifinis Nees *ex* Steud.

Syn: *Bromus inermis* var., *confines* Leyss.

Family: Poaceae.

Part Used : Plant.

Utilization : Used as fodder.

References : Dwivedi (1993), Sharma & Rana (2005).

Broussonetia papyrifera Vent.

Family: Moraceae.

Vern. Names: Karyun, Paper Mulbury.

English, Hindi and Regional Names:

Eng.	:	Paper Mulbury;
Hindi	:	Jangali Toot;
Kan.	:	Kaagada.

Distribution	:	Punjab, A. P. & Meghalaya upto 1,000 m.
Part Used	:	Bark.
Utilization	:	Bark fibre good for making high grade leathery paper, paper handkerchiefs, umbrella covers and lens paper.
References	:	Anonymous (1948), Hooker (1890), Raju (1998).

Buchanania lanzan Spreng.

Syn: *Buchanania latifolia* Roxb.
Family: Anacardiaceae.

Vern. Name: Achav.

English, Hindi, Sanskrit and Regional Names:

Eng.	:	Almondette Tree;
Hindi	:	Charoli, Chironji;
Sans.	:	Chara, Tapasya-priya;
Beng	:	Chironji, Pival;
Guj.	:	Charoli, Chironji;
Kan.	:	Nurkul;
Mal.	:	Munga pera, Nuruvi;
Mar.	:	Charoli;
Ori.	:	Charee, Charu;
Pub.	:	Chiroli, Chironji;
Tam.	:	Morala;
Tel.	:	Sara.

Distribution	:	Throughout Tropical India upto 900 m.
Part Used	:	Leaves lopped for fodder.

Active Constituents : Leaves rich in kaempferol, gallic acid, quercetin and protein 21.6%, 11.90% ash, 2.42% Ca and 1.6% N.

References : Agarwal (2003), Ambasta (1986), Anonymous (1948), Bhattacharjee (1998), Dwivedi (1993), Prajapati *et al.* (2003), Rastogi & Mehrotra (1995b), Singh (1982).

Buddleia paniculata Wall.

Syn: *Buddleia nepalensis* Colla.

Family: Loganiaceae.

Vern. Name: Shivra.

Distribution : Distributed between 1,000- 2,500 m.

Part Used : Leaves.

Utilization : Browsed by all ruminants.

References : Hooker (1885), Purohit & Samant (1995), Sharma (2004).

Bupleurum candollei Wall.

Family: Apiaceae.

Distribution : Upto 3,000-4,000 m.

Part Used : Plant.

Utilization : Browsed by all ruminants.

Reference : Purohit & Samant (1995).

Butea monosperma (Lamk.) Taub.

Syn: *Butea monosperma* Koen. *ex* Roxb.; *Erythrina monosperma* Lamk.; *Plaso monosperma* Kuntze.

Family: Fabaceae.

Vern. Name: Palash.

English, Hindi, Sanskrit and Regional Names:

Eng. : Bastard Teak, Bengal Kino, Flame of the forest;
Hindi : Dhak, Palas;
Sans. : Bijasneha, Palasa, Vatapotha;
Beng : Kinaka, Palas;

Guj.	:	Khakra, Pallaso;
Kan.	:	Muttuga, Pallasa, Palasu;
Mal.	:	Brahmarikshan, Palasinjamata;
Mar.	:	Kakradha, Phalas;
Ori.	:	Kinjuko, Palaso;
Pub.	:	Chachra, Dhat, Palas;
Tam.	:	Kutumurukku;
Tel.	:	Moduga, Palasamo.

Distribution : Throughout India upto 1,400 m.

Parts Used : Leaves. Stem. Root Bark.

Utilization : Leaves a good fodder for elephants. Root bark yields strong fibre for making ropes and triangular baskets by 'Garasias'.

References : Ambasta (1986), Anonymous (1948), Chakraborty & Bhattacharjee (2003), Dastur (1964), Datt (1996), Deshmukh *et al.* (2000), Diwaker & Ansari (2003), Drury (1978), Jain (1991), Mittre (1981, 1997), Oommachan (1977), Raju (1998), Rao & Henry (1996), Sahni (1998), Sharma & Rana (2005), Singh (1982), Singh & Prakash (1996), Singh & Pandey (1998), Singh & Kumar (2000), Varghese (1996).

Butea parviflora Roxb.

Syn: *Spatholobus roxburghii* Benth.
Family: Fabaceae.

Hindi and Regional Names:

Hindi	:	Bando, Maula;
Kan.	:	Mukkate balli;
Mal.	:	Athambuvalli;
Mar.	:	Phalsan;
Ori.	:	Poraso;
Pub.	:	Mula;
Tam.	:	Pilacchivalli, Plashivalli;
Tel.	:	Bodegateega.

Distribution : Common Throughout India upto 900 m.

Part Used : Bark.

Utilization : Bark fibre for ropes.

References : Ambasta (1986), Parrotta (2001).

Caesalpinia crista Linn.

Syn: Caesalpinia bonduc (Linn.) Roxb.; *C. bonducella* Flem.

Family: Fabaceae.

English, Hindi, Sanskrit and Regional Names:

Eng. : Bonduct Tree, Fever Nut;
Hindi : Karanju;
Sans. : Putikaranja;
Beng : Nata;
Bom. : Gaja;
Guj. : Gajga, Kakcha;
Kan. : Gajagakarkai;
Mal. : Kazanchikkuru;
Mar. : Gajaga;
Ori. : Kotokolija;
Tam. : Kazhichikay;
Tel. : Gacacha- kaya.

Distribution : Bengal. S. India upto 900 m.

Part Used : Root Bark.

Utilization : Fibre used for making ropes.

References : Anonymous (1950), Raju (1998).

Cajanus cajan (Linn.) Millsp.

Syn: *Cajanus indicus* Spreng.; *Cytisus cajan* L.; *C. obcordifolia* Singh.

Family: Fabaceae.

Vern. Names: Arhar, Aarar.

English, Hindi, Sanskrit and Regional Names:

Eng. : Congo pea, Pigeon pea, Red gram;
Hindi : Arhar, Tur;
Sans. : Adhaki, Tuvari, Tivarika;
Beng : Tur;

Kan. : Togare;
Mal. : Thuvara;
Mar. : Arhar, Tur, Tuver;
Tam. : Thovaray;
Tel. : Kandulu.

Distribution : Mainly Grown in U. P., Bihar, Bombay, Madras.

Parts Used : Green Leaves. Stem-Husk. Plant.

Utilization : Green leaves, stem-husk considered as fodder. Plant exploited for ropes and mats.

Active Constituents : Leaves comprise of alkanes (16.6), ester (8.4), alcohols (26.4), lupeol (19.2) and phytosterol.

References : Agarwal (2003), Ambasta (1986), Anonymous (1950), Khanna (2002), Oommachan (1977), Parrotta (2001), Prajapati *et al.* (2003), Rastogi & Mehrotra (1991, 1995a), Sood *et al.* (2005).

Calamus andamanicus Kurtz.

Family: Arecaceae.

Regional Name:

Beng : Bet.

Distribution : Andaman & Nicobar Islands.

Part Used : Stem.

Utilization : Ropes and cordage are made from the stem fibre.

References : Chakrabarty & Balakrishnan (2003), Dagar (2003), Roy *et al.* (1998).

Calamus erectus Roxb. (Pl. 3A)

Family: Arecaceae.

Distribution : E. Himalaya and N. E.India.

Part Used : Stem.

Utilization : Stem fibre valued for making ropes, cordage, bags, baskets.

References : Chakrabarty & Balakrishnan (2003), Islam (2002), Raju (1998).

Calamus tenius Roxb.

Family: Arecaceae.

Hindi and Regional Names:

Hindi : Bet;

Ass. : Jattebet;

Beng : Bhandaribet.

Distribution : Sub-Himalayan Tract from Dehradun to Assam.

Parts Used : Leaves. Stem. Branches.

Utilization : Leaves eaten by buffaloes. Stem and branches used for preparing mats, baskets and various domestic articles.

References : Chakrabarty & Balakrishanan (2003), Drury (1978), Raju (1998), Roy *et al.* (1998).

Callicarpa arborea Roxb.

Syn: *Callicarpa farniosa* Roxb.; *C. tectonaefolia* Wall.; *C. villosa* Roxb.

Family: Verbenaceae.

Hindi and Regional Names:

Hindi : Khoja;

Ass. : Khoja, Ghiwata;

Ori. : Bornnopotari, Gerugombhari.

Distribution : Common in Sikkim and from Kumaon to Assam.

Part Used : Leaves.

Utilization : Leaves fed to cattle.

Active Constituents : These contain maslinic acid, ursolic acid, luperol acetate, β- amyrin acetate and sitosterol acetate.

References : Agarwal (2003), Ambasta (1986), Anonymous (1950), Hooker (1885), Purohit & Samant (1995), Rastogi & Mehrotra (1991, 1995b).

Callicarpa macrophylla Vahl.

Syn: *Callicarpa incana* Roxb.; *C. roxburghii* Wall.
Family: Verbenaceae.

Vern. Names: Dahimal, Nagdhava.

English, Hindi, Sanskrit and Regional Names:

Eng.	:	Perfumed cherry;
Hindi	:	Sumali;
Sans.	:	Kanta, Phalini, Priyangu;
Beng	:	Mathara, Mattaranjo;
Guj.	:	Lata- Priyangu;
Mar.	:	Chimpompil;
Ori.	:	Priyangu;
Pub.	:	Priyangu;
Tam.	:	Gnazhal, Gnazhalpoo;
Tel.	:	Prenkanamu.

Distribution	:	Kashmir to Assam & Arunachal Pradesh: 300 - 2,000 m.
Part Used	:	Leaves.
Utilization	:	Leaves fed to cattle during scarcity in winter season.
Active Constituents	:	Leaves contain flavone and propionic acid.
References	:	Rastogi & Mehrotra (1991, 1995a, 1995b).

Calotropis gigantean (Linn.) R. Br. *ex* Ait.

Syn: *Asclepias gigantean* Willd.
Family: Asclepiadaceae.

Vern. Names: Aakanda, Akura.

English, Hindi, Sanskrit and Regional Names:

Eng.	:	Bowstring Hemp, Madar;
Hindi	:	Ak;
Sans.	:	Akra, Alarka, Mandaru;
Beng	:	Akanda;
Guj.	:	Akado;
Kan.	:	Ekka;

Mal.	:	Erikku;
Mar.	:	Rui;
Ori.	:	Akado, Akando;
Tam.	:	Arkkam;
Tel.	:	Jilledu.

Distribution : Throughout India upto 1,000 m.

Parts Used : Twigs. Stem Bark. Floss.

Utilization : Bark fibre good for fishing nets, bow strings and twines. Floss considered a good stuffing material for pillows, cushions, shawls and handkerchiefs.

References : Agarwal (2003), Ambasta (1986), Anonymous (1950), Jain (1991), Dastur (1964), Hooker (1885), Mittre(1997), Parrotta (2001), Raju (1998), Rao & Henry (1996), Sharma & Rana (2005), Singh & Pandey (1998).

Calotropis procera (Ait.) R. Br.

Syn: *Asclepias procera* Willd.; *Calotropis hamiltonii* Wall.; *C. heterophylla* Wall.; *C. wallichii* Wight.

Family: Asclepiadaceae.

Vern. Names: Ak, Madar.

English, Hindi, Sanskrit and Regional Names:

Eng.	:	Dead Sea Apple, Milk weed, Sodom Apple;
Hindi	:	Akada, Ak, Mamiri;
Sans.	:	Alarka, Arka, Bimbora, Mandara, Vasuka;
Beng	:	Akada;
Guj.	:	Akado, Akdo, Ankado, Nanirui, Nano akado;
Kan.	:	Bili yekkada gida, Yakkigida;
Mal.	:	Erukku;
Mar.	:	Mandara;
Ori.	:	Orko;
Pub.	:	Ak, Shakarulusar;
Tam.	:	Vellerukku;
Tel.	:	Erajilledu.

Distribution : Kashmir to Nepal, ascending to 1,300 m.

Parts Used : Stem.Twigs.

Utilization : Stem yields a fibre for fishing nets, lines, bowstrings and twines. The floss fibre of seed used for stuffing pillows.

References : Agarwal (2003), Ambasta (1986), Anonymous (1950), Dastur (1964), Datt (1996), Jain (1991), Khanna (2002), Parrotta (2001), Raju (1998), Sharma & Rana (2005), Singh (2000), Singh & Pandey (1998).

Canna orientalis Rosc.

Syn: *Canna indica* Linn.
Family: Cannaceae.

English, Hindi, Sanskrit and Regional Names:

Eng. : Indian Shot;
Hindi : Sabbajaya, Sarvajjaya;
Sans. : Devakili, Krishnatamara, Sarvajaya;
Beng : Sarvajaya;
Kan. : Kalahu;
Mal. : Devakeli;
Tam. : Kalvalai, Puvalai;
Tel. : Krishnatamara.

Distribution : Warmer Regions of India.

Part Used : Stem.

Utilization : Stem fibre used as a substitute for jute in the manufacture of rope, sacks and cloth.

References : Agarwal (2003), Ambasta (1986), Anonymous (1950), Jain (1991).

Cannabis sativa Linn.

Syn: *Cannabis gigantea* E.; *C. indica* Lamk.
Family: Cannabinaceae.

Vern. Name: Bhang.

English, Hindi, Sanskrit and Regional Names:

Eng. : Marijuana, Marihuana, Soft – Hemp, True – Hemp;
Hindi : Charas, Siddhi;
Sans. : Bhanga, Ganja, Indrasana, Jaya, Vijaya;

Beng	:	Bhange;
Guj.	:	Bhange, Charas, Ganja;
Kan.	:	Bhangi- Gida;
Mal.	:	Ginjilachiachi, kanchavachetta;
Mar.	:	Bhangachajhada;
Pub.	:	Bhang, Charas, Kas, Sabzi;
Tam.	:	Ganja- chedi, Kalpam;
Tel.	:	Ganzai.

Distribution : Sub-Himalayan Tract. Punjab to Bengal.

Utilization : Stemfibre (hemp) used for making ropes, mats, sacs, cordage, tarpoline and parachute webbing.

References : Agarwal (2003), Ambasta (1986), Anonymous (1950), Dastur (1964), Drury (1978), Hooker (1890), Jain (1991), Kaul (1997), Mittre (1997), Shah (1997b, 2000), Sharma & Rana (2005), Singh (2000), Singh & Kumar (2000), Sood & Thakur (2004), Sood *et al.* (2001).

Capparis sepiaria L.

Syn: *Capparis glauca* Wall. *ex* Hook. f. & Thoms.; *C. retusella* Thw.; *C. sepiaria* var., *incanescens* (DC.) Hook. f. & Thoms.; *C. sepiaria* var., *vulgaris* Hook. f. Thoms.; *C. sepiaria* var., *retusella* (Thw.) Thw.

Family: Capparidaceae.

Vern. Names: Jal, Teet.

English, Hindi, Sanskrit and Regional Names:

Eng.	:	Caper Bush, Indian Caper;
Hindi	:	Kanthari;
Sans.	:	Hinara, Kanthari, Kala, Tundi;
Beng	:	Kaliakara;
Guj.	:	Kantharae;
Kan.	:	Basingi, Volle uppi gidda;
Mar.	:	Kanthar, Mastondi, Waghati;
Ori.	:	Halubhi, Kontoko, Sorolakoli;
Pub.	:	Hingerna, Hins, Hius;
Tam.	:	Karindu, Karunjurai;
Tel.	:	Nalluppi, Puyyi, Uppi.

Distribution : Throughout India.

Part Used : Leaves.

Utilization : Leaves used for fodder purposes.

Active Constituents : Leaves rich in α-amyrin, β-amyrin, taraxasterol and erythrodiol.

References : Hooker (1872), Khan & Khanum (2003), Parrotta (2001), Rastogi & Mehrotra (1991), Singh (2000, 2003), Talbot (1976b).

Capparis zeylanica L.

Syn: *Capparis acuminata* Roxb.; *C. brevispina* DC.; *C. harridan* Linn. f.; *C. wallichiana* W. A.

Family: Cappariadaceae.

English, Hindi, Sanskrit and Regional Names:

Eng. : The Caper Plant;
Hindi : Ardanda
Sans. : Karambha;
Beng : Kalokera;
Bom. : Anti, Tarti, Way;
Kan. : Tuttula;
Mar. : Govindi;
Ori. : Govindi, Lobhytai, Osara;
Pub. : His, Karvila;
Tam. : Adandoi
Tel. : Adonda.

Distribution : Most Parts of India from U.P. eastwards & southwards through Peninsular India.

Parts Used : Leaves. Twigs. Fruits.

Utilization : Leaves and twigs eaten by elephants. Ripe fruits relished by sheep, goats and camels.

Active Constituents : Leaves possess carotene.

References : Anonymous (1950), Hooker (1872), Kirtikar & Basu (1984), Oommachan (1977), Parrotta (2001), Raju (1998), Rastogi & Mehrotra (1995a).

Carallia brachiata (Lour.) Merril.

Family: Rhizophoraceae.

Regional Names:

Ass.	:	Kanthekera;
Beng	:	Kierpa;
Kan.	:	Andipunaru;
Mal.	:	Varanga;
Mar.	:	Pansai;
Tel.	:	Karalli.

Distribution : Assam. W. Bengal. S. India. Andamans.

Part Used : Leaves.

Utilization : Leaves lopped for fodder.

Active Constituents : Alkaloides (+) hygroline present in leaves.

References : Raju (1998), Rastogi & Mehrotra (1990), Roy *et al.* (1998).

Cardaria draba (Linn.) Desv.

Syn: *Lepidium draba* Linn.
Family: Brassicaceae.

Hindi Name:

Hindi	:	Bijindak.

Part Used : Herb.

Utilization : Used as fodder.

Reference : Ambasta (1986).

Carduus nutans Linn.

Syn: *Carduus lucidus* Wall.
Family: Asteraceae.

English and Hindi Names:

Eng.	:	Musk Thistle;
Hindi	:	Kanchari.

Distribution : W. Himalaya: (Kashmir to Shimla). Punjab.

Part Used : Plant.

Utilization : Considered a good fodder grass for camel.

References : Agarwal (2003), Ambasta (1986), Hooker (1882).

Carex cernud Beett

Family: Cyperaceae.

Part Used : Plant.

Utilization : Plant used for fodder.

Reference : Islam (2002).

Carex cruciata Wahlenb.

Family: Cyperaceae.

Part Used : Plant.

Utilization : It is used for fodder.

Reference : Islam (2002).

Careya arborea Roxb.

Family: Lecythidaceae.

English, Hindi and Regional Names:

- **Eng.** : Patana Oak, Wild Guava;
- **Hindi** : Kumbhi, Kinihi, Nabhika, Pilu, Shaundi;
- **Guj.** : Kumbi;
- **Kan.** : Kaval;
- **Mal.** : Alam;
- **Mar.** : Kumbia;
- **Ori.** : Kumbhi;
- **Pub.** : Vakhumba;
- **Tam.** : Ayma;
- **Tel.** : Araya, Doddipa.

Distribution : Throughout India upto 1,300 m.

Parts Used : Stem Bark.

Utilization : Bark used for making ropes and brown paper.

References : Agarwal (2003), Anonymous (1950), Dastur (1964), Dwivedi (1993), Hooker (1879), Jain (1991), Khan & Khanum (2005), Kirtikar & Basu (1984), Mittre (1997), Sharma & Rana (2005), Singh (1982), Varghese (1996).

Carica papaya L.

Family: Caricaceae.

Vern. Names: Papaya, Papita.

English, Hindi and Regional Names:

Eng. : Melon Tree, Papaw Tree, Paw- Paw;
Hindi : Papita;
Beng : Papai;
Bom. : Chibhado, Katha;
Guj. : Papayi, Thadchibadi;
Kan. : Pappaye - phal;
Mal. : Kapallam, Nalikadala;
Mar. : Papaya, Popai;
Ori. : Omrytobhondo, Popoya;
Pub. : Anand - Kharpurna, Popai;
Tam. : Pappali, Pappayi;
Tel. : Bappayi.

Distribution : Commonly Cultivated Throughout India.

Part Used : Stem Pith .

Utilization : Stem pith used as fibre.

References : Agarwal (2003), Anonymous (1950), Parrotta (2001).

Carissa congesta Wight.

Syn: *Carissa carandus* L.
Family: Apocynaceae.

Vern. Names: Garna, Garnae.

English, Hindi, Sanskrit and Regional Names:

Eng. : Bengal Currents;
Hindi : Karaunda, Karrona;

Sans. : Jalipushpa, Krishna Pakaphal;
Beng : Hartund, Karinda;
Guj. : Karamadan, Karamarda;
Mal. : Boranda, Haradundi;
Ori. : Khyirokoli, Korond;
Tam. : Kalakkay, Perangala;
Tel. : Vaka.

Distribution : Throughout India ascending to 1,800 m.

Part Used : Leaves.

Utilization : Leaves used as fodder for sheep and goats.

Active Constituents : Leaves contain carissic acid.

References : Rastogi & Mehrotra (1995a), Sharma (2004).

Carissa opaca Stapf.

Syn: *Carissa spinarum* auct. (non L.)
Family: Apocynaceae.

Sanskrit and Regional Names:

Sans. : Karavanda;
Guj. : Karamarda;
Kan. : Karikayi;
Ori. : Anka, Kol;
Tam. : Kallakkay;
Tel. : Vaka.

Part Used : Leaves.

Utilization : Leaves relished by sheep and goats.

Reference : Raju (1998).

Caryota mitis Lour.

Family: Arecaceae.

Hindi Name:

Hindi : Mari supari.

Distribution : Widely Distributed in the Andaman Islands.

Part Used : Leaf.

Utilization : Leaf fibre used for stuffing mattresses and thatching purposes.

References : Ambasta (1986), Anonymous (1948-1976), Chakrabarty & Balakrishnan (2003).

Caryota urens Linn.

Family: Arecaceae.

English, Hindi, Sanskrit and Regional Names:

Eng. : Kittul, Sago;
Hindi : Mari;
Sans. : Dirga, Mada;
Beng : Bankhajur, Galsago;
Kan. : Anapana, Kundapana, Vazapana;
Mar. : Bherlimar;
Ori. : Salopa, Solopo;
Tam. : Koondalpanai, Jilugu, Mari;
Tel. : Jilugu, Mari.

Distribution : Moist Regions of E. & W. Coasts. C. India. Bihar. Assam.

Parts Used : Leaves. Petioles. Floss.

Utilization : Yield a fibre called 'Kittul Fibre' used for making arrows, brushes, brooms, hairs, cloths, polishing yarns, fishing-lines and strong wiry ropes for tying elephants.

References : Agarwal (2003), Ambasta (1986), Anonymous (1950), Caius (2003), Dastur (1964), Drury (1978), Dwivedi (1993), Jain (1991), Kothari & Rao (2000), Mittre (1997), Prasad *et al.* (2002, 2003), Rao & Henry (1996), Sahni (1998), Saklani & Jain (1994), Talbot (1976a).

Cassia auriculata L.

Syn: *Senna auriculata* Roxb.

Family: Fabaceae.

English, Hindi, Sanskrit and Regional Names:

Eng. : Tanners cassia;

Hindi	:	Tarwar;
Sans.	:	Avartaki, Awartakki, Tarwar;
Beng	:	Tarwar;
Guj.	:	Awal, Tangedu;
Mal.	:	Tangedu;
Mar.	:	Tarwad;
Tam.	:	Avaram, Avarai, Avirai;
Tel.	:	Tangedu.

Distribution : Common in A.P. Tamil Nadu & Karnataka.

Part Used : Bark.

Utilization : Yields cassia fibre of commercial importance.

References : Agarwal (2003), Anonymous (1950), Drury (1978), Hooker (1879), Parrotta (2001).

Cassia floribunda Cav.

Syn: *Cassia laevigata* Willd.

Family: Fabaceae.

Distribution : Often cultivated between 1,000 -2,000 m.

Part Used : Leaves.

Utilization : It is fed to animals when the plant is young and the pods have been removed.

Active Constituents : Leaves contain ombuin, floribundone 1 and floribundone 2.

References : Purohit & Samant (1995), Rastogi & Mehrotra (1993, 1995b).

Castanopsis tribuloides (Sm.) A. DC.

Family: Fagaceae.

Hindi and Regional Names:

Hindi	:	Desi Chesnut;
Ass.	:	Phul hingori.

Distribution : Occurs between 1,000-2,500 m.

Part Used : Leaves.

Utilization : It is fed to animals in spring season.

Active Constituents : Leaves contain crude fibre 30.6% and ash 3.6%.

References : Dwivedi (1993), Hooker (1890), Purohit & Samant (1995).

Cedrela toona Roxb.

Syn: *Cedrela australis* F. Muell.; *C. febrifuga* Blume.; *C. hexandra* Wall.; *C. longifolia* Wall.; *Toona ciliata* Roem.
Family: Meliaceae.

Vern. Name: Tooni.

English, Hindi, Sanskrit and Regional Names:

Eng. : Mountain Cedar, Red Cedar;
Hindi : Toona;
Sans. : Nandivriksha;
Ass. : Ponna;
Beng : Mahanim;
Kan. : Mandurike;
Mal. : Malarveppu;
Mar. : Deodari, Kuruk;
Ori. : Deodari, Kuruk;
Pub. : Birsu, Darab, Khanam, Tun;
Tam. : Santanavenbu;
Tel. : Nandi - chetu.

Distribution : Throughout the Tropical Himalaya. Hilly District of C. & S. India upto 1,000 m.

Part Used : Young Shoots.

Utilization : Young shoots lopped for fodder.

References : Anonymous (1950), Raju (1998).

Ceiba pentandra (Linn.) Gaertn. (Pl. 3B)

Syn: *Bombax pentandrum* Linn.; *Eriodendron anfractuosum* DC.; *E. pentandrum* Kurtz.
Family: Bombacaceae.

English, Hindi, Sanskrit and Regional Names:

Eng. : White Silk Cotton Tree, True Kapok Tree;

Hindi : Katan, Hattian, Safed Simal;
Sans. : Chirayu, Moch, Rochana, Sveta Salmali;
Kan. : Buraga, Bili buraga;
Mal. : Ilavu, Mullilavu;
Mar. : Pandhari, Salmali;
Tam. : Ilavum;
Tel. : Tella burga.

Distribution : Widely Distributed in Hotter Parts of India upto 500 m.

Parts Used : Silky Floss of Seed.

Utilization : Source of kapok fibre for bedding, upholstery industry, thermally insulated and sound proof covers and walls, padded surgical dressing, cardings, mattresses and life saving equipment.

References : Agarwal (2003), Ambasta (1986), Dastur (1964), Dwivedi (1993), Mittre (1997).

Celastrus paniculatus Willd.

Syn: *Celastrus montana* Roth.; *C. multiflorus* Roxb.; *C. nutans* Roxb.
Family: Celasteraceae.

English, Hindi, Sanskrit and Regional Names:

Eng. : Black oil Plant, Climbing Staff Tree, Intellect Tree;
Hindi : Makangni;
Sans. : Jyotismati, Kanguni, Pitatalia;
Guj. : Malkangana, Velo;
Kan. : Kariganne;
Mal. : Paluavam;
Mar. : Dhimarbal, Kanguni, Pigari;
Ori. : Korsana, Korsono;
Pub. : Sankhu;
Tam. : Adibarichum, Kaligam;
Tel. : Mala- erikata.

Distribution : Distributed Throughout India upto 1,800 m.

Part Used : Bark.

Utilization : Bark fibre used for sealing the rims of musical instruments.

References : Parrotta (2001), Singh & Pandey (1998).

Celtis australis Linn.

Syn: *Celtis acata* Ham.; *C. caucasia* Willd.

Family: Ulmaceae.

Vern. Names: Khidak, Khirak.

English, Hindi and Regional Names:

Eng. : European Nettle Tree, The Honey Berry Tree;
Hindi : Kharak;
Gar. : Kharik;
Pub. : Batkar, Biugli, Kharak, Khirik, Roku.

Distribution : Himalaya 1,000 -3,000 m.

Part Used : Leaves.

Utilization : Leaves considered a good fodder for sheep and all classes of livestock in winter.

Active Constituents : Leaves have iron, copper and protein 14%.

References : Agarwal (2003), Ambasta (1986), Dwivedi (1993), Hooker (1890), Kapur (1996), Parker (1973), Purohit & Samant (1995), Sharma (2004), Singh (1982, 2000), Sood & Thakur (2004).

Celtis eriocarpa Decne.

Family: Ulmaceae.

Distribution : Occurs between 1,000 -2,000 m.

Part Used : Leaves.

Utilization : It is lopped for fodder in winter. However, it is reported to reduce the milk.

Reference : Purohit & Samant (1995).

Celtis setigerus Vahl.

Syn: *Celtis biflorus* Roxb.; *C. montanus* Nees.
Family: Poaceae.

Distribution : N. W. India. Gujarat. Haryana. Madras & Punjab.

Part Used : Plant.

Utilization : Useful as fodder grass.

Active Constituents : Leaves rich in a diglycoside of cyaniding.

References : Ambasta (1986), Rastogi & Mehrotra (1991).

Celtis tertrandra Roxb.

Syn: *Celtis glabra* var., nepalensis Planch.; *C. roxburghii* Planch.; *C. sirotina* Planch.; *C. trinervia* Roxb.
Family: Ulmaceae.

Hindi Name:

Hindi : Bhrumaj.

Distribution : Kumaon Hills, Bengal, M. P., & W. Ghats.

Part Used : Plant.

Utilization : It is lopped for animals in winter.

References : Agarwal (2003), Hooker (1890), Purohit & Samant (1995).

Cenchrus barbatus Schum.

Syn: *Cenchrus catharticus* Delile.
Family: Poaceae.

Hindi Name:

Hindi : Bharut.

Distribution : Gangetic Plains. Punjab.

Part Used : Plant.

Utilization : Useful as a fodder grass.

References : Agarwal (2003), Ambasta (1986), Duthie (1888).

Chenopodium album L.

Syn: *Chenopodium giganteum* Don.; *C. nepalense* Linn.; *C. purpurascens* Ham.; *C. vulpinum* Wall.

Family: Chenopodiaceae.

Vern. Names: Bathua, Neu.

English, Hindi, Sanskrit and Regional Names:

Eng.	:	Lambs quarters;
Hindi	:	Bethu, Sag;
Sans.	:	Agarolohita, Chillika, Vastuka , Vastukah;
Beng	:	Chanda betu;
Guj.	:	Cheel, Tanko;
Kan.	:	Chakravati;
Mal.	:	Vastuccira;
Mar.	:	Chakavata, Chakvar, Chivil;
Pub.	:	Chakavata, Chakvar, Chivil;
Tam.	:	Parupukkirai;
Tel.	:	Pappukura.

Distribution : Tropical & Temperate Himalaya; ascending to 4,000 m. Plains of Bengal. The Gangetic Valley. Punjab.

Part Used : Whole Plant.

Utilization : Used for fodder purposes.

References : Hooker (1890), Parrotta (2001), Singh *et al.* (1996).

Chimnobambusa jaunsarensis (Gambel) Bahadur & Naithani.

Family: Poaceae.

Hindi Names:

Hindi	:	Jaunsari, Nigal, Ringal.

Part Used : Plant.

Utilization : It provides good fodder for both wild and domestic animals; also useful as raw material for paper making.

Reference : Raju (1998).

Chloris incompleta Roth.

Syn: Chloris roxburghii Edgew.

Family: Poaceae.

Hindi and Regional Names:

Hindi : Bamna, Hikagadi, Mathaniya;
Kan. : Melamelai hullu;
Tel. : Kanthari gadi.

Distribution : Throughout the Plains of India, from Punjab to Myanmar & Southwards to Sri Lanka.

Part Used : Plant.

Utilization : Plant considered as good fodder before flowering.

References : Ambasta (1986), Hooker (1897).

Chloris inflata Link.

Syn: *Chloris barbata* Sw.

Family: Poaceae.

Hindi and Regional Names:

Hindi : Gandi, Jargi;
Tel. : Uppa gaddi.

Part Used : Plant.

Utilization : Plant considered as good fodder before flowering.

Reference : Ambasta (1986).

Chonemorpha fragrans (Moon) Alston.

Syn: *Chonemorpha macrophylla* (Roxb.) G. Don.

Family: Apocynaceae.

English, Hindi, Sanskrit and Regional Names:

Eng. : Wood Vine;
Hindi : Garbhidaro;
Sans. : Morata, Murva;
Kan. : Massjinaru;
Mal. : Perunkurumpa;
Tam. : Perunkurumpa;
Tel. : Chaga.

Distribution : Throughout India. Moist Forest in Andaman Islands.

Part Used : Bark.

Utilization : Bark fibre used for fishing-nets.

References : Ambasta (1986), Prajapati *et al.* (2003).

Chrysopogon gidarba (Ham. *ex* Hook. f.) Haines.

Syn: *Andropogon gidarba* Ham. *ex* Hook. f.
Family: Poaceae.

Part Used : Leaves.

Utilization : Leaves used as fodder.

Reference : Sharma & Rana (2005).

Chrysopogon lancearius (Hook. f.) Haines.

Syn: *Andropogon lancearius* Hook. f.
Family: Poaceae.

Part Used : Plant.

Utilization : Considered as good fodder.

Reference : Ambasta (1986).

Chrysopogon orientalis A. Camus.

Syn: *Andropogon wightianus* Steud.
Family: Poaceae.

Vern. Names: Bhatindu, Bhindu, Kurupan.

Regional Name:

Tel. : Karappa gaddi.

Part Used : Plant.

Utilization : Used as fodder.

Reference : Ambasta (1986).

Cissampelos pareira L.

Syn: *C. caapeba* Roxb.; *C. convolvulacea* Willd.; *C. triandra* Roxb.
Family: Menispermaceae.

Vern. Names: Bhatindu, Bhindu, Kurupan.

English, Hindi, Sanskrit and Regional Names:

Eng.	:	False Pareira Brava, Velvet Leaf;
Hindi	:	Akauadi;
Sans.	:	Ambashthika, Ruchaishya, Vidhakanni;
Beng	:	Akanadi, Nimuka;
Guj.	:	Karandhis, Venivel;
Kan.	:	Padavali;
Mal.	:	Kattuvalli, Patuvalli;
Mar.	:	Onkanibhindi;
Ori.	:	Padavalli;
Pub.	:	But, Batindu path, Katori;
Tam.	:	Appatta, Vatattirupi;
Tel.	:	Pata.

Distribution : Throughout India upto 2,000 m.

Part Used : Stem.

Utilization : Used for fibre purposes.

References : Anonymous (1950), Jain (1991), Kirtikar & Basu (1935), Oommachan (1977), Sharma & Jain (1994), Talbot (1976b).

Cissus adnata Roxb.

Syn: *Vitis adnata* Wall. *ex* Wight.
Family: Vitaceae.

Regional Names:

Bom.	:	Kalezan;
Tel.	:	Kokkitayaralu.

Distribution : Himalaya. A.P. Assam. W. Ghats.

Part Used : Stem.

Utilization : Stem a source for cordage fibre.

References : Ambasta (1986), Roy *et al.* (1998).

Cissus pallida (W. A.) Plamch.

Syn: *Vitis pallida* W. & A.
Family: Vitaceae.

Part Used : Stem.

Utilization : Stem fibre used for making ropes.

References : Rao & Henry (1996), Sood *et al.* (2005).

Cissus repens Linn.

Syn: *Vitis repens* W.& A.
Family: Vitaceae.

Regional Names:

Ass. : Mei- hur- jarap;
Kan. : Elakom bullaballi;
Ori. : Diboria;
Tel. : Nelaboddu.

Distribution : Sikkim Himalaya. Assam. S. India.

Part Used : Stem.

Utilization : Stem used for making ropes.

References : Anonymous (1950), Sur & Halder (2002).

Clematis buchananiana DC.

Syn: *Clematis buchananiana* Hook. f. *et* Thomas.
Family: Ranunculaceae.

Vern. Names: Bald Khuja, Dhudki Bel.

English Name:

Eng.: Old Man Beard.

Distribution : Throughout Temperate Himalaya between 1,800–3,000 m.

Part Used : Plant.

Utilization : Plant yields fibre for cordage.

Reference : Sharma (2004).

Clematis nepaulensis DC.

Syn: *Clematis montana* Don Prodr.
Family: Ranunculaceae.

Vern. Names: Ghantiali, Iali.

Regional Name:

Pub. : Birri, Pawanne, Wondak.

Distribution : Temperate Himalaya from Garhwal to Bhutan.

Part Used : Aerial Plant Parts.

Utilization : It is extensively fed to all classes of livestock.

References : Kirtikar & Basu (1935), Purohit & Samant (1995).

Clinogyne dichotoma Salib.

Syn: *Donax arundinastrum* Lour.; *Phrynium dichotomum* Roxb.
Family: Marantaceae.

Hindi and Regional Names:

Hindi : Sitaphali;
Ass. : Murta;
Beng : Sitaphali.

Distribution : E. India.

Parts Used : Stem. Leaf.

Utilization : Leaf fibre used for making baskets, hats; and stem fibre for mats.

References : Agarwal (2003), Ambasta (1986), Roy *et al.* (1998).

Coccothrinax argentea (Schult. & Schult. f.) K. Schum.

Syn: *Thrinax argentea* Lodd. *ex* Schult. & f. Schult. f.
Family: Arecaceae.

Part Used : Mature Leaves.

Utilization : Mature leaves used for making brooms, hats, baskets and fancy articles.

Reference : Ambasta (1986).

Coccinia cordifolia (L.)

Syn: *Cephalandra indica* Naud.; *Coccinia indica* W. A.
Family: Cucurbitaceae.

English Name:

Eng. : Ivy Gourd.

Part Used : Stem.

Utilization : Stem fibre for making ropes.

Reference : Agarwal (2003).

Cochlospermum religiosum (Linn.) Alston.

Syn: *Cochlospermum gossypium* DC.
Family: Cochlospermaceae.

Vern. Name: Gabadi.

English, Hindi, Sanskrit and Regional Names:

Eng. : Silk Cotton Tree;
Hindi : Kumbi;
Sans. : Girisalmalika, Silakarpasika;
Beng : Gabdi, Golgol;
Guj. : Ganeri, Ganiar, Gungdi, Venivel;
Mal. : Appakuttakka;
Mar. : Ganeri, Ganglay;
Ori. : Kontopalas
Pub. : Kumbi;
Tam. : Kongilam, Tannaku;
Tel. : Adaviburuga.

Distribution : W. Coasts. Sub-Himalayan Tract upto 900 m. Plains of U. P. Orissa. Maharashtra. Gujarat. Deccan. East of the W. Ghats.

Part Used : Pods.

Utilization : Seed floss on the seeds used for stuffing mattresses, pillows, cushions, cordage and life belts.

References : Agarwal (2003), Ambasta (1986), Dastur (1964), Datt (1996), Dwivedi (1993), Drury (1978), Parrotta (2001),

Prajapati *et al.* (1995), Rastogi & Mehrotra (1995b), Sahni (1998), Talbot (1976a).

Coix aquatica Roxb.

Family: Poaceae.

Distribution : Bihar. Sikkim. Orissa. West Bengal.

Part Used : Plant.

Utilization : Plant relished as good fodder.

References : Agarwal (2003), Ambasta (1986), Anonymous (1952), Roy *et al.* (1998).

Coix lacryma-jobi Linn.

Syn: *Coix agrestis* Lour.; *C. arundiacea* Lamk.; *C. axaltala* Jacq.; *C. ovata* Stokes.; *C. pendula* Salisb.; *C. puellarum* Balans.

Family: Poaceae.

Vern. Name: Gurlu.

English, Hindi, Sanskrit and Regional Names:

Eng. : Job's Tears;
Hindi : Sankru;
Sans. : Gavedhukah, Jargadi;
Beng : Gurgur;
Guj. : Kasai, Ranzondlo;
Mal. : Kattugotampu;
Mar. : Ranmakkai;
Pub. : Sanklee, Sanklu;
Tam. : Netpavalum.

Distribution : Throughout the Hotter Parts of India (wild or cultivated upto 2,000 m).

Part Used : Leaves.

Utilization : Leaves used for fodder purposes and also for making ropes.

References : Ambasta (1986), Anonymous (1952), Das & Agarwal (1991), Dastur (1964), Das *et al.* (1996), Duthie (1988), Hooker (1897), Islam (2002), Jain (1991), Khan & Khanum (2005), Oommachan (1977), Parrotta (2001).

Colquhounia coccinea Wall.
Family: Lamiaceae.

Distribution : Occurs between 1,000- 2,500 m.

Part Used : Plant.

Utilization : It is browsed by sheep and goats during rainy season.

References : Nautiyal *et al.* (2003), Purohit & Samant (1995).

Combretum roxburghii Spreng.
Family: Combretaceae.

Part Used : Leaves.

Utilization : Leaves valued as fodder.

References : Sur & Halder (2002), Varghese (1996).

Commelina benghalensis L.
Family: Commelinaceae.

English, Hindi, Sanskrit and Regional Names:

Eng. : Day Flower, Dew Flower;

Hindi : Kanchara;

Sans. : Kankata, Kanchata;

Beng : Kachradam, Kanchara;

Guj. : Mhotunshushmulian;

Mar. : Kena;

Tam. : Kanavazhar;

Tel. : Nirukassuvu, Vomadivikura.

Distribution : Moist Habitats Throughout Tropical & Sub-Tropical India.

Part Used : Plant.

Utilization : Plant browsed for fodder.

References : Anonymous (1950), Caius (2003), Islam (2002), Parrotta (2001).

Corchorus capsularis Linn.

Family: Tiliaceae.

Vern. Name: Tammar- labangphul.

English, Hindi, Sanskrit and Regional Names:

Eng.	:	White jute;
Hindi	:	Narcha, Patt;
Sans.	:	Kalasaka;
Ass.	:	Titamara;
Beng	:	Chinalitapat, Narcha, Nalita, Titapat;
Guj.	:	Chunchdo;
Mar.	:	Karru chantz;
Tam.	:	Pirattaikeerai;
Tel.	:	Parinta.

Distribution : Cultivated Throughout Hotter Parts of India.

Part Used : Stem.

Utilization : Jute fibre from stem used for gunny bags, coarse cloths, twine, carpets, sacs, tarpoline, packing cloths, tents and wide range of domestic and industrial products in paper industry.

References : Agarwal (2003), Ambasta (1986), Anonymous (1952), Drury (1978), Islam (2002), Prasad *et al.* (2002, 2003), Singh & Pandey (1998).

Corchorus olitorius L.

Family: Tiliaceae.

English, Hindi, Sanskrit and Regional Names:

Eng.	:	Desi Jute, Jew's Mallow, Tossa Jute;
Hindi	:	Koahta;
Sans.	:	Brihachanchu, Patta, Singgika;
Beng	:	Bogi, Desipat, Mithapat;
Guj.	:	Banphal;
Mar.	:	Chunch;
Ori.	:	Jhoto, joto;
Pub.	:	Banphal;

Tam.	:	Parratti;
Tel.	:	Parintakura.

Distribution : Commonly in Bengal, Bihar and Orissa.

Part Used : Stem.

Utilization : Jute fibre employed for making antiseptic surgical dressing, sacs, coarse fabrics, ropes, cordage and paper making.

References : Agarwal (2003), Anonymous (1950), Drury (1978), Parrotta (2001), Singh & Pandey (1998), Watt (1972).

Cordia dichotoma Forst. F.

Syn: *Cordia latifolia* Roxb.; *C. myxa* Roxb. non. Linn.; *Cordia oblique* Willd.

Family: Ehretiaceae.

English, Hindi, Sanskrit and Regional Names:

Eng.	:	Clammy cherry, Indian cherry;
Hindi	:	Lasora;
Sans.	:	Bahuvaraka, Ganghapushpa, Selu;
Beng	:	Bahubara;
Guj.	:	Bargund;
Kan.	:	Chikkachalle;
Mal.	:	Cheruviri,Viri;
Mar.	:	Shelvanti;
Ori.	:	Ambata, Gondi, Guhalo;
Pub.	:	Lasura;
Tam.	:	Naruvili;
Tel.	:	Chinna nakkeru.

Distribution : Distributed in Hotter Parts of India upto 1,000 m.

Parts Used : Stem. Leaves.

Utilization : Leaves lopped for fodder and bast fibre used for ropes, brushes and cordages.

Active Constituents : Leaves contain 12.37-15.13% crude protein, 16.45-25.76% crude fibre, 4.24% Ca, 0.30% P, macrophyllin, sitosterol, linolic acid and olic acid.

References : Anonymous (1950), Dastur (1964), Drury (1978), Dwivedi (1993), Hooker (1885), Parmar & Kaushal (1982), Parrotta (2001), Purohit & Samant (1995), Singh (1982), Singh & Pandey (1998), Rastogi & Mehrotra (1995a).

Cordia rothii Roem. & Schult.

Syn: *Cordia angustifolia* Roxb.; *C. reticulata* Roth Nov.; *C. subopposita* DC.
Family: Boranginaceae.

Hindi, Sanskrit and Regional Names:

Hindi : Gondi;
Sans. : Laghushleshmataka;
Guj. : Gundi;
Kan. : Kirichalle;
Mar. : Gondani;
Tam. : Naruvili;
Tel. : Chinnabotuku.

Distribution : Rajasthan. Gujarat. Karnataka.

Part Used : Bark.

Utilization : Bark fibre used for making ropes.

References : Agarwal (2003), Ambasta (1986), Dwivedi (1993), Hooker (1885), Talbot (1976b).

Cordia vestita (DC.) Hook. f. &Th.

Syn: *Cordia incana* Royle.; *Gynaion vestitum* A. DC.
Family: Boranginaceae.

Hindi and Regional Names:

Hindi : Bairola, Kumpaiman;
Pub. : Karuk, Kumbi;
Tam. : Bairola.

Distribution : Hills between 300-1,500 m.

Part Used : Leaves.

Utilization : Leaves lopped for fodder during winter.

References : Dwivedi (1993), Hooker (1885), Purohit & Samant (1995).

Coriandrium sativum Linn. Coriander.

Syn : *Cuminum cyminum* Wall.

Family : Apiaceae.

English, Hindi, Sanskrit and Regional Names:

Eng.	:	Coriander;
Hindi	:	Dhania;
Sans.	:	Dhanya, Dhanika, Vitunnaka;
Beng	:	Dhane;
Bom.	:	Dhana;
Guj.	:	Konphir, Libdhana;
Kan.	:	Kothambri;
Mal.	:	Kothumpalari, Malli;
Mar.	:	Dhanna, Kothimber;
Ori.	:	Dhania;
Pub.	:	Dhania;
Tam.	:	Dhaniya;
Tel.	:	Dhaniyalu.

Distribution : Throughout India.

Part Used : Seed.

Utilization : Seed cakes used as fodder.

Active Constituents : Rich in vit. A. over 1500 I.U./100mg with riboflavin 350.

References : Agarwal (2003), Ambasta (1986), Anonymous (1950), Hooker (1879), Prajapati *et al.* (2003), Pruthi (1976).

Coriaria nepalensis Wall.

Family: Coriariaceae.

Hindi Name:

Hindi : Masuri.

Distribution : Outer Himalaya. N. W. to Sikkim.

Part Used : Branches.

Utilization : Branches browsed by sheep and goats.

References : Anonymous (1950), Raju (1998).

Cornus capitata Wall.

Syn: *Benthania fragifera* Lindl.
Family: Cornaceae.

Distribution : Occurs in Oak Forest from Kullu, Kumaon to Bhutan.

Part Used : Leaves.

Utilization : Leaves valued for fodder purposes.

Active Constituents : Leaves contain iridoid glucoside, verbenalin and phlorin.

References : Dwivedi (1993), Hooker (1879), Purohit & Samant (1995).

Corypha umbraculifera Linn.

Family: Arecaceae.

English, Hindi, Sanskrit and Regional Names:

Eng. : Fan Palm, Talipot Palm;
Hindi : Karalika;
Sans. : Karalika, Sritalam, Tali;
Beng : Tali, Tarit;
Kan. : Sritale;
Mal. : Kuttapanna, Sitalam, Tallipanna;
Mar. : Bajarbatu, Tali;
Tam. : Tali panai;
Tel. : Sritalamu.

Distribution : N. Malabar. Andaman Islands.

Part Used : Leaves.

Utilization : Leaf stocks afford material for manufacture of wrapping paper.

References : Agarwal (2003), Ambasta (1986), Anonymous (1950), Jain (1991), Sahni (1998).

Corypha utan Lam.

Syn: *Corypha elata* Roxb.
Family: Arecaceae.

English, Hindi and Regional Names:

Eng. : Agle Palm, Buri Palm;
Hindi : Bajur;
Beng : Bajur, Bajur batul.

Distribution : Bengal. Andaman Islands.

Parts Used : Unopened Leaves. Petioles.

Utilization : Petiole fibre used for hats, rough cloths, bags, trays and cigarette cases. Ribs of unopened leaves yield a fibre for making Calasiao and Prototan hats.

References : Agarwal (2003), Ambasta (1986), Anonymous (1950).

Cotoneaster obtusa Wall. *ex* Lindl.

Syn: *Cotoneaster bacillaris* Wall.
Family: Rosaceae.

Distribution : Occurs between 2,000-3,000m.

Part Used : Leaves.

Utilization : Leaves considered a good fodder for animals.

Active Constituents : Leaves yield 0.57 % hydrocyanic acid.

Reference : Purohit & Samant (1995).

Cotula hemispherica (Roxb.) Wall. *ex* Clarke.

Family: Asteraceae.

Part Used : Plant.

Utilization : Plant used for fodder purposes.

Reference : Islam (2002).

Crataeva religiosa Thoms.

Family: Capparidaceae.

English, Hindi, Sanskrit and Regional Names:

Eng.	:	The Barna;
Hindi	:	Barna;
Sans.	:	Varuna;
Beng	:	Barun;
Kan.	:	Bole tumbe;
Mal.	:	Kili;
Mar.	:	Vayavarna;
Ori.	:	Barun;
Pub.	:	Barna;
Tam.	:	Maralingam;
Tel.	:	Ulimari.

Part Used : Leaves.

Utilization : Leaves used as fodder for cattle.

References : Raju (1998), Parker (1973).

Crotalaria burhia Buch.-Ham. Benth.

Family: Fabaceae.

Vern. Name: Sanpat.

Hindi and Regional Names:

Hindi	:	Sissai;
Bom.	:	Sis, Sissai;
Guj.	:	Ghugharo;
Mar.	:	Ghagari;
Pub.	:	Khip, lathia, Polo, Sissai .

Distribution : Common in Arid Parts of Punjab. Rajputana & Gujarat.

Part Used : Stem.

Utilization : Bast fibre employed by the 'Bhil' and 'Garasia' tribals for ropes and cordage.

References : Agarwal (2003), Ambasta (1986), Anonymous (1950), Dwivedi (1993), Singh & Pandey (1998), Sur & Halder (2002), Talbot (1977).

Crotalaria juncea Linn.

Syn: *Crotolaria benghalensis* Lamk.; *C. fenestrata* Bot.; *C. porrecta* Wall.; *C. sericea* Willd.; *C. tenulfolia* Roxb.; *C. viminea* Wall.

Family: Fabaceae.

Vern. Name: Madar.

English, Hindi, Sanskrit and Regional Names:

Eng. : Sann Hemp, Sunn;
Hindi : Masina, San, Sunn;
Sans. : Sana, Sanah, Shanapushpi;
Beng : Shon, Shonpat;
Guj. : San, Sun, Suna;
Kan. : Sanabu;
Mal. : Janapa, Wucko nar;
Mar. : Tag;
Ori. : Choni, Somo;
Tam. : Sanal, Sannappu;
Tel. : Janumu.

Distribution : Cultivated Throughout Warmer India (Mysore, Deccan and S. India).

Parts Used : Stem Bark.

Utilization : Bark fibre used for bags, flax-card, matting, rugs, webbing and wrapping paper, cordage, canvas and fishing-nets.

References : Agarwal (2003), Ambasta (1986), Anonymous (1950), Chakraborty & Bhattacharjee (2003), Deshmukh *et al.* (2000), Drury (1978), Hooker (1879), Jain (1991), Khanna (2002), Parrotta (2001), Sharma & Rana (2005), Talbot (1976b), Varghese (1996), Watt (1972).

Crotalaria mucronata Desv.

Syn: *Crotalaria striata* DC.
Family: Fabaceae.

English, Hindi and Regional Names:

Eng.	:	San Hemp;
Hindi	:	San, Sen;
Ass.	:	Ghant akaram.

Distribution : Throughout India.

Part Used : Plant.

Utilization : Yields a fibre similar to sunn hemp.

References : Ambasta (1986), Roy *et al.* (1998).

Cryptolepsis buchanani Roem. & Schult.

Syn: *Cryptolepsis reticulata* Wall.; *Echites cuspidata* Heyne.; *E. reticulata* Roth.; *Nerium reticulam* Roxb.
Family: Asclepiadaceae.

Vern. Names: Khamb, Thundra.

English, Hindi, Sanskrit and Regional Names:

Eng.	:	Indian Sarsaparilla;
Hindi	:	Dudhi, Karanta;
Sans.	:	Gopavalli, Kashmari, Krishna, Saariva;
Kan.	:	Metaguli hambu;
Mal.	:	Kalipalvalu, Kattupalvalli;
Mar.	:	Kaovel, Kawavel;
Ori.	:	Gopikoniora, Maloti;
Tam.	:	Umi- thekku;
Tel.	:	Adavipalotige.

Distribution : Throughout India, ascending to 1,300 m (Kashmir to Assam. C. and S. India).

Part Used : Stem.

Utilization : 'Bhils' of Dungarpur use the ropes made from stem fibre for tightening the leather hoops of musical instruments and knitting their cots.

References : Anonymous (1950), Hooker (1885), Khare (2004), Parrotta (2001), Rao & Henry (1996), Sharma (2004), Singh & Pandey (1998).

Cryptostegia grandiflora R. Br.

Syn: *Nerium grandiflorum* Roxb.
Family: Periplocaceae.

Regional Names:

Mar. : Vilayati vakhandi;
Tam. : Garudappalai, Palai.

Distribution : Cultivated in Various Parts of Bombay, Madras, Punjab, U. P. & Delhi.

Part Used : Seeds.

Utilization : The floss fibre of seeds used for stuffing the pillows and paper industry.

References : Anonymous (1950), Drury (1978), Hooker (1885), Singh & Pandey (1998).

Cyamopsis tetragonoloba (Linn.) Taub.

Syn: *Cyamopsis psoralioides* DC.
Family: Fabaceae.

Vern. Name: Ganwai.

English, Hindi, Sanskrit and Regional Names:

Eng. : Cluster Bean;
Hindi : Gowar;
Sans. : Bakuchi, Gorani;
Bom. : Gauri, Gawari;
Guj. : Guwar;
Kan. : Gorikayi;
Mar. : Bavachi, Gowar;
Pub. : Guwar;
Tam. : Kothaveray;
Tel. : Gorkikudu.

Distribution : Cultivated Throughout India in U.P. & Tamil Nadu.

Part Used : Pods.

Utilization : Used as green fodder for cattle.

References : Agarwal (2003), Ambasta (1986), Blatter & Hallberg (1984), Khare (2004), Kirtikar & Basu (2003), Sharma (2003).

Cyathula tomentosa (Roth.) Miq.

Syn: *Achyranthes sequax* Wall.; *A. tomentosa* Roth.; *Desmochaeta tomentosa* Roem. & Sch.; *Polyscelis sequax* Wall.

Family: Amarantaceae.

Vern. Name: Kutha.

Distribution : Kashmir to Bhutan: 1,000-2,500 m.

Part Used : Aerial Parts.

Utilization : Plant used as fodder during rainy season.

References : Purohit & Samant (1995), Sharma (2004), Sood & Thakur (2004).

Cynodon dactylon Pers.

Syn: *Agrostis linearis* Retz.; *Cynodon linearis* Willd.; *Panicum dactylon* L.

Family: Poaceae.

English, Hindi, Sanskrit and Regional Names:

Eng. : Bahama Grass, Couch grass;
Hindi : Dub;
Sans. : Durva, Haritali, Ruha, Vijaya;
Beng : Dubh, Dubla;
Guj. : Dhro, Durva, Gharo;
Kan. : Garike hullu, Karkeri hullu;
Mal. : Kuruka;
Mar. : Haryali, Karala;
Ori. : Dubbula, Ghanso;
Pub. : Dhub Khabbal, Talla;
Tam. : Arugam-pullu;
Tel. : Gericha gaddi.

Distribution : Throughout India; ascending to 3,000 m in the Himalaya.

Part Used : Herb.

Utilization : Important pasture plant for cattle, sheep and goats.

Active Constituents : Leaves contain crude proteins 10.47% and fibre 28.17%.

References : Ambasta (1986), Diwaker & Ansari (2003), Dwivedi (1993), Hooker (1897), Khare (2004), Oommachan (1977), Sharma & Rana (2003).

Cyanotis axillaris Roem. & Schult.

Syn: *Tradescantia axillaris* Roxb

Family: Commelinaceae.

Hindi and Regional Names:

Hindi : Baghanulla;
Mar. : Itsaka;
Tam. : Nirpulli.

Distribution : Throughout India in the Plains from Upper Gangetic Valley to Assam Southwards.

Part Used : Plant.

Utilization : Plant used as fodder.

References : Caius (2003), Islam (2002), Parrotta (2001).

Cyperus articulatus Linn.

Family: Cyperaceae.

English Name:

Eng. : Guinea Rush.

Distribution : E. India to S. India.

Part Used : Tuber.

Utilization : Fragrant tubers used for making perfumed clothings.

References : Agarwal (2003), Ambasta (1986).

Cyperus compressus L.

Family: Cyperaceae.

Part Used : Plant.

Utilization : Plant used as fodder.

Reference : Islam (2002).

Cyperus cuspidatus H. B. & K.

Family: Cyperaceae.

Part Used : Plant.

Utilization : Plant used as fodder.

Reference : Islam (2002).

Cyperus difformis L. (Pl. 3C)

Family: Cyperaceae.

Distribution : Throughout India to 2,400 m.

Part Used : Plant.

Utilization : Commonly used for fodder.

Reference : Islam (2002).

Cyperus distans L. f.

Family: Cyperaceae.

Distribution : Throughout Plains of India ascending to 900 m.

Part Used : Plant.

Utilization : Used for fodder.

Reference : Islam (2002).

Cyperus haspans Linn.

Family: Cyperaceae.

Part Used : Leaves.

Utilization : Leaves used as fodder.

Reference : Das *et al.* (1996).

Cyperus iria Linn. (Pl. 3D)

Family: Cyperaceae.

Distribution : Throughout India.

Part Used : Leaves.

Utilization : Leaves given as fodder.

References : Das *et al.* (1996), Parrotta (2001).

Cyperus metzii (Hochst.) Mattf. & Kukenthal.

Syn: *Kyllinga squamulata* Vahl.

Family: Cyperaceae.

Part Used : Plant.

Utilization : Eaten by cattle.

Reference : Ambasta (1986).

Cyperus niveus Retz.

Family: Cyperaceae.

Part Used : Plant.

Utilization : It is commonly used for fodder.

Reference : Islam (2002).

Dactylis glomerata L.

Syn: *Dactylis glaucescens* Willd.; *D. hispanica* Roth.; *Festuca glomerata* All.

Family: Poaceae.

English Names:

Eng. : Cock's Foot Grass, Orchard Grass.

Distribution : Grows in Punjab, Assam, Nilgiris Hills.

Part Used : Plant.

Utilization : Grass relished by cattle, sheep and horses.

Active Constituents : Plant rich in xylose (50.0), arabinose (22.0), glucose (8.0) and galactose (7.0).

References : Agarwal (2003), Ambasta (1986), Anonymous (1952), Hooker (1897), Rastogi & Mehrotra (1990), Sharma & Rana (2005), Watt (1972).

Dalbergia latifolia Roxb.

Syn: *Dalbergia emarginata* Roxb.
Family: Fabaceae.

Vern. Name: Shisham.

English, Hindi, Sanskrit and Regional Names:

Eng.	:	Bombay Black Wood, Rose Wood;
Hindi	:	Shisham;
Sans.	:	Shishapa;
Beng	:	Kiri, Ruti, Siso;
Guj.	:	Shisham;
Kan.	:	Bite, Todagatti;
Mal.	:	Itti, Kar- Itti;
Mar.	:	Shisham, Sisu, Siswa;
Ori.	:	Sisso, Sissu;
Pub.	:	Shisham;
Tam.	:	Itti, Karundorviral;
Tel.	:	Biti.

Distribution : Gangetic Plains. Sikkim. C. W. & S. India.

Part Used : Leaves.

Utilization : Source of fodder.

Active Constituents : Leaves possess isosterol, pinitol, glucosylcaviunin, luteolin and apigenin.

References : Agarwal (2003), Ambasta (1986), Anonymous (1950), Dash & Misra (2000), Dastur (1964), Drury (1978), Dwivedi (1993), Hooker (1879), Parrotta (2001), Rastogi & Mehrotra (1995b), Talbot (1976a).

Dalbergia pinnata (Lour.) Prain.

Syn: *Dalbergia tamarindifolia* Roxb.
Family: Fabaceae.

Hindi and Regional Names:

Hindi	:	Ketikat;
Ass.	:	Keti.

Distribution : E. Himalaya. W. Peninsula. Assam.

Part Used : Leaves.

Utilization : Leaves lopped for fodder.

References : Agarwal (2003), Ambasta (1986), Anonymous (1952).

Dalbergia sericea G. Don.

Syn: *Dalbergia hireina* Wall. *ex* Benth.; *D. stenocarpa* Kurz.
Family: Fabaceae.

Distribution : Occurs between 500-1,500 m.

Part Used : Leaves.

Active Constituents : Leaves contain quercetin and hyperin.

References : Purohit & Samant (1995), Rastogi & Mehrotra (1991).

Dalbergia sissoo Roxb.

Syn: *Dalbergia pendula* Tenore.
Family: Fabaceae.

Vern. Name: Shisham.

English, Hindi, Sanskrit and Regional Names:

Eng. : Red Wood, Sisso, South Wood;
Hindi : Shisham, Sissoo, Shishu;
Sans. : Agaru, Dhira, Kapila, Shinshapa;
Ass. : Sissu;
Beng : Shishu;
Guj. : Shisham, Tanch;
Kan. : Agarau, Birid;
Mal. : Iruvil;
Mar. : Sisu, Sissu;
Ori. : Sibu, Simsapa;
Pub. : Shisham, Tali;
Tam. : Itti, Sisu;
Tel. : Etta, Sisso, Sinsupa.

Distribution : W. Himalaya. Tarai of N. India. Sikkim. Assam.

Part Used : Leaves.Twigs.

Utilization : Leaves and twigs browsed by cattle and camels.

Active Constituents : Leaves contain biochanin A-7-glucoside and isoflavin.

References : Agarwal (2003), Ambasta (1986), Anonymous (1952), Drury (1978), Dwivedi (1993), Hooker (1879), Parrotta (2001), Rastogi & Mehrotra (1990), Sharma & Rana (2005), Sharma & Thakur (2004), Singh (1982, 2002, 2003), Talbot (1976b), Watt (1972).

Dalhousiea papyracea Wall. *ex* Steud.

Syn: *Dalhousiea cannabina* Wall.

Family: Fabaceae.

Hindi and Regional Names:

Hindi : Satpura, Setburwa;
Pub. : Niggi jeku.

Part Used : Bark.

Utilization : Bark fibre good for cordage.

Reference : Ambasta (1986).

Daphne cannabina Hook. f.

Syn: *Daphne odora & bholua* Don.; *D. papyracea* Wall. *ex* Steud.; *D. papyrifera* Ham. MS.

Family: Thymelaeaceae.

English, Hindi and Regional Names:

Eng. : Nepal Paper Plant;
Hindi : Satpura, Setburosa;
Pub. : Jeku.

Distribution : Temperate Himalaya (Chamba to Bhutan). Khasia & Naga Hills. Manipur.

Part Used : Bark.

Utilization : It used in the manufacture of folk paper 'Paharay Kagach' and believed to be suitable for making cartridges. Bark fibre also good for tying purpose.

References : Anonymous (1952), Dam & Hajra (1981), Dhiman (2003), Drury (1978), Hooker (1890), Saklani & Jain (1994), Singh & Kumar (2000), Watt (1972), Yonzone *et al.* (1996).

Daphne mucronata Royle

Syn: *Daphne mucronata* Royle
Family: Thymelaeaceae.

Distribution : W. Himalaya from Chamba in H. P. to Bhutan.

Part Used : Stem Bark.

Utilization : Bark fibre excellent for making paper.

Reference : Agarwal (2003).

Debregeasia hypoleuca Wedd.

Syn: *Boehmaria hypoleuca* Hochst.; *B. silicifolia* Don.; *Debregeasia bicolor* Wedd.; *Missiessya hypoleuca* Wedd.; *Monocarpus silicifolius* Blume; *Urtica bicolor* Roxb.
Family: Urticaceae.

Vern. Names: Siaru, Shyaaru.

Hindi and Regional Names:

Hindi : Sansaru, Siaru;
Pub. : Pinchu, Sansaru.

Distribution : W. Himalaya from Kashmir to Kamaun.

Parts Used : Bark. Leaves.

Utilization : Stem excellent for making ropes, cordage, twines and fishing lines. Leaves lopped for fodder.

References : Agarwal (2003), Ambasta (1986), Anonymous (1952), Hooker (1890), Parker (1973), Raju (1998), Sood & Thakur (2004), Watt (1972), Yoganarasimhan *et al.* (1981).

Debregeasia longifolia Wedd.

Syn: *Debregeasia velutina* Gaud.
Family: Urticaceae.

English, Hindi and Regional Names:

Eng. : Wild Rhea;

Hindi	:	Sansaru;
Kan.	:	Kepasi;
Tam.	:	Kattunochi;
Tel.	:	Kerang.

Distribution : C. & E. Himalaya. W. Ghats & Nilgiris: 600-1,800 m.

Parts Used : Leaves. Bark.

Utilization : Leaves fed to all ruminants, especially sheep, goats and cattle during winter and summer. Bark fibre used for cordage, ropes and fishing nets.

References : Agarwal (2003), Ambasta (1986), Anonymous (1952), Bennet *et al.* (1991), Geetha *et al.* (1996), Kumar *et al.* (2000), Purohit & Samant (1995), Rao & Henry (1996), Saklani & Jain (1994), Sharma & Rana (2005), Singh & Kumar (2000).

Decalepis hamiltonii W. & A.

Family: Asclepiadaceae.

Sanskrit and Regional Names:

Sans.	:	Sariba, Svetasariva;
Kan.	:	Makali beru;
Tam.	:	Kizhangu, Mahali, Peru;
Tel.	:	Maredu - geddalu, Neemam theega.

Distribution : Deccan Peninsula. Forest of W. Ghats upto 1,500 m.

Part Used : Plant.

Utilization : Used for making ropes.

References : Anonymous (1952), Parrotta (2001), Rao & Henry (1996).

Deeringia amaranthoides Merrill.

Syn: *Achyranthes amaranthoides* Lamk.; *Deeringia celosioides* R. Br.

Family: Amarantaceae.

Hindi and Regional Names:

Hindi	:	Lemon;
Ass.	:	Rangolilota;
Beng	:	Gola - mohani.

Distribution : Sub-Himalayan Tract from Chenab to Bhutan. Bihar. Bengal. Assam.

Part Used : Leaves.

Utilization : Leaves much valued as a fodder and said to enhance lactation in cows.

References : Anonymous (1952), Raju (1998), Sood *et al.* (2005).

Dendrocalamus hamiltonii Nees *et* Arn. *ex* Munro.

Syn: *Bambusa falconeri* Munro.; *B. maxima* Ham.; *B. monogyna* Griff.

Family: Poaceae.

Vern. Names: Baingh, Maggar.

Hindi and Regional Names:

Hindi : Kyazi - bans;
Ass. : Kokua;
Beng : Kokwa;
Bom. : Pecha.

Distribution : Tropical Himalaya. Meghalaya. Garwhal to Sikkim. Khasia Hills.

Part Used : Stem.

Utilization : Stem used for making ropes.

References : Dwivedi (1993), Hooker (1897), Saklani & Jain (1995), Sharma & Rana (2005), Watt (1972).

Dendrocalamus strictus Nees.

Syn: *Bambusa pubescens* Lodd.; *B. stricta* Roxb.; B. *verticillata* Rottl. *ex* Munro.; *B. tanaea* Ham *ex* Wall.

Family: Poaceae.

Vern. Name: Bans.

English, Hindi, Sanskrit and Regional Names:

Eng. : Mall Bambo, Solid Bamboo;
Hindi : Bans kaban, Bans khurd, Nar bans;
Sans. : Vansha;
Beng : Karail;

Guj.	:	Nakor bans;
Kan.	:	Kiri Bidiru;
Mal.	:	Kal, Mungil;
Mar.	:	Bhariyal, Bhovarli;
Ori.	:	Salia bhanso, Salimbo banaso;
Tam.	:	Kalmungil;
Tel.	:	Sadanapa veduru.

Distribution : Deciduous Forest all over India. Abundant in the Plains & Lowers Hills of N. & C. India.

Parts Used : Leaves. Stem.

Utilization : Leaves lopped for fodder for cattle, horses, elephants and buffaloes. Stem used for making mats, cordage and paper.

References : Agarwal (2003), Ambasta (1986), Anonymous (1952), Dash & Misra (2000), Dastur (1964), Datt (1996), Drury (1978), Dwivedi (1993), Hooker (1897), Jain (1991), Maheshwari *et al.* (1997), Oommachan (1977), Purohit & Samant (1995), Rao & Henry (1996), Sahni (1998), Saklani & Jain (1994), Sharma & Rana (2005), Singh & Kumar (2000), Watt (1972).

Dendrocnide sinuta (Bl.) Chew.

Syn: *Leportea crenulata* Gaud.

Family: Urticaceae.

Hindi and Regional Names:

Hindi	:	Utigun;
Ass.	:	Sorot gach, Tarash;
Beng	:	Charpata;
Mal.	:	Annachoriyan;
Ori.	:	Ottapilavu .

Distribution : Usually upto 1,500 m in the Himalaya. S. India.

Part Used : Stem.

Utilization : Stem fibre good for making ropes.

References : Bennet *et al.* (1991), Saklani & Jain (1994).

Depatratium junceum (Roxb.) Buch -Ham.

Family: Scrophulariaceae.

Part used : Plant.

Utilization : Plant used as fodder.

Reference : Islam (2002).

Derris trifoliata Lour. (Pl. 4A)

Syn: *Derris uliginosa* Benth.; *Robinia uliginosa* Willd.

Family: Fabaceae.

Vern. Name: Panlata.

Hindi and Regional Names:

Hindi : Panlata;
Beng : Panlata;
Tel. : Chirathila thige.

Distribution : Coastal Forest of India & Andamans. E. Himalaya. Assam.

Part Used : Leaves.

Utilization : Leaves lopped for fodder for cattle and sheep.

References : Agarwal (2003), Ambasta (1986), Anonymous (1952), Awasthi & Goel (2000), Dagar & Dagar (2000), Das *et al.* (1996).

Desmodium elegans DC.

Syn: *Desmodium tiliaefolium* G. Don.

Family: Fabaceae.

Distribution : Occurs between 1,500 - 2,000 m.

Parts Used : Leaves. Stem Bark.

Utilization : Leaves relished by sheep and goats. Stem bark used for making ropes.

Active Constituents : Leaves contain n- aliphatic alcohols, wax ester and sitosterol.

References : Nautiyal *et al.* (2003), Purohit & Samant (1995), Rastogi & Mehrotra (1990,1995a), Singh & Kumar (2000).

Desmodium gangeticum DC.

Syn: *Desmodium latifolium* Wight.;
Hedysarum collium Roxb.; *H. gangeticum* L.
Family: Fabaceae.

English, Hindi, Sanskrit and Regional Names:

Eng.	:	Salpan;
Hindi	:	Sarivan;
Sans.	:	Sala parani;
Beng	:	Salpani;
Bom.	:	Daye, Salparni, Salwan;
Guj.	:	Salvan;
Mal.	:	Pullati;
Mar.	:	Darh, Ranbhal, Salparni, Salwan;
Ori.	:	Soloporini;
Pub.	:	Shal purni;
Tam.	:	Pulladi;
Tel.	:	Gita naram.

Distribution : Throughout in the C. & Outer Ranges upto 2,000 m.

Parts Used : Aerial Parts.

Utilization : Fed to all classes of livestock during rainy season.

References : Agarwal (2003), Hooker (1879), Khare (2004), Parrotta (2001), Purohit & Samant (1995), Rastogi & Mehrotra (1990), Watt (1972).

Desmodium microphyllum DC.

Syn: *Desmodium parvifolium* DC.
Family: Fabaceae.

Part Used : Leaves.

Utilization : Leaves used as fodder.

References : Ambasta (1986), Dwivedi (1993), Singh & Kumar (2002).

Desmodium motorium Merrill.

Syn: *Desmodium gyrans* DC.
Family: Fabaceae.

English and Hindi Names:

Eng. : Telegraph Plant;
Hindi : Tarpot.

Part Used : Leaves.

Utilization : Leaves lopped for fodder.

Active Constituents : Leaves contain indole alkaloids, bufotenine and choline.

References : Agarwal (2003), Ambasta (1986), Rastogi & Mehrotra (1991).

Desmodium pannosus Stapf.

Syn: *Unona pannosa* Dalz.
Family: Fabaceae.

Part Used : Inner Bark.

Utilization : Inner-bark useful as source of fibre.

Reference : Ambasta (1986).

Desmostachya bipinnata (L.) Staph.

Syn: *Eragrostis cynosuroides* Beauv.
Family: Poaceae.

English, Hindi, Sanskrit and Regional Names:

Eng. : Sacrificial Grass;
Hindi : Dab, Durva;
Sans. : Darbha, Kusha, Suuchyagra;
Beng : Durbha, Kusha;
Guj. : Dabha, Darabha;
Mal. : Darbha;
Mar. : Durva;
Pub. : Dab, Dhab, Dib, Drab , drabh;
Tam. : Darbhaipul;
Tel. : Darbha, Kushdarbh.

Distribution : Throughout the Plain of India & in Sandy Desert Area.

Part Used : Leaves.

Utilization : Leaves used by tribals for making ropes, cordage. Fibre also used as raw material for paper making.

References : Anonymous (1952), Khare (2004), Kirtikar & Basu (1984), Prajapati *et al.* (2003), Singh & Pandey (1998).

Dichrocephala integrifolia (Ait.) Kuntze.

Syn: *Cotula bicolor* Roth.; *Dichrocephala bicolor* (Roth.) Sch.; *D. latifolia* DC.; *Ethula integrifolia* Don.; *Grangea latifolia* Lamk.; *Hippa integrifolia* Ait.

Family: Asteraceae.

Distribution : Himalaya, U. P. to Sikkim.

Part Used : Plant.

Utilization : Leaves given as fodder to ruminants.

Active Constituents : Leaves contain three isomeric caffic acid ester-3, 4-dicaffeoyl, kaempferol and quercetin.

References : Rastogi & Mehrotra (1995a), Sood & Thakur (2004).

Dichrostachys cinerea (L.) W. A.

Syn: *Acacia cinerea* Linn.; *A. dales* Desv.; *Desmanthus cinereus* Willd.; *Mimosa cinerea* Linn.

Family: Fabaceae.

Hindi, Sanskrit and Regional Names:

Hindi : Vurtuli;
Sans. : Viravriksha;
Bom. : Vurtuli;
Guj. : Marud;
Kan. : Odatare;
Mal. : Vitattal;
Mar. : Segam kati;
Ori. : Khoiridya;
Tam. : Vadatalla, Vadatara;
Tel. : Veltu, Veturu.

Distribution : W. & C. India. Rajputana. Tamil Nadu.

Part Used : Leaves.

Utilization : Leaves as excellent fodder for cattle and horses.

Active Constituents : Leaves yield hentriacontanol and β- amyrin.

References : Anonymous (1952), Dwivedi (1993), Hooker (1879), Kirtikar & Basu (1984), Prajapati *et al.* (2003), Rastogi & Mehrotra (1991), Singh & Pandey (1998), Watt (1972).

Diectomis fastigiata H. B. &K.

Syn: *Andropogon fastigiatus* Sw.
Family: Poaceae.

Distribution : Bihar. Orissa.

Part Used : Plant.

Utilization : Used as fodder in young stage.

References : Ambasta (1986), Anonymous (1952).

Digitaria adscendens (H. B. & K.) Henr.

Syn: *Digitaria marginata* Link.; *Panicum adscendens* H. B.& K.
Family: Poaceae.

English, Hindi and Regional Names:

Eng. : Fig Grass;
Hindi : Takri, Takria;
Bom. : Tara;
Kan. : Hennu Akkibu hullu;
Tam. : Arisipillu.

Distribution : Sub-Tropical Region of India.

Part Used : Plant.

Utilization : Used as fodder.

Active Constituents : Rich in proteins, Ca, P and fibre.

References : Agarwal (2003), Ambasta (1986), Anonymous (1952).

Digitaria ciliaris (Retz.) Koeler.

Syn: *Panicum ciliare* Retz.; *Paspalum sanguinale* Hook. f. in part, var., *ciliare* Hook. f.
Family: Poaceae.

Part Used : Plant.

Utilization : Used as a fodder for cattle.

References : Ambasta (1986), Diwaker & Ansari (2003), Sharma & Rana (2005).

Digitaria jubata (Griseb.) Henr.

Syn: *Paspalum jubatum* Griseb.
Family: Poaceae.

Part Used : Plant.

Utilization : Used as fodder grass.

Reference : Ambasta (1986).

Digitaria longifolia (Retz.) Pers.

Syn: *Paspalum longiflorum* Retz.
Family: Poaceae.

Hindi and Regional Names:

Hindi : Kanta juriya;
Kan. : Tapari hullu;
Tel. : Pakuru gaddi.

Part Used : Plant.

Utilization : Used as fodder.

References : Ambasta (1986), Sharma & Rana (2005).

Dillenia pentagyna Roxb.

Syn: *Colbertia augusta* Wall.; *C. coromandeliana* DC.; *Dillenia augusta* Roxb.; *D. pillosa* Roxb.
Family: Dilleniaceae.

English, Hindi, Sanskrit and Regional Names:

Eng. : Dog Teak;

Hindi	:	Aggai, Karmal;
Sans.	:	Aksikiphala, Akshikiphala;
Ass.	:	Akshi;
Beng	:	Karkotta;
Bom.	:	Karamala;
Kan.	:	Kanigala;
Mal.	:	Kada punna, Punna;
Mar.	:	Karmal;
Ori.	:	Rai;
Tam.	:	Naytekku;
Tel.	:	Chinnakalinga.

Distribution : Bengal, Assam, C. S. & W. India.

Parts Used : Bark. Leaves.

Utilization : Bark fibre used for cordage and leaves as fodder.

References : Ambasta (1986), Anonymous (1952), Dastur (1964), Khan & Khanum (2005), Mittre (1997), Singh (1982), Watt (1972).

Dinebra retroflexa Panz.

Syn: *Dinebra arabica* Jack.

Family: Poaceae.

Regional Names:

Bom.	:	Kali kauli, Kharia;
Kan.	:	Nari balada;
Tel.	:	Wadaka toka gaddi.

Distribution : Cultivated in M. P. W. India. C. & S. District of Tamil Nadu.

Part Used : Green Grass.

Utilization : An excellent fodder for buffaloes to increase the milk production.

References : Ambasta (1986), Anonymous (1952), Oommachan (1977).

Diospyros melanoxylon Roxb.

Syn: *Diospyros dubia* Wall.; *D. tomantosa* Roxb.; *D. tupra* Buch.-Ham.; *D. wightiana* Wall.
Family: Ebnaceae.

Vern. Names: Kendu, Tendu.

English, Hindi, Sanskrit and Regional Names:

Eng.	:	Coromandel Ebnoy Persimmon;
Hindi	:	Tendu, Timburni;
Sans.	:	Dirghapatraka , Kenduka;
Beng	:	Kena, Kendu, Kiu;
Guj.	:	Tamrug;
Kan.	:	Abanasi, Bale;
Mal.	:	Kari;
Mar.	:	Tendu, Tumri;
Ori.	:	Kendu;
Tam.	:	Karai;
Tel.	:	Manchigata, tumku.

Distribution	:	Deccan Peninsula, extending Northwards to Bihar. M. P. Bombay.
Part Used	:	Leaves.
Utilization	:	Leaves used as a source of fodder.
Active Constituents	:	Leaves contain ceryl alcohol, lupeol, betulin, sitosterol and a new triterpene carboxylic acid- diospyric, crude proteins, 25.28% crude fibre, 1.79% Ca and 0.16% P.
References	:	Anonymous (1952), Chatterjee & Pakrashi (1997), Dwivedi (1993), Hooker (1885), Parrotta (2001), Rastogi & Mehrotra (1990, 1991), Singh (1982, 2002, 2003).

Diploknema buttracea (Roxb.) Lamk.

Syn: *Bassia butyracea* Roxb.
Family: Sapotaceae.

English, Hindi and Regional Names:

Eng.	:	Indian butter tree;
Hindi	:	Phulwra;
Beng	:	Gopal.

Distribution : Kumaon Eastwards to Sikkim. Andaman Islands.

Part Used : Leaves.

Utilization : Leaves lopped for fodder during winter.

References : Agarwal (2003), Anonymous (1952), Dwivedi (1993), Purohit & Samant (1995).

Dipsacus fullonum Linn.

Syn: *Dipsacus mitis* Don.
Family: Dipsacaceae.

English, Hindi and Regional Names:

Eng. : Fuller's Teasel;
Hindi : Burash;
Pub. : Burash.

Distribution : Cultivated in Tropical Parts of India. Kashmir to Dehradun. H. P.

Part Used : Dry Flower Heads.

Utilization : Used in the woollen industry for high grade cloths.

References : Agarwal (2003), Ambasta (1986), Anonymous (1952).

Dolichandrone spathacea K. Schum.

Syn: *Dolichandrone rheedii* Seem.
Family: Bignoniaceae.

Hindi and Regional Names:

Hindi : Gorshingian;
Beng : Gorshingiah;
Mal. : Nirpponalyam;
Tam. : Kanballai, Virbadiri.

Distribution : Malabar, Sunderbans & Andaman Islands.

Part Used : Bark.

Utilization : Bark fibre used for fishing nets.

References : Ambasta (1986), Anonymous (1952), Das & Agarwal (1991), Dwivedi (1993).

Dolichos lablab Linn.

Syn: *Dolichos benghalensis* Jack.; *D. cuspidatus* Grah.; *D. curtisii* G. Don Gard.; *D. festivus* Wall.; *D. lignosus* Linn.; *D. perpureus* Linn.; *Lablab cultrutus* DC.; *L. lignosus* Grah.; *L. microcarpus* DC.; *L. vulgaris* Savi.

Family: Fabaceae.

English, Hindi, Sanskrit and Regional Names:

Eng. : Bonavisit Bean, Indian Butter Bean, Lablab Bean;
Hindi : Sem;
Sans. : Madusarkara;
Ass. : Uri, Urshi;
Beng : Shim;
Guj. : Val;
Kan. : Chapparada avare;
Mal. : Avara;
Mar. : Pavta;
Tam. : Ararai;
Tel. : Chikkudu.

Distribution : Cultivated Throughout India: 1,800 - 2,200 m.

Part Used : Pods.

Utilization : Used as fodder.

Active Constituents : Pods rich in proteins 5% , mineral matter in calcium, iron and vit. C.

References : Agarwal (2003), Ambasta (1986), Anonymous (1952), Dwivedi (1993), Oommachan (1977).

Dracaena angustifolia Roxb.

Family: Liliaceae.

Hindi Name:

Hindi : Buckripathi.

Distribution : Himalaya. Khasia Hills & Andaman Islands.

Part Used : Leaves.

Utilization : Used as fodder for sheep and goats.

References : Agarwal (2003), Ambasta (1986), Anonymous(1952), Dwivedi (1993).

Dracocephalum heterophyllum Benth.

Family: Lamiaceae.

Regional Names:

Pub. : Karamm, Shanku, Zanda.

Distribution : N. W. Himalaya. Sikkim.

Part Used : Whole Plant.

Utilization : Leaves browsed by sheep and goats.

References : Anonymous (1952, 1986), Sood *et al.* (2001).

Dregea volubilis (Linn. f.) Benth. *ex* Hook. f.

Syn: *Asclepias volubilis* (L. f.) Staff.;
Maradenia volubilis Cooke.; *Wattakaka volubilis* (L. f.) Staff.

Family: Asclepiadaceae.

Hindi, Sanskrit and Regional Names:

Hindi : Nak – Chhikni;
Sans. : Hemajivanti, Madhumalati;
Beng : Titakunga;
Guj. : Dodi, Malati;
Kan. : Dugdhike;
Mal. : Vattakkakkakkodi;
Mar. : Amri, Hirandodi, Gharphul;
Ori. : Dugdhika;
Tam. : Kodipalai;
Tel. : Dudhipaala.

Distribution : W. Bengal. Assam. Orissa. Bihar. Peninsula.

Part Used : Plant.

Utilization : Plant fibre used for domestic uses.

References : Agarwal (2003), Ambasta (1986), Hooker (1885), Sood *et al.* (2005), Talbot (1976a), Varghese (1996).

Drypetes longifolia Pax & Hoffm.

Syn: *Cyclostemon longifolius* Blume.

Family: Euphorbiaceae.

Distribution : Assam Hills.

Part Used : Plant.

Utilization : Fibre used for paper making.

References : Ambasta (1986), Anonymous (1952).

Eichinochola stagnina Beauv.

Syn: *Panicum stagninum* Retz.

Family: Poaceae.

Hindi and Regional Names:

- **Hindi** : Banti;
- **Beng** : Dul;
- **Bom.** : Banti;
- **Kan.** : Kadu dashai hullu;
- **Tel.** : Bontha oddu.

Distribution : Throughout India.

Part Used : Grass Blades.

Utilization : Used as rich fodder for horses and cattle.

Active Constituents : Rich in sucrose 10 %.

References : Agarwal (2003), Ambasta (1986), Anonymous (1952), Das *et al.* (1996), Islam (2002).

Echinops echinatus Roxb.

Syn: *Echinops griffithianus* Boiss.

Family: Asteraceae.

English, Hindi and Regional Names:

- **Eng.** : Camel's Thistel;
- **Hindi** : Utakanta;
- **Sans.** : Kantual, Rakatpushpa , Uttati, Utakantaka;
- **Guj.** : Shuligo, Utkanto;
- **Mar.** : Kadechubuk, Utani.

Distribution : Upper Gangetic Plains. N.W. Himalaya. Punjab. Banaras Westwards ascending to 1,500 m in Sirmore.

Part Used : Twigs.

Utilization : Used as good fodder for camels.

Active Constituents : Leaves contain amyrin acetates, β- amyrin carporate, betulinic acid , lupeol acetate, ethyl palmitate and β-sitosterol.

References : Anonymous (1952), Hooker (1882), Khan & Khanum (2005), Parrotta (2001), Rastogi & Mehrotra (1995a), Singh & Pandey (1998).

Edgeworthia gardneri (Wall.) Meissn.

Syn: *Daphne gardneri* Wall.; *D. papyrifera* Sieb.; *Edgeworthia papyrifera* Zucc.
Family: Thymelaeaceae.

Distribution : E. Himalaya.

Part Used : Bark.

Utilization : Bark fibre suitable for making ropes, straps and paper.

References : Hooker (1972), Saklani & Jain (1994).

Ehretia acuminata R. Br. var. *serrata* (Roxb.)

Syn: *Ehrethia acuminata* R. Br.
Family: Boraginaceae.

Hindi and Regional Names:

Hindi : Arjun, Koda, Punia;
Ass. : Bual;
Beng : Kula aja;
Gar. : Narra;
Guj. : Punna, Pursan;
Pub. : Punna, Pursan, Sum.

Distribution : Throughout India upto 2,000 m. Assam. Karnataka. Sub-Tropical Himalaya.

Part Used : Leaves.

Utilization : Leaves lopped for fodder.

References : Agarwal (2003), Ambasta (1986), Anonymous (1952), Purohit & Samant (1995), Sharma (2004), Sood & Thakur (2004).

Ehretia laevis Roxb.

Syn: *Ehretia aspera* Willd.; *E. floribunda* Benth *ex* Royle; *E. obtusifolia* Hochst.
Family: Ehretiaceae.

Hindi and Regional Names:

Hindi	:	Koda;
Ass.	:	Hanbouk arong;
Beng	:	Tambolli;
Kan.	:	Kappura;
Mal.	:	Chavandi;
Mar.	:	Daterag, Daterang;
Ori.	:	Mosonea;
Pub.	:	Sakar, Saker;
Tam.	:	Addula;
Tel.	:	Addabukkudu.

Distribution : Throughout Tropical & Sub -Tropical Region.

Part Used : Leaves.

Utilization : Leaves considered as good fodder and fed during winter.

References : Agarwal (2003), Anonymous (1952), Hooker (1885), Parrotta (2001), Purohit & Samant (1995), Sood & Thakur (2004), Watt (1972).

Eichhornia crassipes Solms.

Syn: *Eichhornia speciosa* Kunth.
Family: Pontederiaceae.

English, Sanskrit and Regional Names:

Eng.	:	Water Hyacinth;
Sans.	:	Jalakhumbi;
Beng	:	Dachuripana;
Mal.	:	Kolavaza;
Tam.	:	Akasa – thamarai;
Tel.	:	Gurrapu decca.

Distribution : Tropical Regions of India.

Part Used : Leaves.

Utilization : Leaves eaten by cattle and horses during summer and also a source of cellulose for the manufacture of cover paper.

References : Agarwal (2003), Ambasta (1986), Anonymous (1952), Das *et al.* (1996), Khan & Khanum (2003), Raju (1998), Rastogi & Mehrotra (1993).

Eleocharis artopurpurea (Retz.) Presl.

Family: Cyperaceae.

Part Used : Tubers.

Utilization : Tubers used as fodder for cattle.

Reference : Islam (2002).

Embelia ribes Burm. (Pl. 4B)

Syn: *Calispermum scandens* Lour.; *Embelia glandulifera* Wight.; *E. garciniaefolia* Wall.; *Samara ribes* & *S. floribunda* in part, Kurz in Journ.

Family: Myrcinaceae.

English, Hindi, Sanskrit and Regional Names:

Eng. : Babreng, Embelia;
Hindi : Baberange;
Sans. : Bhasmaka, Ghoshaa, Krimihaa, Vindanga;
Beng : Biranga;
Bom. : Karkannie;
Guj. : Vavading;
Kan. : Vayivalanga;
Mal. : Vishaul;
Mar. : Karkannie, Vavadinga;
Ori. : Baibidanga;
Pub. : Babrung;
Tam. : Vevval;
Tel. : Vaya - vilamgam.

Distribution : Throughout India from C. Himalaya to Sri Lanka.

Part Used : Leaves.

Utilization : Leaves used as fodder.

References : Anonymous (1952), Hooker (1882), Khare (2004), Sharma (2004), Saklani & Jain (1994).

Emblica officinalis Gaertn.

Syn: *Cicca emblica* Kurz.; *Dichelastina nodicaulis* Hance.; *Phyllanthus emblica* Linn.

Family: Euphorbiaceae.

Vern. Name: Amala.

English, Hindi, Sanskrit and Regional Names:

Eng.	:	Emblic myrobalan tree;
Hindi	:	Amla, Amlika;
Sans.	:	Adiphala, Akara, Amlika;
Beng	:	Amla, Amlaki;
Guj.	:	Ambala, Amali;
Kan.	:	Amalaka;
Mal.	:	Amalakam ,Amlaki, Ohalu;
Mar.	:	Anvala;
Ori.	:	Gondhona;
Pub.	:	Amla;
Tam.	:	Amalagam;
Tel.	:	Amalakamu.

Distribution : Commonly Grown in India upto 1,800 m.

Part Used : Leaves.

Utilization : Leaves and fruits used as fodder.

Active Constituents : Fruits rich in corilagin, ellagic acid. Leaves contain 4.48% ash, 1.93% Ca and 47.99% N.

References : Agarwal (2003), Ambasta (1986), Anonymous (1952), Dwivedi (1993), Rastogi & Mehrotra (1990), Singh (1982).

Ensete ventricosum (Welw.) Cheesman.

Syn: *Ensete edule* Horan.; *Musa ensete* J. F. Gmelin.

Family: Musaceae.

Part Used : Plant.

Utilization : Fibre used for cordage and textile purposes.

Reference : Ambasta (1986).

Entada phaseoloides Merrill. (Pl. 4C)

Syn: *Entada pursaetha* DC.; *E. scadens* Benth.; *E. schefferi* Ride.

Family: Fabaceae.

English, Hindi, Sanskrit and Regional Names:

Eng. : Elephant Creper, Garbree Bean, Mackay Bean;
Hindi : Gila;
Sans. : Gilla;
Ass. : Gila lisa;
Beng : Gila;
Kan. : Doddakampi;
Mal. : Kattavalli, Malomanchadi ;
Mar. : Garbe, Garambi;
Ori. : Geredi;
Pub. : Kastorikanam;
Tam. : Chillu, Vattavali;
Tel. : Gilatige.

Distribution : Throughout Sub-Himalayan Tract. Assam. Bihar. Orissa. W. Ghats and Andamans.

Part Used : Stem Bark.

Utilization : Fibre good for cordage, fishing nets and making ropes.

References : Agarwal (2003), Ambasta (1986), Chopra *et al.* (1969), Parrotta (2001), Rao & Henry (1996), Sood *et al.* (2005).

Equisetum arvense L.

Family: Equisetaceae.

English and Hindi Names:

Eng. : Bottle Brush, Horse Tail;
Hindi : Nari, Matti, Totak.

Distribution : Himalaya: 3,000 -4,000 m.

Part Used : Herb.

Utilization : Plant browsed by cattle to increase milk production.

Active Constituents : â- sitosterol and isoquercetin isolated from the leaves.

References : Islam (2002), Prajapati *et al.* (2003), Rastogi & Mehrotra (1990, 1993).

Equisetum debile Roxb.

Family: Equisetaceae.

English and Regional Names:

Eng. : Fox-tailed Asparagus, Horse tail;
Pub. : Banduki, Nari, Trotak.

Distribution : Commonly found in Garhwal Himalaya upto 2,100 m.

Part Used : Plant.

Utilization : Plant relished by cattle.

References : Bennet *et al.* (1998), Islam (2002).

Eragrostiella bifaria (Wight) Bor.

Syn: *Eragrostis bifaria* Wight *ex* Steud.
Family: Poaceae.

Regional Names:

Bom. : Chiraka;
Kan. : Kodimara hullu;
Tam. : Oothu pul;
Tel. : Gubbikal guddi.

Part Used : Grass Blades.

Utilization : Used as a good fooder grass.

Active Constituents : These are rich in carbohydrate, 43.66%, fat 1.77% and proteins 2.50%.

References : Ambasta (1986), Anonymous (1952).

Eragrostis plumosa Link.

Syn: *Eragrostis tenella* Roem. & Schult var., *plumosa* Stapf.
Family: Poaceae.

Hindi and Regional Names:

Hindi : Jhusa;

Kan. : Sanna puralai hullu;
Pub. : Budhan, Palingi;

Part Used : Grass Blades.

Utilazation : A fodder grass eaten by horses and cattle.

Active Constituents : Rich in vit. A, crude protein, 9.91 and crude fibre 33.56.

References : Ambasta (1986), Anonymous (1952), Dwivedi (1993), Watt (1972).

Eragrostis tenella L. Beauv.

Syn: *Poa tenella* Linn.
Family: Poaceae.

Regional Names:

Bom. : Bharbusi;
Mar. : Dhane;
Tel. : Pothika gaddi.

Distribution : Throughtout India from Punjab to Malacca.

Parts Used : Leaves. Twigs.

Utilization : Eaten by cattle both green and as hay.

References : Anonymous (1952), Dwivedi (1993), Hooker (1897), Islam (2002), Roy *et al.* (1998), Sharma & Rana (2005).

Eremopogon foveolatus Stapf.

Syn: *Andropogon foveolatus* Delile.
Family: Poaceae.

Distribution : Warm Regions of India.

Part Used : Plant.

Utilization : Fodder grass relished by cattle and camels.

Active Constituents : Plant contains proteins 2.27, fat 1.32, fibre 37. 36 and ash 7.42.

References : Agarwal (2003), Ambasta (1986), Diwaker & Ansari (2003), Dwivedi (1993), Sharma & Rana (2005).

Erianthus elephantinus HK. f

Family: Poaceae.

Distribution : Assam.

Part Used : Plant.

Utilization : A good fodder relished by sheep and goats.

References : Hooker (1897), Islam (2002).

Erianthus munja Jes.

Syn: *Eeianthus ciliaris* Jesw.; *E. sara* Rumke.; *Saccharaum arundinaceum* Hook. f. in part, non Retz.; *S. benghalensis* Retz.; *S. ciliare* Anderss.; *S. munjà* Roxb.; *S. sara* Roxb.

Family: Poaceae.

English, Hindi, Sanskrit Names:

Eng. : Munja Grass;

Hindi : Munj, Sara;

Sans. : Sarkanda.

Part Used : Leaves.

Utilization : Leaves provide good fodder to buffaloes and horses; and fibre for cordage, ropes and paper.

References : Dastur (1964), Raju (1998).

Erigeron alpinus L.

Family: Asteraceae.

Distribution : Alpine & Temperate Himalaya. Khasia Hills. Nilgiris.

Part Used : Aerial Parts.

Utilization : Plant relished by cattle.

Active Constituents : Stigmasterol, spinasterol and amyrins isolated from the leaves.

References : Hooker (1882), Rastogi & Mehrotra (1995b), Singh & Kumar (2000).

Erinocarpus nimmonii J. Grah.

Family: Tiliaceae.

Regional Names:

Kan. : Kadu bende;
Mar. : Chira, Haladi.

Part Used : Bark.

Utilization : Bark fibre good for making ropes and cordage.

References : Ambasta (1986), Anonymous (1952), Jain (1991), Yoganarasimhan *et al.* (1981).

Eriochloa procera C. E. Hubbard.

Syn: *Erichola annulata* Kunth.; *E. polystachya* H. B. & K.; *E. ramosa* Kuntze.

Family: Poaceae.

Regional Names:

Tam. : Karungani pillu, Mathanka pillu.

Distribution : Plains of India.

Part Used : Plant.

Utilization : A good fodder relished by cattle.

References : Ambasta (1986), Anonymous (1952), Islam (2002).

Eruca sativa Lamk.

Syn: *Brassica eruca* Linn.; *B. erucoides* Roxb.; *Eruca sativa* Mill.

Family: Brassicaceae.

Vern. Name: Taarameera.

English, Hindi, Sanskrit and Regional Names:

Eng. : Rocket, Taramira;
Hindi : Taramira;
Sans. : Bimala, Daradharha, Siddartha;
Beng : Sheet sursha;
Mar. : Jambho;
Pub. : Assu, Jamnia,Tara, Usan.

Distribution : Extensively Cultivated as a cold weather crop in N.W. India.

Part Used : Seed.

Utilization : Seeds given to increase lactation in cattle.

Active Constituents : Seeds rich in glucoerucin.

References : Anonymous (1952), Hooker (1872), Rastogi & Mehrotra (1991), Sood & Thakur (2004), Watt (1972).

Erythrina suberosa Roxb.

Syn: *Erythrina alba.;* Roxb.; *E. reniformis* Ham.; *Cropteryx suberosa* Walp.

Family: Fabaceae.

Vern. Names: Paryara.

English, Hindi, Sanskrit and Regional Names:

Eng.	:	Indian Coral Tree;
Hindi	:	Dauldhak, Madar, Mandara;
Sans.	:	Bahupushpa, Kantaki;
Ass.	:	Mandal;
Beng	:	Madar;
Guj.	:	Bangaro, Janghario;
Kan.	:	Kaduparivala;
Mal.	:	kalyanamurikku;
Mar.	:	Mandar, Pangara;
Ori.	:	Piri;
Pub.	:	Pariaria, Thab;
Tam.	:	Kavir;
Tel.	:	Barijama, Barjapu.

Distribution : Himalaya to Sri Lanka on Dry Region.

Part Used : Stem Bark.

Utilization : Bark fibre good for cordage, lining work, insulating mats and making ropes.

References : Agarwal (2003), Ambasta (1986), Anonymous (1952), Dastur (1964), Dwivedi (1993), Drury (1978), Hooker

(1879), Jain (1991), Mittre (1997), Sharma & Rana (2005), Sood & Thakur (2004), Watt (1972).

Erythropsis colorata (Roxb.) Burkill.

Syn: *Firmania colorata* R. Br.; *Sterculia colorata* Roxb.

Family: Sterculiaceae.

Hindi and Regional Names:

Hindi	:	Bodula, Samerri, Walena;
Ass.	:	Jhari, Kath, Udal;
Beng	:	Mula;
Kan.	:	Bilisulige;
Mal.	:	Malam parathii;
Ori.	:	Kadalo, Panico Dal;
Tam.	:	Malam Baruttii;
Tel.	:	Karu, Karaka.

Distribution : Throughout India.

Parts Used : Leaves.Twigs.Bark.

Utilization : Leaves and twigs used as cattle fodder. Bark fibre good for ropes.

References : Agarwal (2003), Ambasta (1986), Anonymous (1952).

Erythropsis pallens Ridley.

Syn: *Sterculia pallens* Wall.

Family: Sterculiaceae.

Distribution : Throughout India.

Part Used : Bark.

Utilization : Bark yields coarse fibre for ropes.

References : Ambasta (1986), Anonymous (1952).

Ethulia conyzoides L. (Pl. 4D)

Syn: *E. angustifolia* Bojer.; *E. gracilis* Delile.

Family: Asteraceae.

Part Used : Plant.

Utilization : Plant used for fodder.

References : Islam (2002), Islam & Hasin (2003), Hooker (1882).

Euchlaena maxicana Schrad.

Syn: *Euchlaena luxurians* Dur. & Aschers.
Family: Poaceae.

English, Hindi and Regional Names:

Eng. : Teosinte;
Hindi : Makay bajara;
Pub. : Mukchari.

Distribution : Mostly in Warmer Parts of India.

Part used : Plant.

Utilization : Fodder suitable for irrigated areas for cattle and horses. It is believed to stimulate milk production when fed to cows.

References : Agarwal (2003), Ambasta (1986).

Eugenia operculata L.

Syn: *Calyptranthes costata* Ham in Wall.; *C. grandis* & *C. tatna* Ham.; *C. mangiferifolia* Hance in Walp.
Family: Myrtaceae.

Hindi, Sanskrit and Regional Names:

Hindi : Paiman, Raj -Jaman;
Sans. : Bhumijamun, Rhasva, Rhasvaphala;
Mal. : Naral;
Tam. : Monisiajamo.

Distribution : Sub-Himalayan ascending to 3,000 m.

Part Used : Leaves.

Utilization : Leaves considered an excellent fodder for livestock.

References : Hooker (1879), Singh (2000), Watt (1972).

Eulalia cumingii A. Camus.

Syn: *Pollinia cumingii* Nees.
Family: Poaceae.

Distribution : Gangetic Plain, Kumaon, Dehradun, Sikkim Terai, Khasia & Jantia Hills.

Part Used : Plant.

Utilization : Used as fodder.

Active Constituents : Plant contains crude proteins 6.90, fibre 38.67, total ash 8.49, CaO 0.89, MgO 0.49 and K2O 1.72 %.

References : Ambasta (1986), Anonymous (1952).

Eulaliopsis binata (Retz.) C. E.

Syn: *Ischaemum angustifolium* Hack.; *Pollinidum angustifolium* Haines.; *P. binatum* (Retz.) C. E. Hubbard.
Family: Poaceae.

Vern. Name: Bhabber.

English, Hindi and Regional Names:

Eng. : Sabai Gras;
Hindi : Baib, Bhabar, Sabai;
Ori. : Babuia, Bagali.

Distribution : Common in Bihar, Orissa, Bengal, C. India & Punjab.

Part Used : Leaf Blade.

Utilization : Fibre used for making ropes, baskets, brooms and stringing traditional beds called 'charpais. Whole plant string used for tying and stiching roofs.

References : Anonymous (1952), Chakraborty & Bhattacharjee (2003), Kaul *et al.* (1996), Maheshwari *et al.* (1997), Rao & Henry (1996).

Euphorbia geniculata Orteg.

Family: Euphorbiaceae.

Hindi and Regional Names:

Hindi : Choota - dudhi;
Beng : Swet - Keryi.

Distribution : Naturalised in Many Parts of India upto 800 m.

Utilization : Plant used as fodder for milk yielding cattle.

Active Constituents : Alanine, cystine, serine, aspartic acid, proline, glutamic acid glucose, rhamnose and glactose present in leaves.

References : Diwaker & Ansari (2003), Rastogi & Mehrotra (1991), Roy *et al.* (1998), Singh & Kumar (2000).

Euphorbia royleana Boiss.

Syn: *Euphorbia pentagona* Royle

Family: Euphorbiaceae.

Vern. Names: Chuii, Chhun, Danda, Thor.

Hindi and Regional Names:

Hindi : Sulla, Thor;

Pub. : Dandathor.

Distribution : N. W. Himalaya : 1,800 m.

Part Used : Leaves.

Utilization : Leaves used as fodder and browsed by sheep and goats during monsoon season in the desert.

References : Anonymous (1952), Hooker (1890), Raju (1998), Watt (1972).

Eurya acuminata DC.

Family: Theaceae.

Hindi and Regional Names:

Hindi : Diyar;

Bom. : Baunra, Deura, Gonta.

Distribution : Occurs N. E. Himalaya from Kumaon to Bhutan.

Part Used : Leaves.

Utilization : Leaves fed to goats during winter season.

References : Hooker (1872), Purohit & Samant (1995), Watt (1972).

Fagopyrum esculentum Moench.

Syn: *Fagopyrum emarginatum* Meissn.; *Polygonum diocum* Ham.; *P. emarginatum* Roth.; *P. fagopyrum*.

Family: Polygonaceae.

Vern. Name: Tyat.

English, Hindi and Regional Names:

- **Eng.** : Black Wheet, Buck Wheat;
- **Hindi** : Kottu, Kullun;
- **Ass.** : Doron;
- **Gar.** : Kotu;
- **Pub.** : Darau, Obal, Phulan.

Distribution : Kashmir to Khasia Hills. Manipur. Nilgiri Hills.

Part Used : Green Straw.

Utilization : Used as fodder.

Active Constituents : Kaemferol-3-rhamnosyl glucoside, feruloylquinic acid are the main constituents of leaves.

References : Agarwal (2003), Ambasta (1986), Anonymous (1952), Dwivedi (1993), Hooker (1890), Rastogi & Mehrotra (1991), Singh *et al.* (1996), Watt (1972).

Feronia limonia (Linn.) Swingle.

Syn: *Feronia elephantum* Correa.
Family: Rutaceae.

English, Hindi, Sanskrit and Regional Names:

- **Eng.** : Elephant Apple, Wood Apple;
- **Hindi** : Bilin, Kait, Kavitha;
- **Sans.** : Danta, Kappitha, Kapripriya, Satha;
- **Beng** : Kait, Katbel;
- **Guj.** : Kavit, Kotha, Kothi, Kothun;
- **Kan.** : Bela;
- **Mal.** : Vila, Villatti;
- **Mar.** : Kavatha, Kavith, Kovit;
- **Pub.** : Kainth;
- **Tam.** : Vilanga;
- **Tel.** : Velaga.

Distribution : Throughout the Plains of India. W. Himalaya. The Deccan Thanna District of Bombay & Chota Nagpur.

Part Used : Leaves.

Utilization : Tree lopped for fodder in U. P. and Bombay.

Active Constituents : Leaves contain stigmasterol (0.22%), bergaptin (0.10%), anisic acid, vitexin and saponarin.

References : Ambasta (1986), Anonymous (1952), Chopra *et al.* (1969), Rastogi & Mehrotra (1991, 1995a).

Ficus auriculata Lour.

Syn: *Ficus macrophylla* Roxb.; *F. roxburghii* Wall.
Family: Moraceae.

Hindi and Regional Names:

Hindi : Timla, Trimal;
Pub. : Barbaru.

Distribution : Punjab to Assam. Bengal. Orissa.

Parts Used : Leaves. Bark.

Utilization : Leaves lopped for fodder. Bark fibre for stuffing and cordage.

References : Agarwal (2003), Ambasta (1986), Anonymous (1956), Bennet *et al.* (1991), Jain (1991), Saklani & Jain (1994), Singh & Kumar (2000).

Ficus benghalensis Linn.

Syn: *Ficus indica* L.; *Urostigma benghalense* Gasp.
Family: Moraceae.

Vern. Names: Bad, Bare.

English, Hindi, Sanskrit and Regional Names:

Eng. : Banyan Tree;
Hindi : Bargad;
Sans. : Bahupada, Chira, Nandi, Nila, Roihina, Vata;
Ass. : Bot;
Beng : Barbut;
Bom. : Barghat, Wad, War;
Guj. : Vad, Vadlo, Vor;
Kan. : Ala, Alada, Alava;
Mal. : Ala, Vatam;
Mar. : Vada, War;
Ori. : Boro, Boto;

Pub.	:	Bargad, Bera, Bohar, Bohir;
Tam.	:	Al, Alam;
Tel.	:	Mari, Mari peddi.

Distribution : Throughout India; wild only in the Sub-Himalayan Forest.

Parts Used : Leaves. Aerial Roots. Stem Bark.

Utilization : Leaves and twigs relished by elephants and cattle. Fibre from aerial roots made into coarse ropes, cordage and brushes. Paper is also prepared from bark in Assam.

Active Constituents : Protein 9.63, Crude fibre 26.84, friedelin and β-sitosterol are the chief constituents of leaves.

References : Agarwal (2003), Ambasta (1986), Anonymous (1956), Chatterjee & Pakrashi (1991), Dastur (1964), Drury (1978), Dwivedi (1993), Hooker (1890), Jain (1991), Mittre (1997), Parrotta (2001), Rastogi & Mehrotra (1990), Sharma & Rana (2005), Singh (1982), Singh & Pandey (1998), Watt (1972).

Ficus cunia Buch-Ham.

Syn: *Covellia cunia, conglomerata & inaequiloba* Miq.; *Ficus conglomerata* Roxb.

Family: Moraceae.

Hindi, Sanskrit and Regional Names:

Hindi	:	Jharphali, Khain;
Sans.	:	Kharapatra;
Beng	:	Dumbur;
Kan.	:	Garagasa;
Mal.	:	Perina, Porog, Teregam;
Mar.	:	Porodumar;
Ori.	:	Godima, Porishkry;
Pub.	:	Kathjular, Kuri, Trumbal;
Tam.	:	Taragaddi.

Distribution : Sub-Himalayan Tract from W. India to Chota Nagpur, Khasia Hills.

Part Used : Leaves.

Utilization : Leaves lopped for fodder in Bengal, Assam and U. P.

References : Agarwal (2003), Ambasta (1986), Anonymous (1956), Drury (1978), Dwivedi (1993), Hooker (1890), Singh (1982).

Ficus glomerata Roxb.

Syn: *Covellia glomerata* Miq.; *C. mollis* Miq.; *Ficus goolereea* Roxb.; *F. mollis* Miq.; *F. racemosa* Linn.

Family: Moraceae.

Vern. Names: Gular, Umara.

English, Hindi, Sanskrit and Regional Names:

Eng. : Cluster Tree;
Hindi : Gular;
Sans. : Jantuphala, Suprastish - thida
Bom. : Jagya dumar;
Guj. : Gular, Umar;
Kan. : Athi, Santal;
Mal. : Athi, Athimarum;
Mar. : Umar;
Ori. : Dimri;
Pub. : Batwar, Kakallam, Kathagular;
Tam. : Athi, Attimaram;
Tel. : Atti, Medi.

Distribution : Outer Himalaya, Plain & Low Hills of India upto 1,000 m.

Parts Used : Leaves. Fruits.

Utilization : Leaves fed to animals in drought and scarcity conditions. Fruits and leaves greedily eaten by cattle and elephants.

Active Constituents : Main constituents of leaves are crude proteins 9.63%, CaO 2.53%, 3.84% Ca, 2.10% N and total digestible nutrients 51.05%.

References : Ambasta (1986), Anonymous (1956), Dastur (1964), Dwivedi (1993), Hooker (1890), Oommachan (1977), Purohit & Samant (1995), Singh (1982), Watt (1972).

Ficus hispida L. f.

Syn: *Covellia daemonam* Miq.; *C. setulosa* Miq.; *Ficus daemona* Koen. *ex* Vahl.; Koen.; *F. mollis* Willd.; *F. oppostifolia* Roxb; *F. promineus* Wall.; *Sycomorpha ruxburghii* Miq.

Family: Moraceae.

English, Hindi, Sanskrit and Regional Names:

Eng.	:	Devil's Tree;
Hindi	:	Dagurin, Gobla, Kagsha;
Sans.	:	Kakodumaru, Malapa, Phalgu;
Beng	:	Dumoor;
Guj.	:	Dhedaumaro, Jangliangir;
Kan.	:	Adavi attu;
Mal.	:	Peyati;
Mar.	:	Bhokada, Bokria, Kalumbar;
Ori.	:	Bhaidimiri, Korotosano;
Pub.	:	Daduri, Degar;
Tam.	:	Peyatti, Sonattii;
Tel.	:	Bodamamidi.

Distribution : Outer Himalaya from Chenab eastwards to W. Bengal. Assam. Andaman Islands upto 1,200 m.

Part Used : Leaves.

Utilization : Leaves given as excellent fodder for cattle and elephants.

References : Agarwal (2003), Ambasta (1986), Dwivedi (1982), Hooker (1890), Khare (2004), Parrotta (2001), Singh (1982), Watt (1972).

Ficus nemoralis Wall.

Syn: *Ficus binata* Wall.; *F. densa* Miq .; *F. fieldingii* Miq.; *F. gemella* Wall.; *F. trilepis* Miq.

Family: Moraceae.

Vern. Name: Dudil, Thelka.

Distribution : Common in Khasia Hills & Assam: 1,000- 2,500 m.

Part Used : Leaves.

Utilization : Leaves considered a very nutritive fodder and fed to all classes of cattle in winter.

Active Constituents : Leaves contain saponin.

References : Ambasta (1986), Dwivedi (1993), Hooker (1890), Parker (1973), Purohit & Samant (1995), Watt (1972).

Ficus palmata Forsk.

Syn: *Ficus caricoides* Roxb.; *F. pseudosycamorus* Decne.; *F. virgata* Roxb; Brandis.

Family: Moraceae.

Vern. Name: Phaegda.

English, Hindi and Regional Names:

Eng. : Wild Fig;
Hindi : Anjiri;
Guj. : Pepri;
Pub. : Dadhalia, Jamir, Kak;
Tel. : Mangimedi.

Distribution : N. W. India. U. P. Himalaya upto 1,000 m.

Part Used : Leaves.

Utilization : It is considered to be good fodder for milch cows.

Active Constituents : Leaves comprise of β- sitosterol and bergapten.

References : Agarwal (2003), Ambasta (1986), Anonymous (1956), Hooker (1890), Purohit & Samant (1995), Rastogi & Mehrotra (1991), Sharma (2004), Sharma & Rana (2005), Singh (1982), Singh & Kumar (2000), Sood & Thakur (2004), Watt (1972).

Ficus religiosa Linn.

Syn: *Ficus affinior* Griff.; *Urostigma affine* Miq.; *U. religiosum* Gasp.

Family: Moraceae.

Vern. Name: Peeple.

English, Hindi, Sanskrit and Regional Names:

Eng.	:	Peeple Pipule Tree;
Hindi	:	Piple, Pipli;
Sans.	:	Ashvatha Pipala;
Beng	:	Asud, Aswat;
Bom.	:	Pimpal, Piplo;
Guj.	:	Jari, Pipul, Pipro;
Kan.	:	Basri, Ragi, Rangi;
Mal.	:	Arachu, Arayal;
Mar.	:	Pimpala, Pishvatha;
Ori.	:	Pippolo;
Pub.	:	Bhor, Pipal;
Tam.	:	Aswattham, Pimpala.

Distribution : Throughout India upto 1,000 m.

Part Used : Leaves. Branches. Bark.

Utilization : Leaves lopped for fodder for cattle, sheep, goats and elephants. Bark fibre good for ropes.

Active Constituents : Leaves contain crude proteins 13.99, crude fibre 22.36, total ash 15.06, CaO 4.64 and P 0.52%.

References : Agarwal (2003), Ambasta (1986), Anonymous (1956), Dastur (1964), Diwaker & Ansari (2003), Drury (1978), Hooker (1890), Jain (1991), Kalpash *et al.* (2004), Khare (2004), Khanna (2002), Mittre (1981, 1997), Oommachan (1977), Purohit & Samant (1995), Singh (1982), Varghese (1996), Watt (1972).

Ficus retusa Linn.

Syn: *Ficus dilatata* Miq.; *F. benjamina* Willd.; *F. littoralis* Blume.; *F. microcarpa* Linn.; *F. nitida* Thunb.; *F. rubra* Roth.; *Urostigma ovoideum* & *pisiferum* Miq.; *U. retusm* & *nitidum* Miq.

Family: Moraceae.

English, Hindi, Sanskrit and Regional Names:

Eng.	:	Peeple Tree;
Hindi	:	Kamrup, Sir;
Sans.	:	Kuni, Mandiriksha, Tunna;
Beng	:	Pilala;
Kan.	:	Pilala, Pinval;

Mal. : Fili;

Mar. : Nandruk

Tam. : Ichi, Kalllichii, Malaiyitchii;

Tel. : Yerrajujvvi.

Distribution : Commonly in S. India, E. India. M.P. Deccan Peninsula. Outer Himalaya from Indus to Bhutan.

Part Used : Leaves.

Utilization : Leaves lopped for fodder.

Active Constituents : Leaves rich in lupenyl acetate, friedelin, glutinol, taxaxeral, celeanolic acid and two pentacyclic triterpenoides.

References : Agarwal (2003), Ambasta (1986), Hooker (1890), Rastogi & Mehrotra (1995a), Watt (1972).

Ficus rumphii Blume.

Syn: *F. cordifili* Roxb.; *Urostigma cordifolium* Miq.; *U. rumphii* Miq.

Family: Moraceae.

Vern. Name: Badarein.

Hindi, Sanskrit and Regional Names:

Hindi : Gajna, Gayjaira, Kabar, Pakar, Pilkhan;

Sans. : Asvatth;

Ass. : Jakri, Prap;

Distribution : Throughout the Hills upto 1,000 m. Grows in Punjab, Assam, M. P. & S. India.

Parts Used : Leaves. Twigs. Bark.

Utilization : Leaves lopped for fodder and fed to all classes of livestock during winter and summer. Bark fibre valued for coarse blankets, bedding and cordage.

References : Agarwal (2003), Ambasta (1986), Anonymous (1956), Hooker (1890), Purohit & Samant (1995), Rao & Shanpru (1997), Saklani & Jain (1994), Singh (1982), Watt (1972).

Ficus sarmentosa Buch.-Ham.

Syn: *Ficus foveolata* Wall.
Family: Moraceae.

Hindi Names:

Hindi : Beduli.

Distribution : Rock in Shady Moist Ravines between 500-2,000 m.

Part Used : Leaves.

Utilization : It is considered to be a very good nutritive fodder for all milch animals during winter when grasses are scarce.

Reference : Purohit & Samant (1995).

Ficus scandens Roxb.

Syn: *Ficus fructicosa* & *hederacea* Roxb.
Family: Moraceae.

Distribution : Tropical Himalaya. Assam. Khasia Hills. Andaman Islands.

Part Used : Leaves.

Utilization : Leaves fed to all classes of livestock throughout the year.

References : Hooker (1890), Purohit & Samant (1995), Singh & Kumar (2000).

Ficus squamosa Roxb.

Family: Moraceae.

Part Used : Leaves.

Utilization : Leaves considered as good fodder to all livestock throughout the year.

Reference : Purohit & Samant (1995).

Ficus subincisa Buch.-Ham. *ex* J. E. Sm.

Syn: *Ficus clavata* Wall. *ex* Miq.
Family: Moraceae.

Distribution : Throughout the Hills between 500 – 1,000 m.

Part Used : Leaves.

Utilization : Leaves useful as a good fodder and fed to ruminants throughout the year.

Reference : Purohit & Samant (1995).

Ficus tsiela Roxb.

Syn: *Ficus amplissima* Smith in Rees Cycleop.; *F. benjamina* Wall.; *F. indica* Linn.; *Urostigma pseudo- benjamina* Miq.; *U. pseudo-tjiela* Miq.

Family: Moraceae.

Hindi, Sanskrit and Regional Names:

Hindi : Jari;
Sans. : Kanninika, Plaksha;
Bom. : Pimperi;
Guj. : Pipar, Pipari, Pipli;
Mal. : Koyali;
Mar. : Pipar;
Ori. : Jori;
Tam. : Ichi;
Tel. : Pedda juvvi.

Distribution : The Deccan Peninsula from Konkan southwards.

Part Used : Bark.

Utilization : Fibre used for making ropes.

References : Agarwal (2003), Dwivedi (1993), Hooker (1890), Watt (1972).

Ficus virens Ait.

Syn: *Ficus infectoria* Roxb.

Family: Moraceae.

Regional Names:

Ass. : Tengabar;
Gar. : Kobra, Pilkharu.

Distribution : Distributed between 1,000- 2,000 m.

Part Used : Leaves.

Utilization : Leaves lopped for fodder and fed to all ruminants in winter.

Active Constituents : Leaves possess flavone glycoside–infectorin.

References : Bennet *et al.* (1991), Khanna (2002), Purohit & Samant (1995), Rastogi & Mehrotra (1995b).

Flacourtia indica (Burm. F.) Merrill.

Syn: *Flacourtia latifolia* Cooke.; *F. occidentalis* Blatter.; *F. ramontchi* L'Herit.; *F. ramontchi* var. *sapida* (Roxb.) Hook. f. & Thoms.; *F. sapida* Roxb.; *F. sepiara* Roxb.

Family: Moraceae.

English, Hindi, Sanskrit and Regional Names:

Eng. : Madagascar Plum;
Hindi : Bhanbher, Bilangra, Kandai;
Sans. : Swadukantaka;
Beng : Baichi, Benchi, Binja;
Bom. : Kakad;
Guj. : Galgugal, Kankad, Katamla, Rankdi;
Kan. : Gabajira, Gajale, Hanmunki;
Mal. : Aghori;
Mar. : Bhekal, Binka, Kaker, Kuki, Paker;
Ori. : Balibhaincho, Boincha, Boniso;
Pub. : Kakoa, Kandai, Kangu, Kukai;
Tam. : Chothakilai, Kakukali, Sottaikala.

Distribution : Found in India upto 1,000 m.

Part Used : Leaves.

Utilization : Leaves serve as good fodder for cattle.

Active Constituents : Leaves rich in arabinose, glucose, beta sitosterol and its D- glucopyranoside, 2.21% Ca, 46.25% carbon and 2.42% N.

References : Anonymous (1956), Dwivedi (1993), Drury (1978), Parrotta (2001), Prajapati *et al.* (2003), Raju (1998), Singh (1982), Singh & Pandey (1998), Watt (1972).

Flemingia fructiculosa (L.) DC.

Family : Fabaceae.

Part Used : Leaves.

Utilization : Leaves used as fodder.

Active Constituents : A novel chalcone-flomiculosin isolated from the leaves.

References : Rastogi & Mehrotra (1993), Singh & Kumar (2000).

Flemingia macrophylla (Wall.) O. Ktze *ex* Merrill.

Syn: *Flemingia congesta* Ait.; *F. rhodocarpa* Baker.

Family: Fabaceae.

Vern. Name: Bada Salphan.

Hindi and Regional Names:

Hindi : Bhalia, Kusnut, Suptu;
Beng : Bara salpan, Bhalia;
Bom. : Dowdowla;
Mal. : Kamaterri;
Mar. : Dowdowla;
Ori. : Bonokandulo.

Distribution : Common Throughout the Hotter Parts of India.

Part Used : Leaves.

Utilization : Leaves considered as good fodder.

Active Constituents : Leaves contain vicenin-2, reynoutrin, rutin, myricitrin and robinin.

References : Anonymous (1956), Rastogi & Mehrotra (1991, 1995), Singh & Kumar (2000), Watt (1972).

Flemingia semialata Roxb.

Syn: *Flemingia prostrata* Roxb.; *F. stricta* Wall.

Family: Fabaceae.

Distribution : C. Himalaya, ascending to 1,800 m.

Part Used : Pods.

Utilization : Used as fodder.

Reference : Watt (1972).

Floscopa scandens Lour.

Family: Commelinaceae.

Regional Names:

Ass. : Kana himlu;
Mar. : Padathipullu;
Tam. : Vazhaparathi pillu;
Tel. : Konda amadikada.

Part Used : Plant.

Utilization : Used as fodder.

References : Anonymous (1956), Islam (2002).

Fragaria indica Andr.

Syn: *Duchesnea chrysantha* Miq.; *D. fragiformis* Don.; *D. iragarioides* Sm.; *Fragaria arguta* Lindl.; *F. malayana* Roxb.; *F. nilgirica* Zenker.; *F. roxburghii* W. A.; *Potentilla denticulosa, wallichiana* Ser.; *P. durandii* Torr.; *P. fragariaefolia* Klotzsch.; *P. trifida* Lehm.

Family: Rosaceae.

English and Regional Names:

Eng. : The Indian Strawberry, Moek strawberry;
Pub. : Bana - Phal, Bunun, Kansars, Paljor.

Distribution : Temperate & Sub-Tropical Parts along the Foot Hills from the Punjab to Assam. Khasia Hills. Nilgiris acending to 2,400 m.

Part Used : Plant.

Utilization : Plant also used as cattle fodder.

References : Hooker (1879), Islam (2002), Watt (1972).

Funtumia elastica Stapf.

Family: Apocynaceae.

English Names:

Eng. : Ire Rubber, Logos Silk Rubber.

Distribution : Nilgiri Hills.

Part Used : Floss.

Utilization : Floss covering the seed used to stuff cushions.

References : Ambasta (1986), Anonymous (1956).

Furcarea foetida Haw.

Family: Amaryllidaceae.

English and Hindi Names:

Eng. : Mauritius Hemp;
Hindi : Sissal, Vilayati gheequar.

Distribution : Grown in Indian Gardens as an ornamental plant.

Part Used : Leaves.

Utilization : Leaf fibre good for cordage, mats, ropes and netting the cots.

References : Agarwal (2003), Geetha *et al.* (1996).

Furcraea hexapetala Urban.

Syn: *Furcraea cubensis* Vent.
Family: Amaryllidaceae.

Part Used : Leaves.

Utilization : Leaves are a source of Cuba Hemp fibre and valuable for twines, mats, halters and other domestic articles.

Reference : Ambasta (1986).

Furcraea macrophylla Baker.

Family: Amaryllidaceae.

Part Used : Leaves.

Utilization : Fique fibre extracted from leaves used for making ropes, twines, hammocks, halters, sacks and sandal- girths.

References : Ambasta (1986), Anonymous (1948-1976).

Galinsoga parviflora Cav.

Family: Asteraceae.

English Name:

Eng. : Galinsoga.

Distribution : Himalaya; Quite common upto 2,500 m.

Part Used : Whole Plant.

Utilization : Valued in india as fodder. Crude fibre used for stuffing.

Active Constituents : Plant rich in proteins 10.93, fat 0.76, carbohydrates 34.04, crude fibre 38.44 and ash 15.82.

References : Agarwal (2003), Ambasta (1986), Anonymous (1956), Hooker (1882), Roy *et al.* (1998), Sood & Thakur (2004).

Garuga Pinnata Roxb.

Syn: *Garuga madagascariensis* DC.

Family: Burseraceae.

English, Hindi, Sanskrit and Regional Names:

Eng. : Garuga;
Hindi : Kaikar, Kharpat;
Sans. : Golika, Karnikara, Kinkarata, Paranki;
Ass. : Gendeli poma, Rohimala;
Beng : Dabdabe, Nilbhadi;
Guj. : Khusimb;
Kan. : Aranelli;
Mal. : Kattukalinjan , Kosramba;
Mar. : Kakad, Kudak, Kurak;
Ori. : Amru, Mohi, Sompotri;
Pub. : Karpat, Kathula, Kilmira;
Tam. : Arunelli, Karre vembu;
Tel. : Gargu, Garugu.

Distribution : Throughout India from the Jammu Eastwards to Myanmar & Southwards to the Nilgiri Hills.

Parts Used : Leaves. Fruits.

Utilization : Leaves and fruits used as fodder for cattle.

Active Constituents : Main consituents of leaves are amentoflone, 3.59% Ca, 44.15% carbons and 1.94% N.

References : Agarwal (2003), Ambasta (1986), Anonymous (1956), Dwivedi (1993), Hooker (1872), Kirtikar & Basu (1935), Kumar & Jain (1998), Parrotta (2001), Rastogi & Mehrotra (1991), Saklani & Jain (1994), Singh (1982).

Gaultheria hookeri Cl.

Family: Ericaceae.

Distribution : E. Himalaya: 2,000 -4,000 m.

Part Used : Leaves.

Utilization : Leaves relished as fodder.

References : Hooker (1882), Saklani & Jain (1994).

Gerbera lanuginosa Benth.

Syn: *Chaptalia gossypiana* Royle.; *Gerbera gossypiana* (Royle) Beauv.; *G. langinosa* Benth.; *Oreoseris lanuginosa* DC.

Family: Asteraceae.

Vern. Names: Himsaeli, Kaffi, Kapasi.

English and Regional Names:

Eng. : Transvaal Daisy;

Gar. : Gauni;

Pub. : Bur, Buril, Kafi, Kapasi, Tsar.

Distribution : Himalaya from Kashmir to Nepal upto 3,000 m.

Part Used : Leaves.

Utilization : Leaves twisted into a yarn for blankets, sacks, bags, etc.

References : Agarwal (2003), Ambasta (1986), Hooker (1882), Sharma (2003).

Girardinia heterophylla Decne.

Syn: *Girardinia diversifolia* Friis.; *Girardinia heterophylla* var. *palmata* Hook. f.; *G. leschenautiana* Decne.; *Urtica diversifolia* Link.; *U. hetrophylla* Vahl.; *U. horrida* Link.; *U. palmata* Forsk.

Family: Urticaceae.

Vern. Name: Badi- kogsi.

English, Hindi and Regional Names:

Eng.	:	Himalayan Nettle, Nilgiri Nettle;
Hindi	:	Alla, Awa, Bichua, Chikki;
Ass.	:	Horru, Surat;
Mal.	:	Ana schorigenain;
Mar.	:	Moti khajati;
Pub.	:	Anjan, karala, Sanoli.

Distribution : Temperate & Sub-Tropical Himalaya; Assam, Silhet & Southwards to Myanmar & C. India.

Part Used : Stem Bark.

Utilization : Stem-bark valued for ropes, twines and rough cloths.

References : Agarwal (2003), Ambasta (1986), Anonymous (1956), Dastur (1964), Drury (1978), Hooker (1890), Jain (1991), Mittre (1997), Sharma & Rana (2005), Singh (2000), Singh & Kumar (2000), Sood & Thakur (2004), Watt (1972), Yonzone *et al.* (1996).

Glinus lotoides L.

Syn: *Mollugo hirta* Thunb.; *M. lotoides* Kuntze.
Family: Molluginaceae.

Sanskrit and Regional Names:

Sans.	:	Bhissata, Okharadi;
Guj.	:	Golo okhard, Meetho orkhad;
Mar.	:	Kotrak;
Pub.	:	Gandi butti, Porparang;
Tam.	:	Sirooseroopadi.

Distribution : Throughout the Plains of India.

Part Used : Plant.

Utilization : Commonly used as fodder.

References : Anonymous (1956), Islam (2002).

Gliricidia sepium (Jacq.) Walp

Syn: *Gliricidia maculata* (H. B.& K.) Steud.
Family: Fabaceae.

English and Regional Names:

Eng. : The Spotted Gliricidia;
Tam. : Kona, Maram;
Tel. : Madri.

Distribution : Parts of Madras, Mysore, Bombay, Tamil Nadu. Kerala and Maharashtra upto 1,000 m.

Part Used : Leaves.

Utilization : Leaves lopped for fodder.

Active Constituents : Leaves rich in proteins, kaempferol-3-O-rhamnoglucoside.

References : Raju (1998), Rastogi & Mehrotra (1990).

Glochidon calocarpum Kurz.

Family: Euphorbiaceae.

Distribution : Andaman Nicobar Islands.

Part Used : Leaves.

Utilization : Leaves valued as fodder.

References : Dagar (2003), Hooker (1890).

Glochidion velutinum Wight.

Syn: *Bradlcia ovata* Wall.;
Phyllanthus nepalensis Muell.; *P. velutinus* Muell.
Family: Euphorbiaceae.

Regional Names:

Ass. : Dolpuduli, Uding- that;
Kan. : Salai- mara;
Mal. : Kayara;
Mar. : Paritza, Shoura;
Pub. : Amblu, Gol, Kamila, Sama;

Distribution : Himalaya: Kashmir to Myanmar.

Part Used : Leaves.

Utilization : Leaves given as cattle fodder.

References : Anonymous (1956), Rastogi & Mehrotra (1991), Sood & Thakur (2004), Watt (1972).

Glyceria tonglensis C. B. Clarke.

Syn: *G. fluitans* Duthie Grass.; *G. rivularis* Jacq.

Family: Poaceae.

English Names:

Eng. : Floating Grass, Monna Grass;

Distribution : Temperate Himalaya from Kashmir to Sikkim: 2,000-4,000 m

Part 1Used : Foliage.

Utilization : Fodder for cattle, sheep and goats.

Active Constituents : Leaves rich in proteins 9.69, fat 0.43, starch, sugar 75.06 and fibre 0.21.

References : Anonymous (1956), Hooker (1897), Sharma & Rana (2005).

Gnetum latifolium Blume.

Syn: *Gnetum funiculare* Blume.; *G. indicum* (Lour) Merril in Part.; *G. macropodum* Kurz.

Family: Gnetaceae.

Distribution : Andaman & Nicobar Islands.

Part Used : Bark.

Utilization : Bark fibre used for making ropes, paper and net.

References : Ambasta (1986), Anonymous (1956), Jain (1991).

Goniothalamus sequipedalis Hk. f. &Th.

Family: Annonaceae.

Distribution : N. E. India.

Part Used : Bark.

Utilization : Bark used for crude cordage.

Reference : Saklani & Jain (1994).

Gonostegia pentendra (Roxb.) Miq.

Syn: *Pouzalzia pentandra* Benn.
Family: Sapotaceae.

Part Used : Plant.

Utilization : Yields a fibre for cordage.

Reference : Ambasta (1986).

Gossypium arboreum Linn.

Syn: *Gossypium cernuum, Gossypium obtusifolium* Roxb. in part.; *G. indicum* Tod.; *G. intermedium* Tod.; *G. nanking* Meyen.; *G. neglectum* Tod.; *G. sanguineim* Hassk.
Family: Malvaceae.

Vern. Name: Kapah.

English, Hindi, Sanskrit and Regional Names:

Eng. : Tree cotton;
Hindi : Kapas;
Sans. : Karpasamu;
Bom. : Deokapas;
Guj. : Kapas, Rui;
Kan. : Hathi;
Mal. : Pangai, Paruthi;
Mar. : Kapas, Rui;
Ori. : Karpaso, Kopa;
Pub. : Kapas, Rui, Tula;
Tam. : Pangai, Paruthi;
Tel. : Patti.

Distribution : Widely Cultivated Throughout the Warmer Parts of India.

Part Used : Seed.

Utilization : Cotton fibre much valued for fabrics, carpets and making sacred thread 'Janau'.

References : Agarwal (2003), Ambasta (1986), Anonymous (1956), Drury (1978), Hooker (1872), Jain (1991), Saklani & Jain (1994), Watt (1972).

Gossypium barbadense L.

Syn: *Gossypium nigrum* (in part) W. & A.
Family: Malvaceae.

English, Hindi, Sanskrit and Regional Names:

Eng. : Egyptian Cotton, Sea Land Cotton;
Hindi : Janghrui, Maghani;
Sans. : Maghani, Purvam;
Mal. : Chemparutti;
Ori. : Rongokopa;
Tam. : Arattam, Mayiliyam, Sembanju;
Tel. : Ettapatti, Paidipatti.

Distribution : Throughout India.

Part Used : Seed.

Utilization : Fibre soft and valued for hosiery yarns.

References : Agarwal (2003), Anonymous (1956), Caius (2003), Hooker (1872), Watt (1972).

Gossypium herbaceum L.

Syn: *G. album* Lam.; *G. indicum* Lam.; *G. micranthum* DC.
Family: Malvaceae.

English, Hindi, Sanskrit and Regional Names:

Eng. : Asiatic Cotton, Levant Cotton;
Hindi : Karpasi , Rui;
Sans. : Karpas, Karpasi;
Beng : Kapas Tula;
Bom. : Kapas Rui;
Guj. : Kapas, Ru;
Mal. : Badara, Karpas, Karpasi, Pangini, Parutti;
Mar. : Kapus, Ru;

Ori.	:	Karpaso, Kopa, Korpaso, Rui;
Pub.	:	Rui;
Tam.	:	Parutti;
Tel.	:	Edudi, Paratti, Pauti.

Distribution : Throughout the Plains of India.

Part Used : Root Bark. Seed.

Utilization : Fibre good for making cloths and stuffing quiltz, pillows etc.

References : Agarwal (2003), Anonymous (1956), Drury (1978), Hooker (1872), Caius (2003), Prasad *et al.* (2002, 2003), Watt (1972).

Grewia asiatica L.

Syn: *Grewia elastica* Royle; *G. subinaequalis* DC.

Family: Tiliaceae.

Vern. Names: Dhaman, Phalsa.

English, Hindi, Sanskrit and Regional Names:

Eng.	:	Asiatic Grewia;
Hindi	:	Phalsa;
Sans.	:	Porusha;
Beng	:	Phalsa, Shakri;
Bom.	:	Phalasi;
Guj.	:	Phalsa;
Mal.	:	Chadicha;
Mar.	:	Phalsi;
Ori.	:	Pharosakoli;
Pub.	:	Phalsa;
Tam.	:	Tadachi;
Tel.	:	Phutiki, Putiki;

Distribution : Cultivated Throughout India; generelly in the Gangetic Plains & E. Bengal.

Part Used : Bark.

Utilization : Bark fibre for making ropes used for tying purpose.

References : Agarwal (2003), Caius (2003), Dastur (1964), Hooker (1872), Khare (2004), Talbot (1976b), Watt (1972).

Grewia flavescens Juss.

Syn: *Grewia carpinifolia* Mast. non Juss.; *G. pilosa* Mast. in part.
Family: Tiliaceae.

Distribution : U.P. Rajasthan. Bihar. C. and S. India.

Parts Used : Leaves. Branches.

Utilization : Lopped for fodder.

References : Ambasta (1986), Dwivedi (1993), Singh (2002, 2003).

Grewia glabra Blume.

Syn: *Grewia disperma* Drummond, non Rottl.;
G. laevigata Mast. Non Vahl.
Family: Tiliaceae.

Hindi and Regional Names:

Hindi : Bhimul, Kakki, Kath bewal;
Ass. : Senam logda;
Beng : Kath Bimla;
Bom. : Kaori, Gullgollop;
Ori. : Kulokathi;
Tam. : Naraittai, Pirunnu;
Tel. : Allipayaru, Potirike.

Distribution : Found Throughout Greater Parts of India.

Part Used : Bark.

Utilization : Bark yields a cordage fibre for ropes.

References : Agarwal (2003), Ambasta (1986), Anonymous (1956), Awasthi & Goel (2000).

Grewia maderaspatana (L.) Poir.

Syn: *Artemissa maderaspatana* Roxb.; *Cotula maderaspatana* Willd.; *C. sphaeranthus* Link.; *C. sphaeranthus* Koch.
Family: Asteraceae.

English, Hindi, Sanskrit and Regional Names:

Eng. : Madras Warm Weed;

Hindi : Mukhatari, Mustaru;
Sans. : Mastaru;
Beng : Namuti;
Guj. : Jinkimundi;
Kan. : Dovana;
Mar. : Mashipatri;
Tam. : Mashipatri;
Tel. : Save.

Distribution : Throughout India.

Part Used : Twigs.

Utilization : Used as fodder.

References : Islam (2002), Khan & Khanum (2005), Watt (1972).

Grewia multiflora Juss.

Syn: *Grewia coriacea* Garcke.; *G. sepiaria* Roxb.; *G. serrulata* DC.
Family: Tiliaceae.

Distribution : E. Bengal, Assam, Sikkim Himalaya, Khasia & Nilgiri Hills.

Part Used : Stem Bark.

Utilization : Stem-bark used for making ropes.

References : Saklani & Jain (1994), Watt (1972).

Grewia oppositifolia Buch.-Ham.

Syn: *Grewia optiva* Drumm. *ex* Burret.
Family: Tiliaceae.

Vern. Name: Beul.

Hindi and Regional Names:

Hindi : Bahul, Biul, Biung;
Kan. : Thidsal.

Distribution : N. W. Himalaya ascending to 2,400 m.

Parts Used : Leaves. Bark.

Utilization : Leaves lopped for fodder in winter for cattle and goats. Bark fibre used for making ropes, net, strings and paper.

Active Constituents : Leaves contain crude proteins 10.1%, fat 6.8%, crude fibre 14.1%, ash 14.2%, P 0.25% and tannin.

References : Agarwal (2003), Hooker (1872), Parker (1973), Purohit & Samant (1995), Sharma (2004), Sood & Thakur (2004), Watt (1972).

Grewia rothii DC.

Syn: *Grewia excelsa* Mast. in Part.
Family: Tiliaceae.

Regional Names:

Ori. : Homola- poto;
Tam. : Angolam;
Tel. : Kolu Pu, Putiki, Siriana.

Part Used : Bark.

Utilization : Bark fibre good for making ropes.

Reference : Ambasta (1986).

Guazuma ulmifolia Lam.

Syn: *Guazuma tomentosa* H. B. & K.
Family: Sterculiaceae.

English, Hindi and Regional Names:

Eng. : Bastard Cedar, Honey Fruit Tree, Musket tree;
Hindi : Nipaltunth;
Kan. : Rudrakshi;
Mal. : Rudrakshm;
Ori. : Debodari;
Tam. : Rudrasam, Thenmaram, Tubakki;
Tel. : Rudraksha, Udrikpatta.

Distribution : Warmer Parts of India; often cultivated as a shade tree.

Parts Used : Leaves. Stem.

Utilization : Leaves lopped for fodder. Stem fibre considered good for making ropes.

Active Constituents : Leaves comprise of friedelinol-3-acetate and fridelan-3-β-sitosterol.

References : Agarwal (2003), Ambasta (1986), Anonymous (1956), Dastur (1964), Drury (1978), Parrotta (2001), Rastogi & Mehrotra (1991).

Gyrinops walla Gertn.

Family: Thymelaeaceae.

Distribution : S. India.

Part Used : Bark.

Utilization : Bark fibre employed for making ropes, hats, mats and cigar pouches.

Reference : Agarwal (2003).

Haloptelea integrifolia (Roxb.) Planch.

Syn: *Ulmus integrifolia* Roxb.

Family: Ulmaceae.

Vern. Names: Bastendun, Papri.

English, Sanskrit and Regional Names:

Eng. : Indian Elm, Jungle Cork Tree

Sans. : Chirabilvah, Puikaranjah;

Guj. : Kanjho, Waola;

Mal. : Aval, Nettavel;

Mar. : Papara, Vavala, Vavli;

Ori. : Daurango, Dauranja, Taruda;

Pub. : Arjan, Khulan, Papri, Rajaun;

Tam. : Aya, Ayil, Kanchi, Vellaga.

Distribution : Commonly in Deciduous Forest upto 600 m in the Sub-Himalayan Hills. Bihar. W. & S. Parts of Peninsular India.

Part Used : Leaves.

Utilization : Leaves used for fodder purposes.

References : Parrotta (2001), Singh (2002).

Hedera helix auct (non L.) Clarke.

Syn: *Hedera himalaica* Tobler.; *H. nepalensis* K. Koch.; *H. rhombea* Sieb. & Zucc.

Family: Araliaceae.

Vern. Name: Dakari.

English, Hindi and Regional Names:

Eng.	:	Barren Ivy, Beene Tree, Bent Wood, Black Ivy, Wood Bind;
Hindi	:	Lab-Lab;
Ass.	:	Mej- Peosree;
Kum.	:	Banda.

Distribution	:	Himalayan Region.
Part Used	:	Leaves.
Utilization	:	Leaves excellent fodder for goats.
Active Constituents	:	Leaves contain α-hederin, β- elemene, elixene, germacrene, vitamıns, carotene, rutin and caffeic acid.
References	:	Anonymous (1959), Chopra *et al.* (1969), Hooker (1879), Rastogi & Mehrotra (1991, 1995a), Sharma (2004), Singh & Kumar (2000), Sood *et al.* (2005), Watt (1972).

Hedychium spicatum Buch.-Ham. *ex* Smith

Syn: *Hedychium acuminatum* Rosc.

Family: Zingiberaceae.

Vern. Name: Banadrak-safed.

English, Hindi, Sanskrit and Regional Names:

Eng.	:	Lesser Galangal Silked Ginger Lily;
Hindi	:	Gandhapalashi, Sit- ruti;
Sans.	:	Amlaharidra, Durva, Karpura, Palashasathi, Tuni;
Beng	:	Ada, Arna;
Bom.	:	Sir, Sutti;
Guj.	:	Kapurkachari;
Mar.	:	Kapurkachari, Sanatakka;
Pub.	:	Banhaldi, Khor, Saki;
Tam.	:	Simaikkichilikkilhangu.

Distribution : Sub-Tropical Himalaya, C. & W. Himalaya upto 2,000 m. E. India and Hills of S. India.

Part Used : Leaves.

Utilization : Green fodder fed to livestock to enhance lactation. Fibre used for making ropes.

References : Anonymous (1959), Dastur (1964), Dhiman (2003), Khare (2004), Sood & Thakur (2004), Watt (1972).

Helianthus annuus Linn.

Family: Asteraceae.

English, Hindi, Sanskrit and Regional Names:

Eng. : Common Sun Flower, Merigold of Peru;
Hindi : Surajmukhi;
Sans. : Surya - mukhi;
Beng : Surajmuki;
Bom. : Surajmaki;
Guj. : Surajmukhi;
Kan. : Suryakanti;
Mal. : Suryakanti;
Mar. : Surajmakha, Suryaphul;
Pub. : Surajmukhi;
Tam. : Suryakanti;
Tel. : Aditya- bhaktichettu.

Distribution : Throughout India.

Parts Used : Oil Cake. Leaves.

Utilization : Leaves and oil cake used as fodder for poultry and cattle feed.

Active Constituents : Seeds rich in palmitic (10.0), stearic (5.7), linolenic (46.0%) acid. Leaves contain ascorbic acid, citric and malic acid.

References : Agarwal (2003), Ambasta (1986), Anonymous (1959), Caius (2003), Chopra *et al.* (1969), Parrotta (2001), Rastogi & Mehrotra (1991), Singh *et al.* (1996), Watt (1972).

Helicteres isora L. (Pl. 5A)

Syn: *Helicteres chrvsocalyx* Miq.; *H. roxburghii* G. Don. Gen.
Family: Sterculiaceae.

Vern. Name: Mavarphali.

English, Hindi, Sanskrit and Regional Names:

Eng.	:	Indian Screw Tree;
Hindi	:	Jonkaphal, Morarphali;
Sans.	:	Avartani, Mriga- shinga;
Beng	:	Atmora;
Bom.	:	Kevana, Kewan, Khiran;
Guj.	:	Murdasing;
Kan.	:	Kavargi, Yedamuri;
Mal.	:	Isvarmuri, Kavum;
Mar.	:	Kevani, Kewan, Varkati;
Ori.	:	Murmuria, Murimuri;
Pub.	:	Jonkaphal, Maror phalli;
Tam.	:	Kaiva, Valampiri;
Tel.	:	Kavanchi, Nuliti.

Distribution : C. & W. India.

Parts Used : Leaves. Twigs. Stem Bark.

Utilization : Leaves and twigs lopped for fodder. Bark fibre used for sacks, canvas and cordage whereas stem fibre valued for tying purposes, string, cattle harness and container bags.

Active Constituents : Leaves contain pro-vitamin A, crude proteins 13 -25 and CaO 3.15.

References : Agarwal (2003), Ambasta (1986), Anonymous (1959), Bhattacharjee (1998), Dash & Mishra (2000), Dastur (1964), Hooker (1872), Hosagoudar & Henry (1996), Jain (1991), Kothari & Rao (2000), Mittre (1997), Oommachan (1977), Parker (1973), Prasad *et al.* (2002), Raju (1998), Rastogi & Mehrotra (1995b), Singh & Pandey (1998), Sur & Halder (2002), Varghese (1996), Watt (1972).

Hibiscus cannabinus L.

Syn: *Hibiscus wightianus* Wall.
Family: Malvaceae.

English, Hindi, Sanskrit and Regional Names:

Eng.	:	Bimli patam jute, Deccan Hemp, Kenaf, Mesta;
Hindi	:	Ambari, Patsan, Pitwa, Pulu;
Sans.	:	Nalita;
Beng	:	Metsa pat;
Bom.	:	Ambari;
Guj.	:	Ambari, Sheria;
Mal.	:	Kanjaru;
Mar.	:	Ambada, Ambadi;
Ori.	:	Kanuriya;
Pub.	:	Sankokla;
Tam.	:	Pulichhai, Pulimanji;
Tel.	:	Gaynara, Gongura.

Distribution : Distributed Throughout India.

Parts Used : Stem. Leaves.

Utilization : Extracted fibre compete with jute in luster but these are coarser, inflexible, harsher and more brittle; suitable for cordage, ropes and manufacturing of paper.

References : Agarwal (2003), Ambasta (1986), Anonymous (1959), Drury (1978), Geetha *et al.* (1996), Hooker (1872), Jain (1991), Parrotta (2001), Patil (1998), Singh & Pandey (1998), Varghese (1996), Watt (1972).

Hibiscus furcatus Roxb.

Syn: *Hibiscus aculeatus* Roxb.; *H. bifurcatus* Roxb.; *H. hispidissimus* Griff.
Family: Malvaceae.

Hindi and Regional Names:

Hindi	:	Gumchi;
Ass.	:	Kiasi;
Kan.	:	Huligowri;
Mal.	:	Naranampupulli, Pachapuli, Suriyamani;

Ori. : Piri pirika;
Tel. : Adavi, Gogu, Konda.

Distribution : Assam. Chota Nagpur. Mysore. A. P. Kerala.

Part Used : Stem.

Utilization : Fibre suitable for cordage and ropes.

References : Agarwal (2003), Ambasta (1986), Anonymous (1959), Hooker (1872), Talbot (1976a), Watt (1972), Yoganarasimhan *et al.* (1981).

Hibiscus macrophyllus Roxb.

Syn: *Hibiscus setosus* Roxb.; *H. vestitus* Griff.
Family: Malvaceae.

Regional Name:

Beng : Khashia udal.

Distribution : Distributed in N. E. India. Assam. Khasia. Jaintia. Aka & Lushai Hills.

Part Used : Stem Bark.

Utilization : Stem fibre valued for cordage, ropes and paper pulp.

References : Agarwal (2003), Ambasta (1986), Anonymous (1959), Watt (1972).

Hibiscus mutabilis Linn.

Syn: *Hibiscus aestuans* Rottler in Herb.
Family: Malvaceae.

English, Hindi, Sanskrit and Regional Names:

Eng. : Chinese Rose, Cotton Rose;
Hindi : Shala Para, Sthal Kamal;
Sans. : Padma Charini, Stalapadma;
Beng : Stalpadma, Thulpadma;
Kan. : Beltada Tavare, Neladavare;
Mal. : Chinnapparatti, Hinaparuti;
Ori. : Sthalopidmo, Tholopodmo;
Pub. : Gul-i- ajaib;
Tam. : Irratai – vellaichembarattom.

Distribution : All over India from Punjab to Madras.

Part Used : Bark.

Utilization : Bark yields a strong fibre for cordage and ropes.

References : Agarwal (1986), Ambasta (2003), Anonymous (1959), Hooker (1872), Watt (1972).

Hibiscus rosa-sinensis L.

Family: Malvaceae.

Vern. Names: China Rose, Shoe Flower.

English, Hindi, Sanskrit and Regional Names:

Eng. : Shoe Flower;
Hindi : Gulhar, Jasut;
Sans. : Japa, Java, Pusham, Rudra;
Ass. : Joba;
Beng : Jiwa;
Bom. : Jasavanda;
Guj. : Jasuva;
Kan. : Dasavala;
Mal. : Chembarathi;
Mar. : Dasindacha phulla, Jasavanda;
Ori. : Mondaro;
Pub. : Jasum;
Tam. : Semparuthi.

Distribution : Grows for its beautiful flowers upto 500 m.

Part Used : Stem Bark.

Utilization : Bark yields a fibre for cordage.

References : Agarwal (2003), Anonymous (1959), Hooker (1872), Watt (1972).

Hibiscus sabdariffa L.

Syn: *Hibiscus sanguineus* Griff.

Family: Malvaceae.

English, Hindi, Sanskrit and Regional Names:

Eng. : Jamaica Sorrel, Red Sorrel;
Hindi : Lal anbari;

Sans.	:	Ambasthaki;
Ass.	:	Chukiar;
Beng	:	Chukar, Lal mista, Patwa;
Bom.	:	Lal ambari;
Guj.	:	Lalsheria;
Kan.	:	Pulachakiri;
Mal.	:	Polechi, Pulichchai;
Mar.	:	Lal ambadi, Patwa;
Tam.	:	Gogu, Kerai, Pulichchai;
Tel.	:	Yarra gogu.

Distribution : Cultivated Throughout the Hotter Parts of India and Sri Lanka.

Part Used : Stem.

Utilization : Stem fibre valued for cordage, sacks, and mats and as raw material for paper industry.

References : Agarwal (2003), Anonymous (1959), Caius (2003), Khanna (2002), Maheshwari *et al.* (1997), Parrotta (2001), Watt (1972).

Hibiscus tilaceus Linn.

Syn: *Hibiscus tortuosus* Roxb.; *Partium tiliaceus* W. & A.
Family: Malvaceae.

English, Hindi, Sanskrit and Regional Names:

Eng.	:	Coast cotton tree, Yellow mallow tree;
Hindi	:	Bola, Chelva;
Sans.	:	Bala;
Beng	:	Bola, Chelva;
Mal.	:	Attuparathi, Nirparathi;
Mar.	:	Belapata;
Ori.	:	Baniah;
Tam.	:	Attuparathi, Nirparathi;
Tel.	:	Eta gogu.

Distribution : E. & W. Coast of India. Andaman & Sunderbans.

Part Used : Bark.

Utilization : Bark fibre good for cordage, ropes, mats, tow, sails, fishing nets, coarse bags and wrapping paper.

References : Agarwal (2003), Ambasta (1986), Awasthi & Goel (2000), Dagar & Dagar (1996), Das *et al.* (1996), Dastur (1964), Dwivedi (1993), Hooker (1872), Mittre (1997), Watt (1972).

Hiptage benghalensis (Linn.) Kurz. (Pl. 5B)

Syn: *Banisteria benghalensis* Linn.; *Hiptage medablota* Gaertn.
Family: Malpighiaceae.

English, Hindi, Sanskrit and Regional Names:

Eng. : Clustered hiptage;
Hindi : Madhvi lata;
Sans. : Atimukta, Madhari;
Beng : Madhavi lata, Madhubi lata;
Guj. : Madhavi, Rakatpiti;
Kan. : Madhavi, Vasantaduti;
Mal. : Sitampu;
Mar. : Madhavi, Haladvel;
Ori. : Oromali, Gorunda;
Pub. : Benkar, Chabuk, Chopar, Endra;
Tam. : Madhavi.

Distribution : Throughout India upto 1,800 m.

Part Used : Leaves.

Utilization : Leaves fed to all classes of livestock throughout the year.

Active Constituents : Leaves contain iso- flavonoids, quercetin, kaempferol and leucopelargonidin.

References : Agarwal (2003), Anonymous (1959), Dhiman (2003), Khare (2004), Parrotta (2001), Prajapati *et al.* (2003), Purohit & Samant (1995).

Holboellia angustifolia Wall.

Syn: *Holboellia angustifolia* Wall.; *Stauntonia angustifolia* Wall.
Family: Lardizabalaceae.

Distribution : Himalaya from Kumaon Eastwards to Bhutan. Khasia Hills. Upper Assam.

Part Used : Leaves.

Utilization : It is fed to all classes of livestock throughout the year.

References : Anonymous (1959), Purohit & Samant (1995), Watt (1972).

Holostemma annularis K. Schum.

Syn: *Holostemma rheedei* Wall.; *H. rheedianum* auct.
Family: Asclepiadaceae.

Hindi, Sanskrit and Regional Names:

Hindi : Chhirvel, Jivanti;
Sans. : Arkapushpi, Jivanti;
Guj. : Kharnar, Khiravel;
Mal. : Ada kodien, Ada modien;
Mar. : Dudurli, Shidoli, Tulatule;
Tam. : Palay kirai;
Tel. : Dudipalatige, Palatura.

Distribution : Most Parts of India.

Parts Used : Leaves. Flowers.

Utilization : Leaves and flowers eaten by cattle.

References : Agarwal (2003), Ambasta (1986), Anonymous (1959), Talbot (1976b).

Hordeum vulgare Linn.

Syn: *Hordeum aegiceras, H. coeleste, H. gymnodistichum* of Watt.; *H. dificiens* Steud.; *H. distichon* Linn.; *H. hexastichon* Linn.; *H. intermedium* Kornioke.; *H. sativum* Jessen.; *H. zeocriton* Linn.
Family: Poaceae.

English, Hindi, Sanskrit and Regional Names:

Eng. : Barley;
Hindi : Jau, jav;
Sans. : Akshata, Divga, Kanchuki, Yaya;
Beng : Jab, Jau;
Bom. : Jav, Satu;
Guj. : Jau, Jav, Ymvah;

Kan.	:	Jave godhi;
Mal.	:	Yavam;
Mar.	:	Java;
Pub.	:	Jaon;
Tam.	:	Barliya- risi;
Tel.	:	Yaya.

Distribution : Cultivated Throughout Dry Parts of India upto 4,000 m. M.P. U.P. Punjab.

Part Used : Straw.

Utilization : Straw used as fodder for livestock.

Active Constituents : Rich in vit. B, protein 11.5, fibre 3.9, carbohydrates 69.3, CaO 0.03 and Fe 3.7 mg/100g.

References : Agarwal (2003), Ambasta (1986), Anonymous (1959), Hooker (1897), Khare (2004), Prajapati *et al.* (2003), Sood *et al.* (2005), Watt (1972).

Hydrolea zeylanica (L.)Vahl.

Syn: *Hydrolea javanica* Blume Bijd.; *Nama zeylanica* Linn.
Family: Hydrophyllaceae.

Sanskrit and Regional Names:

Sans.	:	Langali;
Beng	:	Isha-langulya;
Mal.	:	Cheruvallel.

Distribution : Throughout India, ascending to 1,300 m in wet places.

Part Used : Plant.

Utilization : Commonly used for fodder purposes.

References : Anonymous (1959), Hooker (1885), Islam (2002), Watt (1972).

Hygrophila phlomoides Nees.

Family: Acanthaceae.

Distribution : Plains of W. Bengal. Assam.

Part Used : Plant.

Utilization : Used as fodder.

Reference : Islam (2002).

Hymenachne amplexicaulis (Rudge) Nees.

Syn: *Hymenachne myurus* Beauv.; *H. pseudointerrupa* C. Muell.; *Panicum myurus* H. B.& K.

Family: Poaceae.

English, Hindi and Regional Names:

Eng. : Dal Grass;
Hindi : Dhamsiria;
Ass. : Bhatdal, Dhopdal, Karanga dal, Tattu;
Bom. : Pokalia.

Distribution : Moist, Warm Parts of India (chiefly in E. India & Deccan).

Part Used : Grass Blades.

Utilization : Used as both green and hay and silage.

Active Constituents : Rich in proteins.

References : Agarwal (2003), Ambasta (1986), Anonymous (1959).

Hypserpa cuspidata (Wall.) Miers.

Syn: *Limacia cuspidata* Hook. f. & Thoms.

Family: Menispermaceae.

Distribution : Parts of Assam, Sikkim, Madras & Andaman Islands.

Part Used : Stem.

Utilization : Stem fibre used for bow-string.

References : Ambasta (1986), Anonymous (1959).

Ichnanthus vicinus Merrill.

Syn: *Ichnanthus pallens* Hook. f., non Munro.

Family: Poaceae.

Distribution : Sikkim Himalaya. Assam.

Part Used : Branches.

Utilization : Grass readily eaten by cattle.

References : Ambasta (1986), Anonymous (1959).

Ichnocarpus frutescens (L.) R. Br.

Syn: *Apocynum frutiscens* Linn.; *Echites frutescens* Wall.; *Ichnocarpus dasycalax* Miquel.; *I. leptodictyus* F. Muell.; *I. radicans* Wall.

Family: Apocynaceae.

English, Hindi, Sanskrit and Regional Names:

Eng. : Black creeper;
Hindi : Dudhi, Siama-lata;
Sans. : Palindi, Paravalli, Sariva;
Ass. : Lamkandol, Paharukibandan;
Beng : Shyama, Syama;
Kan. : Karehambu;
Mal. : Paalvally;
Mar. : Kantebhouri, Krishnasarwa;
Ori. : Madhobi, Soyamnoi;
Tam. : Paravalli, Udargodi;
Tel. : Munta Gajjanamu.

Distribution : W. Himalaya. Upper Gangetic Plains from Delhi to Bengal. Assam.

Part Used : Stem.

Utilization : Stem fibre used for fishing nets, ropes, cordage and baskets.

References : Anonymous (1959), Drury (1978), Hooker (1882), Rao & Henry (1996), Varghese (1996), Watt (1972).

Ilex dipyrena Wall.

Family: Aquifoliaceae.

Vern. Name: Charkha.

English and Regional Names:

Eng. : Himalayan Holly;
Pub. : Dinsa, Diusa, Dodru, Kandlar, Kanjru, Karelu.

Distribution : Common in Oak Forest from Shimla (1,000-2,400 m) to Sikkim (2,000-3,000 m).

Part Used : Leaves.

Utilization : Fed to animals during winter.

References : Dwivedi (1993), Hooker (1872), Purohit & Samant (1995), Sharma (2004), Watt (1972).

Ilysanthes parviflora Benth.

Syn: *Bonnaya parviflora* Benth.

Family: Scrophulariaceae.

Distribution : Wet Places from Punjab to Assam. Bengal. Deccan Peninsula.

Part Used : Plant.

Utilization : Used as cattle fodder.

References : Hooker (1885), Islam (2002).

Impatiens tripetala Roxb.

Syn: *Impatiens multiflora* Wall.

Family: Balsaminaceae.

Distribution : Tropical Himalaya Sikkim, 1,000-2,000 m. Kashmir, 1,500- 2,400 m.

Part Used : Seeds.

Utilization : Seeds fed to cattle.

References : Anonymous (1959), Hooker (1872), Islam (2002).

Imperata cylindrica (L.) P. Beauv.

Syn: *Imperata arundinacea* Cyr.

Family: Poaceae.

English, Hindi, Sanskrit and Regional Names:

Eng. : Alang -alang, Bedding Grass, Thatch Grass;

Hindi : Dabh, Siru, Ulu;

Sans. : Bahir, Darbha, Yagnika

Ass. : Batta, Khair, Ullu;

Beng	:	Ooloo, Ulu;
Kan.	:	Dabbai, Hullu, Sanna;
Mal.	:	Vidulam;
Pub.	:	Dab, Kusa, Sil, Sir;
Tam.	:	Darbai pul, Nanal, Varli pillu;
Tel.	:	Balbajamu, Barhisam.

Distribution : Common Weed in the Hilly Areas; Assam, Sikkim, Orissa & W. Bengal.

Parts Used : Leaves. Stem.

Utilization : Leaves beaten and twisted to form cordage which are used variously. Stem used for paper making. Also, a number of plants are twisted together for making ropes by the tribals for carrying fodder and fire wood from the forest.

References : Anonymous (1959), Dastur (1964), Islam (2002), Mittre (1997), Prajapati *et al.* (2003), Sharma & Rana (2005), Singh & Pandey (1998).

Indigofera artopurpurea Buch.-Ham.

Syn: *Indigofera hamiltonii* Garh.

Family: Fabaceae.

Hindi Names:

Hindi : Bankati, Kala, Sakuna.

Distribution : Throughout the Himalaya from Hazara to the Khasia Hills & also on Shiwalik Hills.

Part Used : Plant.

Utilization : Plant valued for its high quality fodder.

References : Purohit & Samant (1995), Watt (1972).

Indigofera cylindracea Wall.

Family: Fabaceae.

Distribution : Nepal.

Part Used : Plant.

Utilization : It is browsed by animals.

References : Hooker (1879), Purohit & Samant (1995).

Indigofera gerardiana Wall.

Syn: *I. dosua* Wall.; *Indigofera heterantha* Wall. *ex* Brandis.; *I. polyphylla* DC.; *I. rubroviolacea* Dunn.; *I. virgata* Roxb.

Family: Fabaceae.

Vern. Names: Kali - kathi, Kathura.

English, Hindi and Regional Names:

Eng. : Himalayan Indigo;
Hindi : Khenti;
Pub. : Kati, Khenti, Kutz, Mattu.

Distribution : Temperate Sub-Tropical Himalaya from Kashmir to Bhutan & Khasia Hills: 500-3,000 m.

Parts Used : Leaves. Twigs.

Utilization : Leaves relished by all ruminants especially sheep and goats. Twigs used for rope making.

References : Agarwal (2003), Anonymous (1959), Hooker (1879), Purohit & Samant (1995), Singh & Kumar (2000).

Indigofera nummularifolia (Linn.) Iivera.

Syn: *Indigofera echinata* Willd.

Family: Fabaceae.

Part Used : Leaves.

Utilization : Used as fodder plant.

Reference : Ambasta (1986).

Indigofera pulchella Roxb.

Syn: *Indigofera arborea* Roxb.; *I. cassoides* Rottl. *ex* DC.; *I. elliptica* Roxb.; *I. gibsonii* Grah.; *I. glaucescens* Grah.; *I. jirahulia*.; Hamilt.; *I. purpurascens* Roxb.; *I. violacea* Roxb.

Family: Fabaceae.

Vern. Name: Kathi.

Hindi and Regional Names:

Hindi	:	Sakena;
Guj.	:	Gali, Gari, Nil;
Kan.	:	Goggl;
Mal.	:	Manali, Nilam;
Mar.	:	Baroli;
Ori.	:	Girrili;
Pub.	:	Kainthi, Kathi;
Tam.	:	Nairnji;
Tel.	:	Siralli, Vuyye.

Distribution : Throughout the Himalayan Tract.

Part Used : Leaves.

Utilization : Considered as a good fodder plant and fed in rainy season.

References : Agarwal (2003), Anonymous (1959), Hooker (1879), Purohit & Samant (1995).

Inula cappa DC. Prodr.

Syn: *Conyza cappa* Ham.; *C. lanuginosa, eriphora and argentea* Wall.; *Duhaldea chinensis* DC.; *Inula pseudo- cappa and eriophora* Wall.; *I. salviodora* Schultz - Bip.
Family: Asteraceae.

Distribution : Kumaon to Bhutan: 1,000- 2,000 m.

Part Used : Plant.

Utilization : Used as fodder plant.

Reference : Hooker (1882).

Ipomoea aquatica Forsk.

Syn: *Ipomoea reptans* Poir.
Family: Convolvulaceae.

English, Hindi, Sanskrit and Regional Names:

Eng.	:	Swamp cabbage;
Hindi	:	Kalmi sag, Karmi, Patua sag;
Sans.	:	Kalamb;
Beng	:	Kalmi sag;

Bom. : Nalichi baji;
Guj. : Nalanibhaji;
Mar. : Nadishaka;
Pub. : Ganthian, Nali, Nari;
Tam. : Villaikeerai;
Tel. : Tuti- Kura.

Distribution : Abundantly in the Plains of W. Bengal. Orissa. Assam.

Part Used : Leaves.

Utilization : Green fodder of high nutritive value; relished by cattle and pigs.

Active Constituents : Rich in minerals, vit.B–complex, vit .C, E 1200mg/ 100gm and mineral ash above 2% /100gm.

References : Agarwal (2003), Ambasta (1986), Anonymous (1959), Chatterjee & Pakrashi (1997), Maya *et al.* (1997), Watt (1972).

Ipomoea digitata Linn.

Syn: *Ipomoea paniculata* R. Br., non Burm.
Family: Convolvulaceae.

English, Hindi, Sanskrit and Regional Names:

Eng. : Milky yam;
Hindi : Bilaikand;
Sans. : Bhumi kushmanda, Vidari;
Beng : Belai kand, Bhumi kumara;
Bom. : Bhui kohala;
Kan. : Bhumichekrigadde, Bhuja gumbala;
Mal. : Mutalakkanta, Palmutakku;
Mar. : Bhuikobola, Vidarikand;
Ori. : Bhuinkoharu;
Tam. : Palmudangi;
Tel. : Bhuchakragadda.

Distribution : E. India. Karnataka. Bihar. Orissa. W. Bengal. Assam. Deccan. W. Coast.

Parts Used : Stem. Leaves.

Utilization : Used as cattle fodder.

Active Constituents : Fresh leaves contain 6.3mg/100g of carbohydrates and carotene.

References : Agarwal (2003), Ambasta (1986), Anonymous (1959), Khare (2004), Watt (1972).

Ipomoea eriocarpa R. Br.

Syn: *Convolvulus sphoerocephalus* Roxb.; *Ipomoea hispida* Roth.; *I. sessiliflora* Roth.

Family: Convolvulaceae.

Vern. Name: Safed Ghardan.

Hindi, Sanskrit and Regional Names:

Hindi : Ghiabato;
Sans. : Nakhari;
Ass. : Kalman;
Mal. : Pulichevidu;
Ori. : Panonoi;
Pub. : Bhanwar;
Tel. : Purititige.

Distribution : Throughout India. Deccan Peninsula. Bengal.

Utilization : Plant yields good fodder for milking cattle and fibre for making ropes.

Part Used : Plant.

Active Constituents : Plant contains crude proteins, 15.56, carbohydrates, 35.12, crude fibre, 20-78 and mineral matter, 8.88.

References : Agarwal (2003), Ambasta (1986), Anonymous (1959), Hooker (1885), Sharma (2004), Hooker (1885).

Iris ensata Thunb

Syn: *Iris moorcroftiana* Wall.

Family: Iridaceae.

Hindi and Regional Names:

Hindi : Irsia;
Kash. : Marjal.

Distribution : W. Himalaya upto 3,000 m.

Part Used : Leaves.

Utilization : Leaves used as fodder. Also its fibre used for matting, basket work, thatching and making ropes.

References : Agarwal (2003), Ambasta (1986), Kapur (1996), Roy *et al.* (1998), Sharma & Rana (2005), Watt (1972).

Iris lactea Pallas.

Family: Iridaceae.

Regional Names:

Ladh. : Tesma mantok.

Part Used : Plant.

Utilization : Used as a good fodder and enhance lactation.

Reference : Chaurasia *et al.* (2000).

Iseilema laxum Hack.

Syn: *Iseilema prostrata* Anders.

Family: Poaceae.

English and Regional Names:

Eng. : Mushan, Mushel;
Bom. : Chamgi, Dangers, Gandhi, Shata;
Ori. : Pandasual;
Pub. : Chatt, Champ, Gandi;
Tam. : Thenga nari pillu;
Tel. : Errachengali gaddi.

Distribution : Gangetic Plains. N. W. India on low lying land.

Part Used : Plant.

Utilization : Considered as a best fodder grass in India and relished by cattle both as green and hay; more nutritious before flowering.

Active Constituents : Rich in crude proteins 5.67, crude fibre 28.03, free N 51.78 and Cao 1.30.

References : Agarwal (2003), Ambasta (1986), Anonymous (1959), Diwaker & Ansari (2003), Dwivedi (1993), Hooker (1897), Watt (1972).

Iseilema prostratum Anders.

Syn: *Iseilema wightii* Anders.
Family: Poaceae.

Regional Names:

Bom. : Achi grass, Mabil, Sona, Tambit;
Pub. : Gauni;
Tel. : Yerra kala kasuvu.

Distribution : Throughout India in Damp Areas.

Part Used : Grass Blades.

Utilization : Used as fodder for cattle in the pre-flowering and flowering stages.

References : Ambasta (1986), Anonymous (1959).

Ixora arborea Roxb.

Syn: *Ixora parviflora* Vahl.
Family: Rubiaceae.

English, Hindi, Sanskrit and Regional Names:

Eng. : Torch wood Ixora;
Hindi : Kota gandhal, Nevar;
Sans. : Iswara, Nevali;
Beng : Rangan;
Guj. : Nevari;
Kan. : Kansuragi, Gorabikattige;
Mar. : Likandi, Mekadi, Nevali;
Ori. : Kilakriya, Telokrya;
Tam. : Shulundu - Kora, Korivi;
Tel. : Kachipala, Korivipala, Gorrivi, Puttapala;

Distribution : Gangetic Plains of E. India & S. India.

Part Used : Leaves.

Utilization : Leaves used as fodder for buffaloes.

Active Constituents : Contains ixoral and sitosterol.

References : Agarwal (2003), Ambasta (1986), Anonymous (1959), Chatterjee & Pakrashi (1997), Jain (1991), Varghese (1996).

Jasminum multiflorum (Burm. f.) Andr.

Syn: *Jasminum bracteatum* Wight.; *J. congestum* Wall.; *J. hirsutum* Willd.; *J. multiflorum* Roth.; *J. pubescena* (Retz.) Willd.; *Nyctanthes multiflora* Burm. f.; *N. pubescens* Reth.; *Mogorium pubescens* Lamk.

Family: Oleaceae.

Vern. Name: Chameli.

English, Hindi, Sanskrit and Regional Names:

- **Eng.** : Downy Jasmine, Musk Jasmine;
- **Hindi** : Chameli, Kunda, Kundo, Kund phul;
- **Sans.** : Attacha, Atimukha, Damand, Kunda, Madhya, Tarapushpa, Vanahasa;
- **Mal.** : Kundam, Kuruna, Gujari;
- **Mar.** : Mogra;
- **Ori.** : Kontabelo;
- **Tam.** : Magarandam, Malligai;
- **Tel.** : Adavimalli, Gujari, Kundamu, Molla.

Distribution : Common in the Sub-Himalayan Tract upto 1,000 m & W. Ghats.

Part Used : Leaves.

Utilization : Leaves used as fodder.

Active Constituents : Leaves comprise of kriedelin, lupeol, amyrin, betulinic and ursolic.

References : Anonymous (1959), Drury (1978), Rastogi & Mehrotra (1995a), Singh & Kumar (2000).

Juglans regia Linn.

Syn: *Juglans arguta* Wall.; *J.regia* L. var. *kumaonica* Cas.

Family: Juglandaceae.

Vern. Name: Akhrot.

English, Hindi, Sanskrit and Regional Names:

Eng.	:	European Walnut, Himalayan Walnut;
Hindi	:	Akhrot;
Sans.	:	Akhoda, Akschota, Aksotah;
Ass.	:	Kabsingh;
Beng	:	Akhrot, Dkrut;
Guj.	:	Akhrot;
Kan.	:	Ak, Akhor, Kharat;
Kash.	:	Ak, Akhor, Akhrot, Kharot;
Kum.	:	Akhor, Kharot;
Mal.	:	Akrottu;
Mar.	:	Akroda;
Ori.	:	Akhrot;
Pub.	:	Akhori, Charmagz, Kabotang;
Tam.	:	Akrottu;
Tel.	:	Akrotu.

Distribution : Throughout Himalaya 1,800-3,000 m.

Part Used : Twigs.

Utilization : Twigs fed to animals.

Active Constituents : These are 21, monoterpenes, 2 sesquterpenes, 23 hydrocarbons along with eugenol, geranic acid, proteinss about 16%, mineral like Na 2.75, K 6.85, Ca 6.1%, Mg 13%, Fe 2%, I 2.8% and vit. B & C.

References : Agarwal (2003), Anonymous (1959), Dwivedi (1993), Hooker (1890), Prajapati *et al.* (2003), Purohit & Samant (1995), Rastogi & Mehrotra (1993), Watt (1972).

Juncus effusus Linn.

Syn: *Juncus communis* E. Mey.

Family: Juncaceae.

English Names:

Eng. : Maltting Rush, Soft common rush.

Distribution : Sikkim Himalaya (1,800-3,000 m). Khasia Hills (1,500-1,800 m).

Part Used : Herb.

Utilization : Used for making mats, baskets and chair bottoms. A fibre which can be spun into thread is obtained from alkali digested pulp of the herb. Fine straw from plant good for stuffing.

References : Agarwal (2003), Ambasta (1986).

Kleinhovia hospita Linn.

Family: Sterculiaceae.

Hindi and Regional Names:

Hindi : Bola;
Beng : Bola;
Tam. : Panaitteku.

Distribution : E. & W. Peninsula of India.

Part Used : Bark.

Utilization : Bark fibre used for ropes.

References : Agarwal (2003), Ambasta (1986).

Kobresia duthiei Clarke.

Family :Cyperaceae.

Distribution : Distributed from 3,500-4,000 m.

Part Used : Twigs.

Utilization : Used as a fodder plant in W. Himalaya.

Reference : Purohit & Samant (1995).

Kochia indica Wight.

Syn: *Kochia griffithii* Bunge in Boiss.; *Panderia pilosa* Herb.

Family: Chenopodiaceae.

Hindi and Regional Names:

Hindi : Bui;
Pub. : Bui, Kaura -ro, Ui – chotti.

Distribution : Delhi Westwards. Deccan Peninsula. Saline soils of Coimbatore district. S. India.

Parts Used : Leaves. Twigs.

Utilization : Relished by cattle, camels and mules.

References : Agarwal (2003), Ambasta (1986), Anonymous (1959), Hooker (1890), Watt (1972).

Kochia scoparia Schrad.

Syn: *Chenopodium scoparia* Linn.

Family: Chenopodiaceae.

English Name:

Eng. : Belvedere.

Distribution : N. W. India, Naturalized in Fields & Mountains.

Part Used : Leaves

Utilization : Used as fodder for cattle.

Active Constituents : Rich in proteins, carbohydrates, saponin, triterpenoides and betanin.

References : Ambasta (1986), Anonymous (1959), Hooker (1890), Prajapati *et al.* (2003).

Koeleria cristata Pers.

Syn: *Aria cristata* Linn.

Family: Poaceae.

English and Hindi Names:

Eng. : Crested Hair Grass;
Hindi : Mahinghas.

Distribution : Grows on high altitude from Nepal to Kashmir.

Part Used : Grass Blades.

Utilization : Yields good fodder.

Active Constituents : These are crude proteins 10.96, crude fibre 24.73, carbohydrates 41.68 and CaO 0.87.

References : Agarwal (2003), Ambasta (1986), Anonymous (1959), Watt (1972).

Kydia calycina Roxb.

Syn: *Kydia fraterna* Roxb.; *K. pulverulenta* Ham.; *K. roxburghiana* Wight.
Family: Malvaceae.

Vern. Name: Pula.

English, Hindi and Regional Names:

Eng. : Pula;
Hindi : Choupultea;
Ass. : Bankopah;
Beng : Bonkopas, Pola;
Bom. : Motipotari, Varanga, Varung;
Gar. : Pillo, Pulao;
Guj. : Motihirwani, Nihotilirwani;
Kan. : Bellatta, Nayebende;
Mal. : Nedunar, Velluku, Venta;
Mar. : Bhoti, Potari, Warung;
Ori. : Bankopasia, Bharimo;
Pub. : Pola, Pula;
Tam. : Vendai;
Tel. : Konda patri, Potri.

Distribution : Sub-Tropical Regions from Peninsular India. N. Maharashtra.

Parts Used : Bark. Twigs.

Utilization : Twigs used as fodder. Bark fibre used for coarse ropes and strings.

Active Constituents : Leaves contain crude proteins 11.09-13.61%, crude fibre 14-23.71%, Ca 3.06% and P 0.31-0.59%.

References : Agarwal (2003), Ambasta (1986), Anonymous (1959), Dastur (1964), Dwivedi (1993), Hooker (1872), Jain (1991), Mittre (1997), Parrotta (2001), Prasad *et al.* (2002), Rao & Shanpru (1981, 1997), Saklani & Jain (1994), Singh (1982), Talbot (1976a), Varghese (1996), Watt (1972).

Kyllinga brevifolia Rottb. Non. Cons.

Syn: *Cyperus brevifolius* (Roth.) Hassk.
Family: Cyperaceae.

Vern. Name: Kyllinga.

Distribution : Common upto 1,000 m.

Utilization : Plant used for fodder.

Active Constituents : These are proteins 8.47, fat 0.94, starch 45.0 and raw fibre 29.86.

References : Anonymous (1959), Islam (2002), Prajapati *et al.* (2003).

Lagerstroemia speciosa Pers.

Syn: *Lagerstromia flasreginae* Retz.; C. B. Clarke in part.
Family: Lythraceae.

Vern. Name: Jarul.

English, Sanskrit and Regional Names:

Eng. : Queen Crap Myrtle;
Sans. : Arjuna, Jarula, Kramuka;
Guj. : Bondaro, Jarul, Moto- bhandara;
Mal. : Manimaruthu;
Mar. : Mota- Bondara, Taman;
Ori. : Patoli;
Pub. : Jarul;
Tam. : Kadali, Pumaruthu.

Distribution : Throughout India in W. Bengal, Assam & Peninsular India to 900 m.

Parts Used : Bark.

Utilization : Bark yields a coarse fibre.

References : Ambasta (1986), Anonymous (1962), Parrotta (2001).

Lannea coromandelica (Houtt.) Merr.

Syn: *Lannea grandis* Engl.; *Odina wodier* Roxb.
Family: Anacardiaceae.

English, Hindi, Sanskrit and Regional Names:

Eng. : Wodier;

Hindi	:	Jinghun, Kaimil, Mohi, Mohin;
Sans.	:	Jingini;
Beng	:	Bhadi, Jeal, Jiol;
Guj.	:	Indramati, Mavadi, Moi;
Kan.	:	Kuratige, Godda, Gopal;
Mal.	:	Anakkaram, Kalasam, Kalayasam;
Mar.	:	Moi, Munidi, Shimat, Shimti;
Ori.	:	Gongoromohi, Indramai, Jiyolo, Mohi, Moi;
Pub.	:	Batrin, Dila, Kemal, sulambra;
Tam.	:	Kalasan, Odi, Udayan, Uthi, Wodier;
Tel.	:	Appriuda, Gumpini.

Distribution : Sub-Himalayan Sal Forest. Greater Parts of India 1,500 m.

Parts Used : Leaves. Shoots.

Utilization : Tender shoots and leaves relished by elephants and cattle.

Active Constituents : Leaves contain D- galactose and complex monobasic acid.

References : Dwivedi (1993), Prajapati *et al.* (2003), Sahni (1998), Singh (1982), Singh & Kumar (2000), Varghese (1996).

Laportea crenulata Gaudich.

Syn: *Laportea gigantean & latifolia* Gaud.; *L. stimulus* Miq.; *Urtica crenulata* Roxb.; *U. gigantea* Poir.; *U. sinuata* Blume.
Family: Urticaceae.

English, Hindi and Regional Names:

Eng.	:	Devil's nettle;
Hindi	:	Chorpata, Utigun;
Ass.	:	Sorot- guch;
Beng	:	Chorpata;
Mal.	:	Anachorian;
Tam.	:	Ottapilavu.

Distribution : E. India. Tamil Nadu. Kerala. Tropical Himalaya. Sikkim eastwards Assam. Khasia Hills.

Part Used : Bark.

Utilization : Bark fibre strong and used for cordage and coarse cloths.

References : Agarwal (2003), Ambasta (1986), Hooker (1890), Watt (1972).

Lasiosiphon eriocephalus Decne.

Syn: *Lasiosiphon metzianus* Miq.; *L. sisparensis* Hugelei & Insularis.; *L. specious* Dene.

Family: Thymelaeaceae.

English, Hindi and Regional Names:

Eng. : Wooly- Headed gnidia;
Hindi : Ramita;
Bom. : Ramatla, Rametha;
Kan. : Rami;
Mal. : Nanca, Nangu;
Mar. : Rami, Ramita, Rametta;
Tam. : Nachinaar.

Distribution : Throughout S. India; W. Ghats from the Konkan Southwards ascending to 2,000 m on the Nilgiris.

Part Used : Bark.

Utilization : Bark fibre useful for tying bundles and paper manufacture.

References : Agarwal (2003), Ambasta (1986), Anonymous (1962), Hooker (1890), Talbot (1976b), Watt (1972).

Lasiurus hirsutus (Forsk.) Boiss.

Syn: *Elionurus hirsutus* Munro *ex* Benth.

Family: Poaceae.

Regional Names:

Pub. : Sewan, Shewar, Sin.

Distribution : Tropics & Sub-Tropical Regions of India.

Part Used : Plant.

Utilization : Yields good nutritive fodder; relished by elephants and sheep.

References : Ambasta (1986), Anonymous (1962), Duthie (1888), Dwivedi (1993).

Lathyrus aphaca Linn.

Family: Fabaceae.

Vern. Names: Mater Phali. Sudu.

English and Regional Names:

Eng. : Yellow Flowered Tree, Yellow Vetchling;
Hindi : Jangali Mater;
Beng : Jangali Mater;
Pub. : Gagla, Rewan, Rewari.

Distribution : Temperate Regions from Punjab to Bengal. Kumaon. Hazara. N. W. Provinces.

Part Used : Whole Plant.

Utilization : Used as fodder.

References : Agarwal (2003), Ambasta (1986), Anonymous (1962), Khanna (2002), Sood & Thakur (2004), Watt (1972).

Lathyrus sativus Linn.

Family: Fabaceae.

English, Hindi, Sanskrit and Regional Names:

Eng. : Chickling Vetch, Grass Pea;
Hindi : Khesari;
Sans. : Sandika,Triputi;
Ass. : Khesari;
Beng : Kassur, Khesari, Teora;
Guj. : Lang;
Mar. : Lakh;
Ori. : Khesra;
Pub. : Chural, Kharas, Karil, Kisari.

Distribution : Practically Throughout India.

Part Used : Leaves.

Utilization : Leaves used as fodder.

Active Constituents : Rich in vit. A and proteins.

References : Agarwal (2003), Ambasta (1986), Anonymous (1986), Oommachan (1977), Singh *et al.* (1996), Watt (1972).

Launaea nudicaulis Hook. f.

Syn: *Brachyrampus obtusus* DC.; *Chondrilla nudicaulis* Linn.; *Lactuca nudicaulis* Murray.; *L. obtusa* Clarke.; *Microrhynchus fallax* Jaub & Spach.; *M. nudicaulis* Lers.; *Prenanthus dichotona* Wall.; *P. obtusa* Ham.; *P. patens* Wall.; *P. procumbens* Roxb.

Family: Asteraceae.

Regional Names:

Bom. : Pathari;
Pub. : Butthal, Dudhlak.

Distribution : Throughout the Plains of India ascending upto 2,400 m in the Himalaya.

Part Used : Whole Plant.

Utilization : Affords fodder for goats.

References : Ambasta (1986), Anonymous (1962).

Leea crispa Willd.

Syn: *Leea asiatica* (L.) Rid.; *L. aspera* Edgew.; *L. edgeworthi* Sant.; *L. pinnata* Andr.

Family: Leeaceae.

Vern. Name: Ghangolae.

Regional Names:

Beng : Ban - Chalita;
Mal. : Naluga, Nellu;
Ori. : Hatikanopotro.

Distribution : Tropical Himalaya. Khasia Hills. W. Ghats. Oudh.

Part Used : Leaves.

Utilization : Leaves used as fodder.

References : Hooker (1872), Kirtikar & Basu (1935), Sood & Thakur (2004), Watt (1972).

Leersia hexandra Sw.

Syn: *Homalocenchrus hexandrus* Kuntze.; *Leersia australis* R. Br.

Family: Poaceae.

English and Regional Names:

Eng. : Rice Grass;
Mal. : Nirvalli Pullu.

Distribution : Lower Himalaya. Assam. Myanmar. Sri Lanka.

Part Used : Whole Plant.

Utilization : Affords good forage when cut early and used both as green feed and hay.

References : Ambasta (1986), Das *et al.* (1996), Duthie (1888), Hooker (1897), Islam (2002), Watt (1972).

Lens culinaris Medic.

Syn: *Ervum lens* Linn.; *Lens esculenta* Moench.

Family: Fabaceae.

English, Hindi, Sanskrit and Regional Names:

Eng. : Lentil;
Hindi : Masser, Masur, Masuri;
Sans. : Masurah;
Ass. : Masurmoha;
Beng : Masser, Masur, Masuri;
Guj. : Masser, Masuri, Mussur;
Kan. : Chanangi, Masur;
Mal. : Athanapparippu, Canampayar, Vattupparippu;
Mar. : Masser, Masur, Masuri;
Pub. : Malka, Masur, Masri;
Tam. : Atanapparuppu, Misur, Parupu;
Tel. : Chirisanagalu, Misurpappu.

Distribution : Throughout Sub-Tropical Regions of India U.P. M.P. Bihar. W. Bengal, smaller extent in Punjab, Rajasthan, Maharashtra & Gujarat.

Parts Used : Leaves. Stalks.

Utilization : Leaves and stalks used as fodder both as green feed and hay.

Active Constituents : Contains protein 30 % and above 2% mineral matter.

References : Agarwal (2003), Ambasta (1986), Anonymous (1962), Prajapati *et al.* (2003).

Lepidium apetalum Willd.

Syn: *Lepidium ruderale* Linn.
Family: Brassicaceae.

English Names:

Eng. : Pepper Grass, Pepper Weed.

Distribution : Cultivated Throughout India & W. Tibet.

Part Used : Whole Plant.

Utilization : Plant used as fodder.

References : Prajapati *et al.* (2003), Singh & Kumar (2000).

Lepidium sativum Linn.

Family: Brassicaceae.

English, Hindi, Sanskrit and Regional Names:

Eng. : Garden Grass, Water Grass;
Hindi : Halim, Huruf;
Sans. : Ashalika, Bhadra, Chandrashura, Nandini;
Beng : Halim;
Bom. : Ahaliva;
Guj. : Asalio, Halim;
Mal. : Asail, Asali;
Mar. : Ahliva;
Pub. : Halim, Shargundei, Tezak;

Tam. : Aliverai;
Tel. : Adalawtulu.

Distribution : Throughout India.

Part Used : Herb.

Utilization : Herb used as fodder for horses and camels.

Active Constituents : These are proteins, fats, carbohydrates, mineral matter, vit. A, thiamine, riboflavin, niacin and ascorbic acid.

References : Agarwal (2003), Ambasta (1986), Anonymous (1962), Hooker (1872), Parrotta (2001), Prajapati *et al.* (2003), Sood & Thakur (2004), Watt (1972).

Leptadenia pyrotechnica (Forsk.) Decne.

Syn: *Leptadenia sportium* Wight.
Family: Asclepiadaceae.

Hindi and Regional Names:

Hindi : Khip;
Guj. : Khip;
Pub. : Khip.

Distribution : Dry, Sandy Regions of N. India & W. India.

Parts Used : Twigs. Plant.

Utilization : Fibre suitable for ropes and paper. Plant used as fodder.

References : Agarwal (2003), Ambasta (1986), Blatter & Hallberg (1984), Singh & Pandey (1998).

Leucaena glauca Benth. (Pl. 5C)

Syn: *Acacia biceps* DC.; *A. caringa* Ham.; *A. frondosa* Willd.; *A. leucocephala* DC.; *Leucaena leucocephala* Lamk de Wit.; *Mimosa leucocephala* Lamk.
Family: Fabaceae.

Vern. Name: Alseenia.

English, Hindi and Regional Names:

Eng. : White popinae;
Hindi : Toria Kadam;

Guj. : Lasobavae;
Mal. : Takarannivam;
Ori. : Rajokasundiri;
Tam. : Tagarai;
Tel. : Kaniti.

Distribution : Throughout Plains of India.

Part Used : Whole Plant.

Utilization : Affords fodder for cattle, sheep and goats.

Active Constituents : Contains carbohydrates, crude protein 24.2%, crude fibre 13.35, Ca 1.98% and P 0.27%.

References : Agarwal (2003), Ambasta (1986), Anonymous (1962), Chopra *et al.* (1969); Dwivedi (1993), Hooker (1879), Parrotta (2001), Prajapati *et al.* (2003), Singh (1982), Sood & Thakur (2004), Watt (1972).

Limnanthemum nymphaeoides Link.

Syn: *Menyanthes nymphaeoides* Linn.; *Villarsia nymphaeoides* Vent.
Family: Gentianaceae.

Regional Name:

Pub. : Khairposh, Kuru.

Distribution : W. Himalaya: 2,000- 3,000 m.

Utilization : Largely used as a galactagogue.

References : Ambasta (1986), Anonymous (1962), Hooker (1885).

Limnophilia heterophylla Benth.

Syn: *Limnophilia heterophylla* (Roxb.) Benth. var. *reflexa* (Benth.) Hook. f.
Family: Scrophulariaceae.

Distribution : Throughout India.

Part Used : Plant.

Utilization : Used for fodder purposes.

Reference : Islam (2002).

Limnophila hirsuta Benth.

Family: Scrophulariaceae.

Part Used : Plant.

Utilization : Used for fodder purposes.

Reference : Islam (2002).

Lindernia crustacea (L.) F. V. Muell.

Syn: *Vendellia crustacea* Benth.

Family: Scrophulariaceae.

Distribution : Throughout India ascending upto 1,600 m in the Himalaya.

Utilization : Used as fodder.

References : Anonymous (1962), Islam (2002).

Lindera pulcherrimum Benth.

Syn: *Daphnidium pulcherrimum* Nees in Wall.

Family: Lauraceae.

Distribution : Temperate Himalaya (Kumaon, Sikkim, Bhutan).

Part Used : Leaves.

Utilization : Leaves fed to cattle throughout the year.

References : Anonymous (1962), Hooker (1890), Purohit & Samant (1995).

Lithocarpus dealbatus (Miq.) Rehder.

Syn: *Quercus dealbata* Miq.

Family: Fagaceae.

Distribution : Small tree or shrubs in E. Himalaya N. E. India. Nilgiris.

Part Used : Fruits.

Utilization : Fruits used as pig fodder.

Reference : Saklani & Jain (1994).

Litsea glutinosa (Lour.) C. B. Robins.

Syn: *Litsea chinensis* Lam.; *L. sebifera* Pers.

Family: Lauraceae.

English, Hindi, Sanskrit and Regional Names:

Eng.	:	Common Tallow Lowrel, Vasa;
Hindi	:	Gar bijaur, Maida lakri;
Sans.	:	Maidaa- lakdi, Medasaka, Vasa;
Beng	:	Gargur, Kukur Chita, Ratun;
Guj.	:	Maeda lakari;
Mar.	:	Maida lakadi;
Ori.	:	Jaisanda;
Pub.	:	Chandra, Meda lakri, Mada sak;
Tam.	:	Elumpurukki, Mushaippeyett, Uralli;
Tel.	:	Narra alagi.

Distribution : Throughout India upto 1,350 m in the Himalayan Region.

Part Used : Leaves.

Utilization : Leaves used as fodder.

References : Agarwal (2003), Ambasta (1986), Anonymous (1962), Chatterjee & Pakrashi (1997), Khan & Khanum (2005), Khare (2004).

Litsea monopetala (Roxb.) Pers.

Syn: *Litsea polyantha* Juss.

Family: Lauraceae.

Vern. Name: Gwanyu.

Hindi, Sanskrit and Regional Names:

Hindi	:	Gwa, Kakuri, Meda, Papria, Sangran;
Sans.	:	Gajapippali;
Ass.	:	Muga hoanlu;
Beng	:	Bara Kakur Chita;
Mar.	:	Ranamba;
Pub.	:	Gwa, Harein, Medalakri, rian;
Tel.	:	Mamudu, Nara, Nara mamidi.

Distribution : Grows Practically Throughout Foot Hills of India.

Part Used : Leaves.

Utilization : Leaves considered as good quality fodder and fed throughout the year.

Active Constituents : Polysaccharides, xylose and arabinose isolated from the leaves.

References : Agarwal (2003), Anonymous (1962), Dwivedi (1993), Purohit & Samant (1995), Rastogi & Mehrotra (1993, 1995b), Sood & Thakur (2004).

Litsea salicifolia (Nees.) Hk. f.

Syn: *Tetranthera attenuata* Wall.; *T. glauca* Wall.; *T. lanceaefolia* Roxb.; *T. laurifolia* Roxb.; *T. salicifolia* Roxb.

Family: Lauraceae.

Regional Names:

Ass. : Diglotti, Sum.

Distribution : N. & N. E. India: Nepal to Sikkim ascending upto 2,000 m.

Part Used : Leaves.

Utilization : Leaves used as fodder.

Active Constituents : Leaves contain icentrione and sanupaphenlr.

References : Anonymous (1962), Rastogi & Mehrotra (1993), Saklani & Jain (1994), Watt (1972).

Litsea umbrosa Nees.

Syn: *Litsea consimilis* Nees.; *L. umbrosa* Nees var. *consimilis* (Nees) Hook. f.; *neolitsea pallens* (D. Don) Momiyana & Hara.; *N. umbrosa* (Nees) Gamble.; *Tetradenia pallens* Don.; *T. umbrosa* Nees.

Family: Lauraceae.

Vern. Name: Pangra.

Regional Names:

Pub. : Chindi, Chirudi.

Distribution : Kashmir. Chamba. Shimla. Kumaon. Khasia Hills. Sikkim Mountains.

Utilization : Commonly browsed by ruminants and fed throughout the year.

References : Hooker (1890), Purohit & Samant (1995), Watt (1972).

Livistonia australis Mart.

Family: Arecaceae.

English and Hindi Names:

Eng. : Australian Fan Palm;
Hindi : Pankhatar.

Distribution : Cultivated in the Indian Gardens.

Part Used : Leaves.

Utilization : Fibre from the unropped leaves used for making hats.

References • : Agarwal (2003), Ambasta (1986), Anonymous (1962).

Livistonia chinensis R. Br.

Syn: *Livistonia mauritiana* Wall.

Family: Arecaceae.

English Name:

Eng. : Chinese fan Palm.

Distribution : Throughout India.

Part Used : Leaf-Stalks.

Utilization : Fibrous sheath of leaf -stalks made into ropes.

References : Ambasta (1986), Anonymous (1962).

Lodoicea maldivica (Poir.) Pers. (Pl. 5D)

Syn: *Lodoicea seychellarum* Labill.

Family: Arecaceae.

English, Hindi, Sanskrit and Regional Names:

Eng. : Double coconut palm, Sea coconut palm;
Hindi : Darya ka gola;
Sans. : Samundra - naarikela;
Guj. : Daryanuariyal;

Kan. : Kadalutengu;
Mal. : Kataltenna;
Mar. : Jahari - naral;
Tam. : Kadalthengai;
Tel. : Samundrapu tentaya.

Distribution : Cultivated in the Bengal Gardens.

Part Used : Leaves.

Utilization : Young leaves suitable for stuffing pillows, mattresses and baskets.

References : Agarwal (2003), Ambasta (1986), Amonymous (1962), Khare (2004).

Lolium multiflorum Lam.

Syn: *Lolium italicum* A. Br.
Family: Poaceae.

English and Hindi Names:

Eng. : Italian Ryegrass;
Hindi : Ryeghas.

Distribution : Assam.

Part Used : Plant.

Utilization : Affords valuable fodder grass for grazing and hay.

Active Constituents : A good source of carotene and vit. B group.

References : Ambasta (1986), Anonymous (1962).

Lolium perenne Linn. (Pl. 6A)

Family: Poaceae.

English Name:

Eng. : Perennial Ryegrass.

Distribution : Shimla. Sikkim, Shilong. Nilgiris & Palni Hill.

Part Used : Plant.

Utilization : Affords fodder for horses, cattle and sheep.

Active Constituents : These are carotene and alkaloid perloline.

References : Ambasta (1986), Anonymous (1962).

Lonicera quinquelocularis Hardw.

Syn: *Lonicera diversifolia* Wall.; *L. royleana* Wall.
Family: Caprifoliaceae.

Vern. Name: Chaunsh.

English and Regional Names:

Eng. : The Himalayan Honey Suckle;
Kash. : Sakhur, Tita bateri;
Pub. : Bakru, Phut.

Distribution : Himalaya: 1,800-3,200 m (Kashmir to Bhutan).

Part Used : Leaves.

Utilization : Leaves used as fodder for goats.

References : Anonymous (1962), Sharma (2004), Singh & Kumar (2000), Watt (1972).

Lotus corniculatus Linn.

Syn: *Lotus bracteatus* Wall.
Family: Fabaceae.

English Name:

Eng. : Birds foot trefoil.

Distribution : W. Himalaya.

Utilization : Fed as silage and hay. As compared to lucerne and white clover it dose not cause bloat in cattle.

Active Constituents : Hay is a good source of carotene, quercetin, lecocyanidin, kaempferol and ferulic acid.

References : Ambasta (1986), Anonymous (1962), Hooker (1879), Rastogi & Mehrotra (1993), Singh & Kumar (2000).

Luffa acutangula (L.) Roxb.

Syn: *Cucumis acutangula* Wall.; *Luffa kleinii* W. & A.; *L. umbellata* (Klein. *ex* Willd) Roem.

Family: Cucurbitaceae.

English, Hindi, Sanskrit and Regional Names:

Eng.	:	Ribbed Group Ridged Group;
Hindi	:	Jinga, Torai, Turi;
Sans.	:	Devasale, Fingakaka, Koshataki;
Beng	:	Finga, Jhinga;
Bom.	:	Gonsali, Jhingha, Sirola, Turai;
Guj.	:	Ghisoda, Turin;
Kan.	:	Hirkaye;
Mal.	:	Puichenggah;
Mar.	:	Shirola;
Ori.	:	Fanhi;
Pub.	:	Jingha, Tori, Turai;
Tam.	:	Pikunkai;
Tel.	:	Bira - kaya, Burkai.

Distribution : N. W. India. Sikkim. Assam. E. Bengal.

Part Used : Mesocarp.

Utilization : Dry fibrous mesocarp of mature fruits is chiefly used as a brush for cleaning the utensils in the tribal areas.

References : Anonymous (1962), Hooker (1879), Parrotta (2001), Singh & Pandey (1998), Sood *et al.* (2005), Watt (1972).

Lycopersicon esculentum Mill.

Syn: *Lycopersicon lycopersicon* (Linn.) Britt. & Brown.; *L. lycopersicon* (Linn.) Farwell.; *L. lycopersicon* (L.) Karsten.; *Solanum humboldtii* Willd.; *S. lycopersicum* L.; *S. pseudolycopersium* Jacq.

Family: Solanaceae.

Vern. Names: Chotta tamater, Olmos.

English, Hindi, Sanskrit and Regional Names:

Eng.	:	Love apple, Tomato;
Hindi	:	Vilayathi boigan;
Sans.	:	Raktamach, Raktavardhak, Ramphala, Shak - stnesdha;

Beng	:	Vilayathi boigan;
Bom.	:	Wal Wangi;
Guj.	:	Vilayati Vengan;
Kan.	:	Chapper - badne;
Mal.	:	Takkali;
Mar.	:	Vel vengi;
Pub.	:	Bhatte;
Tam.	:	Simie - takalie- palams.

Distribution : Cultivated Throughout India.

Part Used : Pressed Cake.

Utilization : Pressed cake suitable as fodder for cattle.

Active Constituents : Rich in mineral matter, Ca 20 mg, Fe 2.4, riboflavin 60 µg and ascorbic acid 30mg/100gm.

References : Agarwal (2003), Ambasta (1986), Anonymous (1962), Parrotta (2001), Prajapati *et al.* (2003), Watt (1972).

Lyonia ovalifolia (Wall.) Drude.

Syn: *Pieris ovalifolia* (Wall.) D. Don.

Family: Ericaceae.

Vern. Name: Bhelar.

Hindi and Regional Names:

Hindi	:	Angiri, Angyar;
Pub.	:	Ailan, Angyar.

Distribution : Throughout Himalaya Nilgiris; Cultivated for Ornamental Purposes.

Part Used : Leaves.

Utilization : Leaves browsed by livestock.

Active Constituents : Leaves rich in putatine, quercetrin, hyperoside and caffic acid.

References : Agarwal (2003), Anonymous (1962), Purohit & Samant (1995), Rastogi & Mehrotra (1991).

Machilus gamblei King *ex* Hook. K.

Family: Lauraceae.

Regional Names:

Ass.	:	Mijli;
Beng	:	Kawla;
Pub.	:	Kharam.

Distribution : Sub-Himalayan Tract: Eastwards in Assam upto 1,500 m.

Part Used : Leaves.

Utilization : Used for fodder.

References : Ambasta (1986), Anonymous (1962).

Madhuca indica J. F. Gmel.

Syn: *Bassia longifolia* Koeing.; *B. Latifolia* Roxb.; *Maduca latifolia* (Roxb.) Mac Bride.; *M. longifolia* (Linn.) Mac Bride.

Family: Sapotaceae.

English, Hindi, Sanskrit and Regional Names:

Eng.	:	Butter Tree, Mahua;
Hindi	:	Mahua, Mohwa;
Sans.	:	Madhuka;
Beng	:	Mahul, Mahwa;
Guj.	:	Mahuda;
Kan.	:	Hippe;
Mal.	:	Ilupa, Poonam;
Mar.	:	Mahwa, Mohwra;
Ori.	:	Madgi, Mahula, Moha;
Pub.	:	Mahua;
Tam.	:	Elupa, Ilupei;
Tel.	:	Ippa.

Distribution : Throughout W. Bengal, Bihar, Orissa, M.P., Punjab, U.P., Karnataka, Kerala & Rajasthan.

Part Used : Leaves.

Utilization : Used as fodder for cattle.

References : Agarwal (2003), Ambasta (1986), Chatterjee & Pakrashi (1997), Dwivedi (1993), Khan & Khanum (2005), Parrotta (2001), Prajapati *et al.* (2003), Rastogi & Mehrotra (1991), Sahni (1998), Singh (1982, 2002).

Maesia indica Wall.

Syn: *Baeoboyrys indica* Roxb.;
B. nemorallus Roxb.; *Maesia montana* W. & A.
Family: Myrsinaceae.

Regional Names:

Ass.	:	Awuas pat;
Beng	:	Ramjani;
Bom.	:	Atki;
Gar.	:	Gadchiana, Jiundali;
Kan.	:	Guddehargi;
Mal.	:	Kirithi;
Mar.	:	Akti;
Tam.	:	Vamarai.

Distribution : Common in the N. E. Himalaya, E. Bengal, Manipur.

Part Used : Leaves.

Utilization : Used as fodder throughout the year.

Active Constituents : A triterpene glycoside isolated from the leaves.

References : Anonymous (1962), Hooker (1882), Purohit & Samant (1995), Rastogi & Mehrotra (1995a), Watt(1972).

Malachra capitata Linn.

Family: Malvaceae.

Hindi and Regional Names:

Hindi	:	Banbhindi;
Beng	:	Ban bhindi;
Bom.	:	Ran or van bHindi

Distribution : Naturalized in Hot & Damp Parts of India.

Part Used : Bast.

Utilization : Bast fibre used as a fine substitute to jute for cordage, gunny bags, coarse textile and paper.

References : Agarwal (2003), Ambasta (1986), Anonymous (1962), Caius (2003), Hooker (1872), Watt (1972).

Mallotus philippensis Muell-Arg.

Syn: *Croton cascarilloides* Reuesch *ex* Steud.; *C. coccineus* Vahl.; *C. distans* Wall.; *C. montanus* Willd.; *C. punctatus* Retzs.; *Rottlera affinis* Hassk.; *R. aurantiaca* Hook & Arn.; *R. mollis* Wall.; *R. montana* Wall.; *R. tintorii* Roxb.

Family: Euphorbiaceae.

Vern. Name: Kambhal.

English, Hindi, Sanskrit and Regional Names:

Eng. : Kamla tree, Money face tree;
Hindi : Kamala, Rohini, Sindur;
Sans. : Kampilla, Kapila;
Ass. : Jorat, Losan;
Beng : Kamala;
Guj. : Kapilo;
Kan. : Kankumadamara;
Mal. : Manjana;
Mar. : Shendri;
Ori. : Kapilogundi, Kumalo, Sinduri;
Pub. : Kambal, Rulya;
Tam. : Kapli, Kapila;
Tel. : Kunkuma, Sinduri.

Distribution : Upto 1,500 m in the Himalaya from Kashmir to Bhutan.

Part Used : Leaves.

Utilization : Leaves used as fodder to cattle.

Active Constituents : Mainly contain rottlerin and isoallorotterin.

References : Agarwal (2003), Ambasta (1986), Anonymous (1962), Dash & Misra (2000), Hooker (1890), Parrotta (2001), Purohit & Samant (1995), Rastogi & Mehrotra (1993), Singh (1982), Singh & Kumar (2000), Watt (1972).

Mallotus tetracoccus (Roxb.) Kurz.

Syn: *Mallotus albus* Muell.-Arg.
Family: Euphorbiaceae.

Distribution : W. Ghats.W. Bengal. Assam.

Part Used : Plant.

Utilization : Lopped for fodder.

Reference : Ambasta (1986).

Malvestrum coromandelianum Garcke.

Syn: *Malva coromandeliana* L.; *Malvestrum tricuspidatum* A. Gray.
Family: Malvaceae.

Hindi Name:

Hindi : Dakshini.

Distribution : Scattered in Warmer Parts of India.

Part Used : Stem.

Utilization : Stem fibre used for making cordage and brooms.

References : Agarwal (2003), Ambasta (1986), Anonymous (1962), Dwivedi (1993), Singh & Kumar (2000).

Manilkara hexandra (Roxb.) Dubard.

Syn: *Mimusops hexandra* Roxb.
Family: Sapotaceae.

English, Hindi, Sanskrit and Regional Names:

Eng. : Obtuse leaved mimusops;
Hindi : Khirini;
Sans. : Kshirika;
Beng : Khirkhejur;
Guj. : Khirini, Rayan;
Kan. : Khirini;
Mal. : Pala, Palamunpala;
Mar. : Raini, Ranjana, Rayan;
Ori. : Khiri, Khirakuli;
Tam. : Palai, Palla;
Tel. : Manjipala.

Distribution : Found in C. India. Deccan; Often Cultivated in N. Drier Regions.

Part Used : Leaves.

Utilization : Leaves used as fodder.

Active Constituents : Rich in vit. A and vit. C 16mg /100mg, crude proteins 23.14%, crude fibre 54.94%, Ca, 1.49% and P 0.35%.

References : Agarwal (2003), Ambasta (1986), Anonymous (1962), Chatterjee & Pakrashi (1997), Prajapati *et al.* (2003), Singh (1982).

Maoutia puya Wedd.

Syn: *Boermaria puya* Hook.;
B. frutescens Don.; *Urtica puya* Ham.
Family: Urticaceae.

Hindi and Regional Names:

Hindi : Pooah, Poi, Puya;
Beng : Yenki;
Kum. : Dhaul- kagshi.

Distribution : Sub-Himalayan Tract from Gharwal to Eastwards.

Part Used : Bark.

Utilization : Bark fibre, 'puya' used as a good substitute to hemp for making nets, bags, game bags, fishing lines and ropes.

References : Agarwal (2003), Ambasta (1986), Anonymous (1962), Hooker (1890), Purohit & Samant (1995), Watt (1972).

Mariscus albscens Gaud.

Syn: *Cyperus canescens* Vah.
Family: Cyperaceae.

Distribution : Sunderbans. Orissa. Konkan. W. Ghats. Deccan.

Part Used : Laeves.

Utilization : Fibre used for mats.

References : Agarwal (2003), Anonymous (1962).

Marsdenia roylei Wight.

Family: Asclepiadaceae.

Hindi and Regional Names:

Hindi : Murkula;
Gar. : Shengori;
Kum. : Murkila;
Sim. : Kurung.

Distribution : Temperate Himalaya; Shimla to Kumaon ascending to 1,600 m.

Part Used : Bark.

Utilization : Fibre used for fishing nets.

References : Agarwal (2003), Ambasta (1986), Anonymous (1962), Hooker (1885), Raju (1998), Singh & Kumar (2000), Watt (1972).

Marsdenia tenacissima Wight & Arn.

Syn: *Asclepias tenacissima* Roxb.;
A. tomentosa Herb. Madr.; *Gymnema tenacissima* Spreng.
Family: Asclepiadaceae.

English, Hindi and Regional Names:

Eng. : Rajmahal hemp;
Hindi : Chiti, Jiti, Tongus;
Beng : Chiti, Jiti;
Bom. : Haba;
Ori. : Gha;
Tam. : Chiti, Fiti;
Tel. : Karudushhiu.

Distribution : Himalaya. Assam. Hills of Bengal.

Parts Used : Bark. Seed. Plant.

Utilization : Bark and seed fibre used for fishing lines, nets, cordage and entire plant used for fastening the bundles of fire wood.

References : Agarwal (2003), Ambasta (1986), Anonymous (1962), Dastur (1964), Drury (1978), Hooker (1885), Jain (1991), Mittre (1981), Singh & Pandey (1998), Singh & Prakash (1996), Sood *et al.* (2005), Watt (1972).

Mastersia assamica Benth.

Family: Fabaceae.

Part Used : Bark.

Utilization : Bark for making fishing nets and ropes.

Reference : Saklani & Jain (1994).

Mazus pumillus (Burm. f.) Van Steenis.

Family: Scrophulariaceae.

Part Used : Herb.

Utilization : Used as fodder.

Reference : Islam (2002).

Medicago hispida Gaertn.

Syn: *Medicago canescens* Grah.; *Medicago denticulata* Willd.; *M. polymorpha* Roxb.

Family: Fabaceae.

Vern. Name: Khokhani.

English and Regional Names:

- **Eng.** : Californian Bur Clover, Toothed Bur Clover, Toothed Mesik;
- **Beng** : Maina;
- **Pub.** : Maina.

Distribution : Found in Punjab, Kumaon, Upper Gangetic Plains, N. Bengal & Madras, ascending to 1,500 m in the Himalaya.

Parts Used : Leaves. Twigs.

Utilization : Relished by cattle, sheep, poultry, horses and camel.

Active Constituents : Leaves rich in ascorbic acid 85mg/100g and carotene 51mg/100g.

References : Ambasta (1986), Anonymous (1962), Chopra *et al.* (1969), Sood & Thakur (2004).

Medicago sativa Linn. (Pl. 6B)

Syn: *Medicago falcata* Stebler & Schroter.
Family: Fabaceae.

English, Hindi and Regional Names:

Eng. : Alfa alfa;
Hindi : Lasunghas, Wilayati - Gawuth;
Guj. : Vilkayati ghas;
Kan. : Vilayati hullu;
Pub. : Lusan.

Distribution : Sub- Tropical Regions of India: Cultivated in Kashmir.

Part Used : Plant.

Utilization : Highly valued in India as a legume fodder for horses; cultivated also for hay and silage.

Active Constituents : Source of vit. A and E, chlorophyll and carotenoids.

References : Agarwal (2003), Ambasta (1986), Chopra *et al.* (1969), Singh *et al.* (1996), Watt (1972).

Melilotus alba Desr.

Syn: *Melilotus altissima* Wall.;
M. leucantha Koch.; *M. vulgaris* Willd.
Family: Fabaceae.

English and Regional Names:

Eng. : Bokhara clover, White sweet clover;
Guj. : Safed banmethi;
Pub. : Safed senji.

Distribution : Throughout India.

Part Used : Plant.

Utilization : Cultivated for fodder in Punjab, U.P. and to a small extent in M. P.

Active Constituents : These are proteins and prenyllated flavanone.

References : Agarwal (2003), Ambasta (1986), Anonymous (1962), Dwivedi (1993), Khanna (2002), Hooker (1879), Rastogi & Mehrotra (1995b).

Melilotus indica All.

Syn: *Melilotus parviflora* Desf.
Family: Fabaceae.

English, Hindi, Sanskrit and Regional Names:

Eng. : Annual yellow sweet clover;
Hindi : Ban methi;
Sans. : Banam methika;
Beng : Ban methi;
Ori. : Ban methi, Huring upu;
Pub. : Ban methi, Metha, Senji.

Distribution : Throughout N. India, Punjab & U.P.

Part Used : Plant.

Utilization : Cultivated for fodder and fed in admixture with 'bhusa'.

References : Agarwal (2003), Ambasta (1986), Anonymous (1962), Dwivedi (1993), Rastogi & Mehrotra (1991).

Meliosma dilleniaefolia Walp.

Family: Sabiaceae.

Hindi and Regional Names:

Hindi : Bativa;
Pub. : Kanna, Karsan.

Distribution : E. Himalaya 1,000- 2,500 m.

Part Used : Leaves.

Utilization : Leaves fed to sheep, goats and other livestock.

References : Agarwal (2003), Anonymous (1962), Dwivedi (1993), Hooker (1879), Purohit & Samant (1995).

Melochia corchorifolia Linn.

Syn: *Melochia affinis* Wall.; *M. pauciflora* Wall.; *M. supina* L.; *M. truncata* Willd.; *Riedleia concatenata* DC.; *R. corchorofolia* DC.; *R. supina* DC.; *R. truncata* DC.; *Visenia corchorifolia* Sprenge.

Family: Sterculiaceae.

Hindi and Regional Names:

Hindi : Bilpat;
Beng : Tikiokra;
Mal. : Ceruvuram;
Ori. : Chyeron, Dasokarotun;
Tam. : Pinnak kuppundu;
Tel. : Ganuga pindikura.

Distribution : Throughout Hotter Parts of India, from Kumaon to Sikkim.

Part Used : Stem Bark.

Utilization : Fibre used for fishing lines and hut roofs.

References : Agarwal (2003), Ambasta (1986), Anonymous (1962).

Melochia umbellata Linn.

Family: Sterculiaceae.

Regional Name:

Bom. : Methuri.

Distribution : Deccan. Konkan. Kerala. Andaman Island.

Part Used : Bark.

Utilization : Bark fibre used for cordage, strings, ropes and turtle nets in Andaman.

References : Ambasta (1986), Anonymous (1962).

Microcos paniculata Linn.

Syn: *Grewia microcos* Linn.

Family: Tiliaceae.

Distribution : Throughout Warmer Parts of India.

Part Used : Leaves.

Utilization : Yields a fibre.

Reference : Ambasta (1986).

Microstegium nudum (Trin.) A. Camus.

Syn: *Pollinia nuda* Trin.

Family: Poaceae.

Part Used : Plant.

Utilization : Used as fodder.

Reference : Ambasta (1986).

Mikania cordata (Burm.) B. L. Robinson.

Syn: *Mikania scandens* Hook. f. non Willd.

Family: Asteraceae.

Parts Used : Stem. Leaves.

Utilization : Stem and leaves given to cattle in times of scarcity.

Active Constituents : Leaves rich in sitosterol, stigmasterol, vit. A & C, fumaric acid and mikanin.

References : Ambasta (1986), Anonymous (1962), Chopra *et al.* (1969), Islam (2002), Rastogi & Mehrotra (1990, 1995a).

Miliusa tomentosa (Roxb.) J. Sinclair.

Syn: *Saccopetalum tomentosum* Hook. f. & Thoms.

Family: Annonaceae.

Hindi and Regional Names:

Hindi : Kari, Kirua, Hoom;
Guj. : Hoom, Humba;
Kan. : Hessare;
Mar. : Hoom, Humba;
Ori. : Gandhap alsa;
Tam. : Periuvay;
Tel. : Chilkadudu.

Distribution : Throughout Sub-Tropical Parts of India.

Part Used : Leaves.

Utilization : Leaves looped for fodder.

References : Agarwal (2003), Ambasta (1986), Anonymous (1962).

Millettia extensa Benth *ex* Baker

Syn: *Milletia auriculata* Baker *ex* Brandis.; *Otosema extensa* Benth.
Family: Fabaceae.

Hindi and Regional Names:

Hindi	:	Gauj;
Mal.	:	Valli muritalli;
Ori.	:	Arkawala, Kissi;
Tel.	:	Kondatongeditige.

Distribution : Foot Hills of the Himalaya Eastwards to Assam & Southwards C. India to N. Andhra Pradesh.

Parts Used : Leaves. Branches.

Utilization : Leaves used as hay to cattle. Branches beaten to obtain coarse fibre for cordage.

Active Constituents : Leaves contain proteins and calcium.

References : Agarwal (2003), Ambasta (1986), Anonymous (1962), Jain (1991), Parrotta (2001), Purohit & Samant (1995), Raju (1998), Rastogi & Mehrotra (1993).

Millettia racemosa Benth.

Syn: *Pongamia racemosa* Grah.; *P. corcor* Grah.; *Robinia racemosa* Roxb.; *Tephrosia racemosa* W. & A.; *Wisteria pallida* Dalz & Gibs.; *W. racemosa* Dalz.
Family: Fabaceae.

Hindi and Regional Names:

Hindi	:	Jungi;
Tel.	:	Galuga

Distribution : Tropical Forest of the Konkan & Orissa.

Part Used : Leaves.

Utilization : Leaves used as fodder.

References : Ambasta (1986), Anonymous (1962), Hooker (1879).

Mirabilis himalaica (Edgew.) Heimerl.

Syn: *Oxybaphus himalaicus* Edgew.
Family: Nyctaginaceae.

Part Used : Twigs. Leaves.

Utilization : Used as a fodder plant.

Reference : Ambasta (1986).

Mitragyana parvifolia (Roxb.) Korth.

Syn: *Stephegyne parvifolia* Korth.
Family: Rubiaceae.

Vern. Name: Mundi.

English, Hindi, Sanskrit and Regional Names:

Eng. : Kaim-Kaime;
Hindi : Kaim, Kailmi;
Sans. : Bhumikadamba, Dhulikadamba;
Beng : Gulikadam;
Guj. : Kadam;
Mal. : Nirkadambu, Vimbu;
Mar. : Kalam, Kuddam;
Ori. : Gudikaima, Mundi, Mur;
Pub. : Kam, Kalam, Keim;
Tam. : Chinna, Kadambu;
Tel. : Nir kadambe.

Distribution : Throughout Greater Parts of India ascending upto 1,200 m in Outer Himalaya.

Parts Used : Bark. Leaves.

Utilization : Plant lopped for fodder. Bark fibre used to drag elephants.

Active Constituents : Leaves afford 4-alkaloids, rotundifoline, isorotundifoline, rhynchophylline, isochynchophylline, 7.73% crude protein, 19.57% crude fibre, 2.27% Ca and 0.24% P.

References : Agarwal (2003), Ambasta (1986), Anonymous (1962), Dwivedi (1993), Jain (1991), Mittre (1997), Parrotta (2001), Prajapati *et al.* (2003), Rastogi & Melhotra (1990), Singh (1982).

Momordica balsamina Linn.

Family: Cucurbitaceae.

English, Hindi and Regional Names:

Eng. : Balsam Apple;
Hindi : Makha;
Guj. : Chhochhiodan.

Distribution : Punjab, U.P. Gujarat. N. W. India.

Parts Used : Leaves. Stem.

Utilization : Leaves and stem used as fodder.

References : Agarwal (2003), Ambasta (1986), Anonymous (1948-1976), Hooker (1897), Watt (1972).

Moringa oleifera Lam.

Syn: *Moringa pterygasperma* Gaertn.
Family: Moringaceae.

English, Hindi, Sanskrit and Regional Names:

Eng. : Drum stick tree, Horse radish tree;
Hindi : Mungna, Sinajana, Shajna;
Sans. : Shobhanjana, Sigruh;
Ass. : Saijana, Sohjna;
Beng : Sajin;
Guj. : Saragavo;
Kan. : Nugge;
Mal. : Murrina, Sigru;
Mar. : Achajhada, Shevgi;
Ori. : Sajina;
Pub. : Sainjana, Soonjana;
Tam. : Murungai;
Tel. : Mulaga, Munaga.

Distribution : Sub - Himalayan Tract. E. India. Warmer Regions of India.

Parts Used : Twigs. Leaves.

Utilization : Twigs lopped for fodder.

Active Constituents : Leaves contain arginine, lysinetryptophan, methonine, lucine and valine, niazinin A and niazinin B. Pods rich in lucine and cystine.

References : Agarwal (2003), Ambasta (1986), Anonymous (1962), Dastur (1964), Dwivedi (1993), Prajapati *et al.* (2003), Rastogi & Mehrotra (1990, 1995b), Sahni (1998), Singh (1982), Singh & Pandey (1998), Varghese (1996).

Morus alba Linn.

Syn: *Morus serrata* Wall.; *M. tetraica* L.
Family: Moraceae.

Vern. Names: Cheemu. Toot.

English, Hindi, Sanskrit and Regional Names:

Eng. : White mulberry;
Hindi : Chinni, Tut, Tutri;
Sans. : Toola, Tula;
Beng : Tut;
Guj. : Shetur;
Mar. : Ambat, Tut;
Ori. : Tuticoli, Tuto;
Pub. : Tut, Tutri;
Tam. : Chedi, Kambeli, Musukette;
Tel. : Reshme chattu.

Distribution : Cultivated Throughout India (N.W. Himalaya).

Parts Used : Leaves. Bark. Twigs.

Utilization : Leaves used as cattle fodder. Bark valued for paper making and textile fibre. Twigs used for making ropes for binding material.

Active Constituents : Leaves contain vit.-C, Ca, astragalin, crude protein 71.20% and crude fibre 54.13%.

References : Agarwal (2003), Ambasta (1986), Anonymous (1962), Dastur (1964), Dwivedi (1993), Jain (1991), Khanna (2002), Rastogi & Mehrotra (1991), Sharma (2004), Singh & Kumar (2000), Singh *et al.* (1996), Sood & Thakur (2003), Varghese (1996), Viswanathan *et al.* (1981), Watt (1972).

Morus australis Poir.

Syn: *Morus indica* Thunb.
Family: Moraceae.

English Name:

Eng. : Common Mulberry.

Distribution : Temperate Himalaya; Cultivated in Meghalaya.

Part Used : Leaves.

Utilization : Leaves lopped for fodder.

References : Panda (1996), Saklani & Jain (1994).

Morus laevigata Wall. *ex* Brandis

Syn: *Morus glabrata* Wall.
Family: Moraceae.

Hindi and Regional Names:

Hindi : Tut;
Ass. : Bola;
Pub. : Tut.

Distribution : Occurs between 300-600 m in Tropical & Sub-Tropical Himalaya.

Part Used : Plant.

Utilization : Fodder plant in W. Himalaya.

References : Agarwal (2003), Ambasta (1986), Hooker (1890).

Morus serrata Roxb.

Syn: *Morus alba* Linn.; *M. pabularia* Dcne.; *M. vicorum* Jacquem.
Family: Moraceae.

English, Hindi and Regional Names:

Eng.	:	Himalayan Mulberry;
Hindi	:	Kimu, Himu;
Pub.	:	Karam, Kimu, Karttit.

Distribution : Colder Region of W. Himalaya & Kumaon Hills.

Part Used : Leaves.

Utilization : Tree lopped for fodder.

Active Constituents : Leaves contain crude protein 11.48%, crude fibre 34.02%, and mineral 1.76.

References : Agarwal (2003), Ambasta (1986), Anonymous (1962), Hooker (1890), Purohit & Samant (1995), Singh (1982), Watt (1972).

Mucuna cochichinnensis Cheval.

Syn: *Mucuna nivea* (Roxb.) DC.; *Stizolobium niveum* Kuntze.
Family: Fabaceae.

English and Regional Names:

Eng.	:	Lyon Bean;
Beng	:	Khamach.

Distribution : Bengal. Bihar.

Part Used : Pods.

Utilization : Affords good fodder.

References : Agarwal (2003), Ambasta (1986), Anonymous (1962).

Mucuna deeringiana (Bort) Merrill.

Syn: *Stizolobium deerimgianum* Bort.
Family: Fabaceae.

English, Hindi and Regional Names:

Eng.	:	Georgia Velvet Beanor Florida;
Hindi	:	Makhmali Sem;
Pub.	:	Makhmali Sem.

Distribution : Cultivated as Kharif Crop in Sub-Tropical Regions of India.

Part Used : Plant.

Utilization : Used as fodder for cattle, sheep and goats and also fed as hay.

References : Agarwal (2003), Ambasta (1986), Anonymous (1948-1976).

Mucuna pruiens Baker, non DC.

Syn: *Carpopogon niveum* Roxb.; *C. pruiens* Roxb.; *M. nivea* (Roxb.) DC.; *Mucuna prurita* Hook.; *Stizalobium pruriens* (L.) Medikus.

Family: Fabaceae.

English, Hindi, Sanskrit and Regional Names:

Eng. : Common Cow witch, Cow hage;
Hindi : Kaunch, Kiwach, Konch;
Sans. : Atmagupta, Kappikachhu;
Beng : Alkushi, Bichehoti;
Guj. : Kavatch, Kivanch;
Kan. : Nasukunni;
Mal. : Corivalli, Naicorna;
Mar. : Lavacha, Kanchkuri, Kuhilli;
Ori. : Kaincho;
Pub. : Gugliu, Kavanvh;
Tam. : Poonaipidukkan, Poonaikalei;
Tel. : Dulagondi.

Distribution : Upto 600- 1,100 m Throughout India & in the Andaman & Nicobar Islands.

Part Used : Pods.

Utilization : Used as a fodder plant in W. Himalaya.

References : Ambasta (1986), Jain (1991), Parrotta (2001), Prajapati *et al.* (2003), Thakur *et al.* (1989).

Mucuna nigricans Steud.

Syn: *M. imbricata DC.*

Family: Fabaceae.

Regional Names:

Ass. : Ghila;
Beng : Kasi.

Distribution : Upto 1,000 m.

Utilization : Used as fodder.

Reference : Ambasta (1986).

Muhelenbergia huegelii Trin.

Syn: *M. viridissima* Nees *ex* Steud.

Family: Poaceae.

Distribution : Temperate Himalaya from Kashmir to Sikkim: 900-2,700 m Khasia Hills.

Part Used : Leaves.

Utilization : Used as a fodder.

Active Constituents : These are proteins 12.8, fats 2.46, carbohydrates 44.8 and fibre 29.7.

References : Ambasta (1986), Anonymous (1962).

Muntinga calabura Linn.

Family: Tiliaceae.

English Name:

Eng. : Japanese Cherry.

Distribution : Sandy Areas of India.

Part Used : Bark.

Utilization : Bark fibre used for cordage and ropes.

References : Agarwal (2003), Ambasta (1986), Anonymous (1962).

Musa paradisiaca Linn.

Syn: *Musa sapientum* Linn.

Family: Musaceae.

English, Hindi, Sanskrit and Regional Names:

Eng. : Edible Banana, Plantain;

Hindi : Kela;

Sans. : Kadali, Rambha;

Beng : Kala;

Bom.	:	Kela;
Guj.	:	Kela;
Kan.	:	Bale, Balehannu;
Mal.	:	Vazha;
Mar.	:	Kadali, Kel;
Ori.	:	Bontolokodoli, Kodoli, Rombhha;
Pub.	:	Kela, Khela, Mus;
Tam.	:	Vezhai;
Tel.	:	Anati, Arati.

Distribution : Throughout India.

Parts Used : Flowers. Stem.

Utilization : Flowers used as fodder. Banana fibre valued for cordage, ropes and coarse paper.

References : Agarwal (2003), Ambasta (1986), Anonymous (1962), Dastur (1964), Datt (1996), Drury (1978), Jain (1991), Kirtikar & Basu (1984), Varghese (1996), Watt (1972).

Musa textiles Nees.

Family: Musaceae.

English, Hindi and Regional Names:

Eng.	:	Abaca, Manila Hemp;
Hindi	:	Naaru;
Kan.	:	Natalbale;
Mal.	:	Naaru vazhai;
Tam.	:	Naaru vazhai.

Distribution : W. Bengal. S. India. Andaman Islands.

Part Used : Stem.

Utilization : Source of excellent fibre, 'Manila Hemp'which is used for cordage, binder twine, bagging, papier mach, wrapping paper, fabrics, mats and lustrous cloths.

References : Agarwal (2003), Ambasta (1986), Anonymous (1962), Watt (1972).

Myrica esculenta Buch.-Ham. *ex* Don. (Pl. 6C)

Syn: *Myrica faguhariana* Wall.; *M. nagi* Thunb.; *M. sapida* Wall.
Family: Myricaceae.

Vern. Name: Kaphal.

English, Hindi, Sanskrit and Regional Names:

Eng.	:	Box Myrtle, Yangmae of China;
Hindi	:	Kaphal, Khaiphal;
Sans.	:	Aranya, Bhadra, Kaphala, Katphala, Nasanu, Purusha;
Beng	:	Katphal;
Bom.	:	Katphal;
Mal.	:	Marula;
Mar.	:	Kayaphala;
Pub.	:	Kaheha, Kahi, Kaphal;
Tam.	:	Marulam;
Tel.	:	Kaidarama.

Distribution : Throughout Himalaya on Lower Levels upto 2,000 m.

Part Used : Leaves.Twigs.

Utilization : Leaves fed to animals in times of scarcity. Twigs browsed by sheep and goats.

References : Agarwal (2003), Anonymous (1962), Purohit & Samant (1995).

Myricaria germanica (Linn.) Desv.

Syn: *Myricaria bracteata* Royle;
M. hoffmeisteri Klotz.; *Tamarix germanica* Linn.
Family: Tamaricaceae.

Regional Name:

Pub. : Bis, Umbu.

Distribution : Temperate & Alpine Himalaya. Sikkim to Kumaon.

Part Used : Leaves.

Utilization : Given as fodder to sheep and goats.

References : Chopra *et al.* (1956), Hooker (1872), Sood *et al.* (2001), Watt (1972).

Nannorrhops ritchieana H. Wendl.

Syn: *Chamaerops ritchieana* Griff.
Family: Poaceae.

English, Hindi and Regional Names:

Eng. : Mazari Palm;
Hindi : Mazari;
Pub. : Kilu, Kaluin, Patha.

Distribution : Found on W. Himalaya. Often Cultivated in N. India.

Part Used : Leaves.

Utilization : Fibre used for hats, sandals, cordage and ropes.

References : Agarwal (2003), Ambasta (1986), Dastur (1964), Watt (1972).

Naravelia zeylanica DC.

Syn: *Atragene zeylanica* L.
Family: Ranunculaceae.

Sanskrit and Regional Names:

Sans. : Dhanavalli;
Beng : Chayal - bati;
Kan. : Neendamalli;
Mal. : Karuppakkoti, Potulalacci, Poytalacci, Vatakkoti;
Tam. : Vathomkolli.

Distribution : Tropical Forest of the Himalaya. Assam. W. Bengal. Bihar. Deccan Peninsula.

Part Used : Stem.

Utilization : Stem fibre good for ropes.

References : Hooker (1872), Saklani & Jain (1994), Prajapati *et al.* (2003), Watt (1972), Yoganarasimhan *et al.* (1981).

Nelumbo nucifera Gaertn.

Syn: *Nelumbium nelumbo* Druce.; *N. speciosum* Willd.
Family: Nymphaeaceae.

English, Hindi, Sanskrit and Regional Names:

Eng. : Chinese Water Lily, Indian Lotus, Sacred Lotus;

Hindi	:	Kamal;
Sans.	:	Ambuja, Kamala, Padma, Pankaja;
Ass.	:	Podum;
Beng	:	Padma;
Guj.	:	Suriyakamal;
Mal.	:	Senthamara, Thamara;
Mar.	:	Kamal;
Ori.	:	Padam;
Pub.	:	Kanval, Pampash;
Tam.	:	Ambal, Thamari.

Distribution : Throughout Warmer Regions of India.

Part Used : Leaf-Stalks.

Utilization : Leaf-stalks yield a fibre.

References : Agarwal (2003), Ambasta (1986), Kirtikar & Basu (1935).

Nepeta floccosa Benth.

Syn: *Marmorotis rotundifolia* Benth.; *Nepeta rotundifolia* Benth.
Family: Lamiaceae.

Distribution : W. Tibet: 2,000-3,500 m.

Part Used : Leaves.

Utilization : Browsed by sheep and goats.

References : Ambasta (1986), Hooker (1885).

Nypa fructicans Wurmb.

Family: Arecaceae.

English, Hindi and Regional Names:

Eng.	:	Nipa Palm;
Hindi	:	Galphal;
Anda.	:	Poothada;
Beng	:	Gabna, Galphal, Gulga;
Guj.	:	Pardesi – tadio.

Distribution : Sunderbans of Bengal. Saurashtra. Andaman Islands.

Part Used : Leaves.

Utilization : Young leaflets used for wrapping cigarettes and mature ones for coarse mats, baskets and bags.

References : Agarwal (2003), Ambasta (1986).

Ochna jabotapia Linn.

Syn: *Ochna squarrosa* Linn.
Family: Ochnaceae.

English, Hindi and Regional Names:

- **Eng.** : Kanakchampa;
- **Hindi** : Kanak champa, Sherundi;
- **Kan.** : Mudah, Narole;
- **Ori.** : Koniari, Nobunisero;
- **Tam.** : Chilanti, Sherundi;
- **Tel.** : Sunari, Yerrajuvi;

Distribution : E. India. Deccan Peninsula.

Part Used : Twig.

Utilization : Plant lopped for fodder.

References : Agarwal (2003), Ambasta (1986), Dwivedi (1993).

Olea ferruginea Royle

Syn: *Olea cuspidata* Wall. *ex* G. Don.
Family: Oleaceae.

English, Hindi and Regional Names:

- **Eng.** : Indian Olive;
- **Hindi** : Kahu;
- **Gar.** : Bairbanj.

Distribution : Kashmir to Kumaon Hills upto 2,500 m.

Part Used : Leaves.

Utilization : Leaves valued for fodder.

References : Agarwal (2003), Ambasta (1986), Singh (1982).

Ophiuros extralatus (Linn.) Kuntze.

Syn: *Ophiuros corymbosus* Gaertn. f. Hook. f. in part.
Family: Poaceae.

Regional Names:

Tam.	:	Kinangu pillu, Sothu alagu pillu;
Tel.	:	Pedda Panuku.

Part Used : Plant.

Utilization : Young grass eaten by cattle.

References : Ambasta (1986), Duthie (1888).

Opomoea aquatica Forsk.

Family: Convolvulaceae.

English, Hindi and Regional Names:

Eng.	:	Swamp Cabbage;
Hindi	:	Kalmisag;
Beng	:	Kalmisak;
Guj.	:	Nalanibhaji;
Mar.	:	Nadishaka, Sarnali;
Raj.	:	Tauasag;
Tam.	:	Vellakerrai;
Tel.	:	Tutikada.

Part Used : Plant.

Utilization : Green fodder nutritious and fed to cattle.

Active Constituents : Carotene.

Reference : Raju (1998).

Opuntia dillenii Haw.

Syn: *Cactus indicus* Roxb.; *Opuntia stricta* (Haw.) Haw. var. *dillenii* Haw.
Family: Cactaceae.

Vern. Names: Chu- chuar. Thor.

English, Hindi, Sanskrit and Regional Names:

Eng.	:	Slopper thorn;
Hindi	:	Hathhathoria;
Sans.	:	Bahudaydhika, Nagaphana;
Beng	:	Nagphana;
Guj.	:	Chulhatalo;
Kan.	:	Papaskalli;
Mal.	:	Palakakkali;
Mar.	:	Chopal;
Ori.	:	Nagaphenia;
Pub.	:	Chittarhohar;
Tam.	:	Nagathali;
Tel.	:	Nagajemundu.

Distribution : Naturalised all over India; Bengal & Madras to Punjab & in the Himalaya.

Parts Used : Stem. Leaves.

Utilization : Stem fibre as a raw material in paper pulp industry. Leaves utilised as fodder for cattle in times of scarcity.

References : Raju (1998), Watt (1972).

Opuntia ficusindica (Linn.) Mill.

Family: Cactaceae.

English Name:

Eng. : Indian Fig.

Part Used : Leaves.

Utilization : Spineless plants browsed by stray cattle.

Active Constituents : Vit. -C.

References : Ambasta (19860, Prajapati *et al.* (2003).

Oreocnida frutiscens (Thunb.) Miq.

Family: Urticaceae.

Distribution : Tropical Himalaya. Meghalaya.

Part Used : Bark.

Utilization : Bark affords fibre.

Reference : Saklani & Jain (1994).

Oreocnida integrifolia (Gaud.) Miq.

Syn: *Villebrunia integrifolia* Gaudich.

Family: Urticaceae.

Regional Name:

Ass. : Banrhea, Lukot, Risa;

Distribution : E. Himalaya & N. E. India. Sikkim. Assam. Andaman & Nicobar Islands.

Part Used : Stem.

Utilization : Fibre used for fishing nets, lines, ropes, cordage nets, canvas and coarse clothings.

References : Agarwal (2003), Ambasta (19860, Chopra *et al.* (1969), Dastur (1964), Jain (1991), Rao & Shanpru (1981), Saklani & Jain (1994).

Oroxylum indicum Vent.

Syn: *Bignonia petandra* Lour.; *B. indica* L.; *Calosanthus indica* Blume.; *Spathodea indica* Pers.

Family: Bignoniaceae.

Vern. Names: Arlu, Tataein.

English, Hindi, Sanskrit and Regional Names:

Eng. : Indian Trumpet Flower;
Hindi : Arlu, Soona, Ullu;
Sans. : Shoshana, Shyonaka, Tuntuka;
Ass. : Bhatgila, Dingari, Kering;
Beng : Nasona, Sona;
Bom. : Phalphura, Tetu;
Guj. : Aralu, Tentu;
Kan. : Bunepale;
Mal. : Palagapaiyani;
Mar. : Tetu;
Ori. : Phapni, Phonphonia;

Pub.	:	Mulin, Tatmorang;
Tam.	:	Achi, Priarlanthei;
Tel.	:	Dundilambu.

Distribution : Throughout Greater Parts of India from Himalaya to Sri Lanka upto 1,200 m.

Part Used : Twigs.

Utilization : Tree lopped for fodder.

Active Constituents : Leaves afford baicalein, its 6-glucuronide & 7-glucuronide, scutellarein and its 7- glucuronide.

References : Agarwal (2003), Ambasta (1986), Hooker (1885), Parrotta (2001), Rastogi & Mehrotra (1991), Watt (1972).

Orthanthera viminea Wight.

Syn: *Apocynea viminea* Wall.

Family: Asclepiadaceae.

Hindi and Regional Names:

Hindi	:	Khip, Muhur;
Pub.	:	Chapkia, Khip, Lanebar, Matti.

Distribution : Sub- Himalayan Tract from Punjab to U.P.

Part Used : Stem.

Utilization : Fibre used for strong, durable ropes and cordage.

References : Agarwal (2003), Ambasta (1986), Hooker (1885), Watt (1972).

Oryza officinalis Wall. *ex* Watt.

Syn: *Oryza latifolia* Hook. f. non Desv.

Family: Poaccac.

Part Used : Plant.

Utilization : Grass eaten by cattle.

Reference : Ambasta (1986).

Oryza rupipogon Griff.

Family: Poaceae.

Part Used : Whole Plant.

Utilization : Plant used for fodder.

Reference : Islam (1998).

Oryza sativa Linn.

Syn: *Oryza fatua* Koen. *ex* Trin.; *O.palustris* Salisb.; *Oryza plena* (Prain.) Chowdhury.; *O. repens* Herb. Ham. *ex* Wall.; *O. sarghoides* Steud.; *O. segetalis* Russ. *ex* Wall.
Family: Poaceae.

English, Hindi, Sanskrit and Regional Names:

Eng. : Paddy, Rice;
Hindi : Chawal, Dhan;
Sans. : Dhanya, Nivara, Syali, Vrithi;
Beng : Chal;
Bom. : Bhatta, Dangar;
Guj. : Choka, Dangar;
Kan. : Bhatta, Nellu;
Mal. : Ari, Nellu;
Mar. : Bhat, Dhan, Tandula;
Ori. : Dhan, Rabana;
Pub. : Dhan, Munjii, Shal, Shali, Tai;
Tam. : Arisi, Nellu;
Tel. : Vadlu, Ullu.

Distribution : Cultivated Throughout India in Warmer Regions.

Part Used : Rice Straw.

Utilization : Rice straw valued as fodder.

Active Constituents : Rich in protein, starch and glucotricin.

References : Agarwal (2003), Ambasta (1986), Duthie (1888), Islam (1998), Rastogi & Mehrotra (1990), Watt (1972).

Ougeinia oogeinensis (Roxb.) Hochr.

Syn: *Ougeinia dalbergioides* Benth.
Family: Fabaceae.

English, Hindi, Sanskrit and Regional Names:

Eng. : Chariot Tree, Punjab Kino, Sandan;

Hindi	:	Panam, Sandan, Tinsa;
Sans.	:	Chitrakrt, Nemi, Vajjala;
Beng	:	Tinis;
Guj.	:	Tanach;
Kan.	:	Kari - hone;
Mal.	:	Malavenna;
Mar.	:	Kalaphulas;
Ori.	:	Bandhona;
Tam.	:	Malavenna;
Tel.	:	Tella motukku.

Distribution : Throughout Outer Himalaya from Jammu to Bhutan. C. India. Deccan.

Parts Used : Stem. Leaves.

Utilization : Blast fibre used for cordage. Leaves fed to cattle.

References : Agarwal (2003), Ambasta (1986), Dastur (1964), Dwivedi (1993), Khare (2004), Oommachan (1977), Prajapati *et al.* (2003), Purohit & Samant (1995), Sahni (1998), Singh (1982).

Oxytropis mollis Royle *ex* Benth.

Syn: *Oxytropis thomsonni* Benth.

Family: Fabaceae.

Distribution : W. Himalaya. Kashmir to Sikkim upto 4,500 m.

Part Used : Plant.

Utilization : Eaten by animals, also important to horses.

Active Constituents : Oxalic acid 4.3% on dry matter basis.

References : Agarwal (2003), Ambasta (1986), Hooker (1879).

Pandanus odoratimus Linn. f. (Pl. 6D)

Syn: *Pandanus amaryllifolius* Roxb.; *P. fascicularis* Lam.; *P. laevis* Kunth.; *P. latifolius* Hask.; *P. tectorius* Sonald *ex* Parkinson.; *P. variegatus* Miq.

Family: Pandanaceae.

English, Hindi, Sanskrit and Regional Names:

Eng.	:	Screw Pine;
Hindi	:	Keora, Ketki, Keura, Kewda;
Sans.	:	Ketaki;
Bom.	:	Kenda;
Guj.	:	Kevoda, Kewoda;
Kan.	:	Kyad-ageida;
Mal.	:	Kaida, Mugali, Thala;
Mar.	:	Keora, Kevada;
Tam.	:	Tezai, Thalay, Tilai;
Tel.	:	Kelaki, Mugali.

Distribution : Found Throughout the Indian Coastal Belt & Andamans. Also in Bengal, Madras.

Parts Used : Leaves. Roots.

Utilization : Leaves are a source of paper sacks for coffee, sugar, grains and material for making umbrellas. Roots commonly used as brushes for painting and white wash.

References : Agarwal (2003), Ambasta (1986), Dastur (1964), Dey *et al.* (1973), Sahni (1998), Talbot (1976a), Watt (1972).

Panicum antidotale Retz.

Syn: *Panicum maximum* Wall.; *P. polygomum* Herb Madr *ex* Wight.; *P. pruinasum* Bernh *ex* Trin.; *P. subalbidum* Kunth Revis.; *P. tenue* Roxb *ex* Wight.

Family: Poaceae.

English, Hindi and Regional Names:

Eng.	:	Blue Panic;
Hindi	:	Gunara;
Bom.	:	Git, Girni, Male;
Guj.	:	Dhusdo, Dhusghas;
Raj.	:	Bangagli, Banwari;
Tam.	:	Nassiam pillu, Pinisu pillu.

Distribution : Distributed in Punjab, Kashmir, Gangetic Plains & S. India.

Part Used : Grass.

Utilization : Grass grazed by cattle only.

References : Agarwal (2003), Ambasta (1986), Dwivedi (1993), Duthie (1888), Hooker (1897), Singh & Pandey (1998), Watt (1972).

Panicum atrosanguineum Hochst. *ex* A. Rich.

Syn: *Panicum hydaspicum* Edgew.

Family: Poaceae.

Part Used : Whole Plant.

Utilization : Excellent fodder for cattle.

Reference : Ambasta (1986).

Panicum colonum Linn.

Syn: *Echinochola colona* Kunth.; *Oplismenus colonus* Kunth.; *Panicum brizoides* Linn.; *P. flavidum* Retz.; *P. flutitans* Retz.

Family: Poaceae.

Regional Names:

Beng : Shama;

Pub. : Jangali Sanwak, Sanwak.

Distribution : Found All Over the Plains.

Part Used : Aerial Parts.

Utilization : It is greedily eaten by all kinds of cattle.

Reference : Duthie (1888).

Panicum incomtum Trin.

Syn: *Panicum sarmentosem* Hook. f. Roxb.

Family: Poaceae.

Part Used : Leaves.

Utilization : Used as fodder.

Reference : Ambasta (1986).

Panicum maximum Jacq. (Pl. 7A)

Syn: *Panicum jumentorum* Pers.
Family: Poaceae.

English, Hindi and Regional Names:

Eng.	:	Guinea Grass;
Hindi	:	Ginighas;
Guj.	:	Ginighas;
Kan.	:	Gini hullu;
Mar.	:	Gini gawat;
Tam.	:	Gini pullu.

Distribution : Dry Regions as a Fodder Grass in India.

Part Used : Grass.

Utilization : Used mainly as green fodder, also makes excellent silage and hay given to cattle to prevent tympanitis.

Active Constituents : Carotene, vit.C, B and tocophrrol.

References : Agarwal (2003), Ambasta (1986), Dwivedi (1993), Duthie (1888), Hooker (1897).

Panicum miliaceum Linn.

Syn: *Milium esculentum* Meth.; *M. ponicum* Mill.; *Panicum asperrimum* fisch.; *P. callosum* Hochst *ex* A.; *P. dansepilosum* Steud.; *P. milium* Pers.; *P. pilosum* Roxb.; *P. striatum* Spreng.
Family: Poaceae.

English, Hindi, Sanskrit and Regional Names:

Eng.	:	Common Millet, Hog Millet, Proso Millet;
Hindi	:	Chena, Chin, China;
Sans.	:	Anu, China, Chini, Rad;
Beng	:	Cheena;
Bom.	:	Chenah, Vari, Wari;
Guj.	:	China, Vari;
Kan.	:	Bilibarragu, Karibaragu, save;
Mar.	:	Vari, Varo;
Ori.	:	Rala;

Pub.	:	Anne, China, Salan, Salar;
Tam.	:	Kadukanni, Panivaragu;
Tel.	:	Varagalu, Variga.

Distribution : Cultivated in Dry Regions of India upto 3,500 m. Himalaya.

Part Used : Whole Plant.

Utilization : Green plant excellent fodder for cattle and horses and also used as hay.

Active Constituents : Leaves rich in proteins 18%.

References : Agarwal (2003), Ambasta (1986), Bajpayee & Dixit (1996), Caius (2003), Duthie (1888), Rastogi & Mehrotra (1990), Watt (1972).

Panicum paludosum Roxb. non Nees.

Syn: *Panicum proliferum* auct; non Lam.

Family: Poaceae.

Distribution : Common Along the Banks of Rivers, Canals etc. & in Low Lying Areas of Rice Fields.

Part used : Plant.

Utilization : Provides a favourite fodder for elephants and buffaloes.

Reference : Ambasta (1986).

Panicum sumatrense Roth. *ex* Roem. & Schult.

Syn: *Panicum miliare* Lam.; *P. psilopodium* Trin.

Family: Poaceae.

English, Hindi, Sanskrit and Regional Names:

Eng.	:	Little Millet;
Hindi	:	Kungu, Kutki, Shavan;
Sans.	:	Syamakah;
Beng	:	Gondula, Gundi;
Guj.	:	Gadro;
Kan.	:	Shame, Save;
Mal.	:	Shama;

Mar.	:	Sava;
Ori.	:	Suniva;
Pub.	:	Kutki;
Tam.	:	Samai;
Tel.	:	Karallu.

Distribution : Cultivated in Outer Parts of N. India upto 2,100 m.

Part Used : Plant.

Utilization : Used as a quick growing fodder.

Active Constituents : Carotene 18.50mg/100g.

References : Agarwal (2003), Ambasta (1986), Duthie (1888).

Panicum virgatum Linn.

Syn: *Panicum glaberrimum* Steud.; *P. pruinosum* Bernh. *ex* Trin.

Family: Poaceae.

Part Used : Grass.

Utilization : Used as a fodder grass.

Reference : Ambasta (1986).

Parkinsonia aculeata Linn.

Family: Fabaceae.

English, Hindi and Regional Names:

Eng.	:	The Jerusalem Thron;
Hindi	:	Vilayati Kikkar;
Beng	:	Balati Kikkar;
Bom.	:	Devi Babul, Vilayti Kikkar;
Guj.	:	Baval, Pardeshi, Ram baval;
Mar.	:	Adanti;
Tel.	:	Sima tumma.

Distribution : Throughoutt Drier Parts of India; Commonly in Gujarat & Rajasthan.

Parts Used : Twigs. Leaves.

Utilization : Tree lopped for fodder.

Active Constituents : Leaves contain 7.5% protein, 29.0% fibre, 4.16% Ca and 0.17 % phosphorus.

References : Agarwal (2003), Ambasta (1986), Dwivedi (1993), Rastogi & Mehrotra (1990), Singh (1982), Watt (1972).

Parrotiopsis jackuemontiana (Retz.) Rehd.

Syn: *Parrotia jackuemontiana* Decne.

Family: Hamamelidaceae.

Hindi Names:

Hindi : Killar, Pahu, Paser, Pishor.

Distribution : W. Himalaya upto 3,000 m. Khasia Hills.

Part Used : Leaves.

Utilization : Used as fodder for cattle and goats.

References : Agarwal (2003), Ambasta (1986), Singh (1982).

Paspalidium geminatum (Forsk.) Stapf.

Syn: *Panicum paspaloides* Pers.

Family: Poaceae.

Part Used : Plant.

Utilization : Used as an excellent fodder.

Reference : Ambasta (1986).

Paspalum kora Willd.

Family: Poaceae.

Hindi and Regional Names:

Hindi : Kodu;
Pub. : Kora.

Part Used : Plant.

Utilization : Cattle and buffaloes eat it readily when young.

Reference : Duthie (1888).

Paspalum scrobiculatum Linn.

Syn: *Paspalum commersonni* Lam.;
P. scrobiculatum var. *frumentaceum* Stapf.
Family: Poaceae.

English, Hindi, Sanskrit and Regional Names:

Eng.	:	Koda Millet;
Hindi	:	Kodo, Kodra;
Sans.	:	Kodrava;
Beng	:	Kodoa Dhan;
Bom.	:	Harik, Kodri, Pakodi, Pakod;
Guj.	:	Kodro, Menya;
Kum.	:	Kodo, Kodra, Kodram;
Mal.	:	Varagu;
Mar.	:	Harik, Kodra;
Ori.	:	Kodus;
Pub.	:	Kodan, Kodra;
Tam.	:	Varagu;
Tel.	:	Arikalu.

Distribution : Throughout India, chiefly in Assam & Andamans.

Part Used : Leaves.

Utilization : Leaves grazed by cattle especially by buffaloes.

Active Constituents : Rich in arginine, histine, lycine, valine and methionine.

References : Agarwal (2003), Ambasta (1986), Duthie (1888), Hooker (1872), Islam (2002), Prajapati *et al.* (2003), Watt (1972).

Paspalum vaginatum Sw.

Family: Poaceae.

Distribution : Upto 1,500 m in Hills.

Part Used : Leaves.

Utilization : Used as fodder.

Reference : Khanna (2002).

Pavonia odorata Willd.

Syn: *Hibiscus odoratus* Roxb.; *P. romborua* Wall.; *P. rosea* Wall.; *Pavonia sidoides* Horn in DC.

Family: Malvaceae.

English, Hindi, Sanskrit and Regional Names:

Eng.	:	Fragrant Sticky Mallows;
Hindi	:	Sugandha bala;
Sans.	:	Ambu, Baala, Jala, Hribera, Nira, Paya, Toya;
Beng	:	Sugandha bala;
Bom.	:	Bala, Kalavala;
Guj.	:	Kalowalo;
Kan.	:	Balarakkasi, Peramutiberu;
Mal.	:	Kurantotti;
Mar.	:	Kalavala;
Ori.	:	Karumuli;
Tam.	:	Avibattam, Peramutti;
Tel.	:	Erra kutti, Chitti benda, Tige benda.

Distribution : W. Peninsula.

Part Used : Plant.

Utilization : Soft plant fibre used for making ropes.

References : Ambasta (1986), Khare (2004), Hooker (1872), Watt (1972).

Pegolettia senegalensis Cass.

Family: Asteraceae.

Hindi Name:

Hindi : Unti chari.

Distribution : Dry Regions of India.

Part Used : Leaves.

Utilization : Useful as fodder for sheep and camels.

References : Agarwal (2003), Ambasta (1986).

Pennisetum cenchroides Rich.

Syn: *Cenchrus ciliaris* Linn.

Family: Poaceae.

Regional Name:

Pub. : Dhamman, Kurkan.

Distribution : Common All Over the Plains of N.W. India.

Part Used : Grass.

Utilization : Excellent fodder grass for lactating cows and horses.

Reference : Duthie (1888).

Pennisetum divisum (Forsk. *ex* Gmel.) Henr.

Syn: *Pennisetum dichotomum* Delile.

Family: Poaceae.

Distribution : Drier Regions.

Part Used : Plant.

Utilization : Relished by horses and donkeys.

Reference : Ambasta (1986).

Pennisetum holcoides Schult.

Syn: *Panicum holcoides* Roxb.

Family: Poaceae.

Regional Name:

Beng : Sivati.

Part Used : Plant.

Utilization : Used as fodder grass.

Reference : Duthie (1888).

Pennisetum purpureum Schum.

Family: Poaceae.

English Names:

Eng. : Elephant Grass, Napier Grass.

Distribution : Cultivated Throughout India.

Part Used : Young Herbage.

Utilization : Readily eaten by cattle.

Active Constituents : Contains 1.0% protein, 7.4% fibre, 3.1% mineral matter, 0.12% Ca and 0.07% P.

References : Ambasta (1986), Dwivedi (1888).

Pennisetum typhoides (Burn. f.) Stapf. & C. E. Hubbard.

Syn: *Pennisetum typhoideum* Rich.
Family: Poaceae.

English, Hindi and Regional Names:

Eng. : Bulrush Millet, Pearl Millet, Spiked millet;
Hindi : Bajra;
Beng : Bajra, Lahra;
Guj. : Bajri;
Kan. : Sajje;
Mar. : Bajri;
Tam. : Kambu;
Tel. : Ganti, Sajja.

Distribution : Crop Plant of Dry Regions in India.

Part Used : Plant.

Utilization : Grown as fodder for cattle.

Active Constituents : Proteins 12-15%.

References : Agarwal (2003), Ambasta (1986), Duthie (1888), Watt (1972).

Pennisetum verticillatum R. Br.

Syn: *Panicum verticillatum* Linn.; *Setaria verticillatum* Beauv.
Family: Poaceae.

Regional Names:

Pub. : Chirchira.

Distribution : Plains of N. W. India upto 2,000 m.

Part Used : Plant.

Utilization : Young plant used as a fodder plant for cattle.

Reference : Duthie (1888).

Perotis indica (Linn.) Kuntze.

Syn: *Perotis latifolia* Ait.

Family: Poaceae.

Hindi and Regional Names:

Hindi : Kuras;

Bom. : Kuras;

Kan. : Jabburu Korlai Hullu, Narimisai Hullu;

Tam. : Narivel Pillu, Thopparai Pillu;

Tel. : Nakka Peethu, Nakka Toka.

Distribution : Found Throughout India.

Part Used : Grass.

Utilization : A good fodder grass relished by stock at all stages.

References : Agarwal (2003), Ambasta (1986).

Persea duthiei (King. *ex* Hook. f.) Kost.

Syn: *Machilus duthiei* King.

Family: Lauraceae.

Distribution : Himalaya: 1,000-3,000 m.

Part Used : Leaves.

Utilization : Leaves fed to sheep, goats and cattle.

References : Purohit & Samant (1995), Raju (1998).

Persea gammieana (King. *ex* Hook. f.) Kost.

Syn: *Machilus gammieana* King.

Family: Lauraceae.

Part Used : Leaves.

Utilization : Leaves used as fodder.

Reference : Saklani & Jain (1994).

Persea odoratissima (Nees) Kost.

Syn: *Machilus odoratissima* Nees.
Family: Lauraceae.

Distribution : Occurs between 1,000-2,000 m.

Part Used : Leaves.

Utilization : Leaves fed to cattle.

Reference : Purohit & Samant (1995).

Persicaria capitata (Buch.-Ham.) H. Gross.

Syn: *Polygonum capitatum* Buch.-Ham.
Family: Polygonaceae.

Part Used : Leaves.

Utilization : Leaves used as fodder.

Reference : Singh & Kumar (2000).

Phanera integrifolia (Roxb.) Benth.

Family: Fabaceae.

Part Used : Stem Bark.

Utilization : Bast fibre widely used for making coloured ropes and cordage by tribals of S.E. Rajasthan.

Reference : Singh & Pandey (1998).

Phararis minor Retz.

Syn: *Phalaris ambigua* Fig. & Notar.; *P. aquatica* Ait.; *P. bulbosa* Desf.; *P. canariensis* Wall.; *P. capensis* Thunb.; *P. nepalensis* Trin.
Family: Poaceae.

Part Used : Plant.

Utilization : Used as a fodder plant.

References : Hooker (1879), Khanna (2002).

Phaseolus lathyroides Linn.

Syn: *Phalaris semierectus* Linn.
Family: Fabaceae.

English Name:

Eng. : Phasemy Bean.

Part Used : Whole Plant.

Utilization : Highly nutritive fodder for pigs and also used as hay.

References : Ambasta (1986), Blatter & Hallberg (1984).

Phaseolus lunatus Linn.

Syn: *Phaseolus inamoenus* Linn.; *P. vulgaris* Wall.
Family: Fabaceae.

English, Hindi and Regional Names:

Eng. : Burma bean, Double Bean, Lima Bean;
Hindi : Lobiya, Sem;
Beng : Banbur Butti;
Bom. : Daful;
Pub. : Lobiya;
Tam. : Khasi Kollu.

Distribution : Throughout India.

Part Used : Green Vines.

Utilization : Green vines used as fodder and also made into hay.

References : Agarwal (2003), Ambasta (1986), Hooker (1879), Rastogi & Mehrotra (1995a), Watt (1972).

Philadelphus coronarius Linn.

Family: Saxifragaceae.

Part Used : Plant.

Utilization : Fibre used for making ropes.

Reference : Raju (1998).

Phoebe lanceolata Nees. (Pl. 7B)

Syn: *Laurus lanceolata* Wall.; *L. lanceolaria* Roxb.; *L. salicifolia* Herb. Hamilt.; *Ocotea lanceolata* Nees.

Family: Lauraceae.

Hindi and Regional Name:

Hindi	:	Haulia;
Ass.	:	Chang Chak, Dieng -Lalong, Moso- Sigaro Baphang;
Gar.	:	Bhardroi, Bhadeur;
Kum.	:	Kanwal;
Pub.	:	Badror.

Distribution : Himalaya. Khasia Hills. Mountains of S. India.

Part Used : Leaves.

Utilization : Leaves used as fodder for all ruminants in winter.

References : Agarwal (2003), Ambasta (1986), Dwivedi (1993), Hooker (1890), Purohit & Samant (1995), Watt (1972).

Phoenix acaulis Roxb.

Syn: *Phoenix acaulis* var. *melanocarpa.*; *P. ouseleyana* Griffith.; *P. pedunculata* Griffith.

Family: Arecaceae.

English, Hindi and Regional Names:

Eng.	:	The Dwarf Date Palm;
Hindi	:	Ban Khajur, Khajur;
Beng	:	Khajur;
Mar.	:	Boichind, Shevra, Shilind;
Ori.	:	Kojiri;
Pub.	:	Pind Khajur;
Tel.	:	Yita.

Distribution : Outer Himalaya on Kumaun to Assam. M.P. Deccan.

Parts Used : Leaves. Roots.

Utilization : Leaves lopped for fodder and also used for thatching, brooms and ropes. Mature roots also used for making ropes.

References : Agarwal (2003), Ambasta (1986), Dwivedi (1993), Jain (1991), Raju (1998), Singh (1982), Varghese (1996), Watt (1972).

Phoenix dactylifera Linn.

Family: Arecaceae.

English, Hindi, Sanskrit and Regional Names:

Eng. : Date Palm, The Edible Date;
Hindi : Khajur;
Sans. : Kharjjuraha;
Beng : Khaji, Khajur;
Bom. : Khajur;
Guj. : Khajur;
Kan. : Kharjura;
Mal. : Ittappazham, Tenitta;
Mar. : Khajur;
Ori. : Khorjurri;
Pub. : Khajur, Khaji;
Tam. : Karchuram, Perichchankay;
Tel. : Ita, Kharjuramu.

Distribution : Cultivated in Some Warmer Parts of India.

Parts Used : Spathes. Petioles.

Utilization : Fibre from green spathes used for ropes and those from petioles used for making insulating boards.

References : Agarwal (2003), Ambasta (1986), Dastur (1964), Sahni (1998), Watt (1972).

Phoenix sylvestris Roxb.

Syn: *Elate sylvestris* Linn.; *Phoenix humilis* Royle

Family: Poaceae.

Vern. Name: Khajara.

English, Hindi, Sanskrit and Regional Names:

Eng. : Wild Date Palm, Date Sugar Palm;
Hindi : Khajur;
Sans. : Bhumikharjurika, Haluka, Kharju;

Beng	:	Khajur, Khajar;
Guj.	:	Kharak, Khakri;
Kan.	:	Ichalu;
Mal.	:	Inta;
Mar.	:	Boichand, Shindi, Sen;
Ori.	:	Khajuri, Kojori;
Pub.	:	Khaji, Khajur;
Tam.	:	Icham, Madal;
Tel.	:	Pedda ita.

Distribution : Throughout India; abundant in Bengal, Bihar, Gujarat, Kumaon, S.W. Punjab.

Parts Used : Leaves. Petioles.

Utilization : Leaves lopped for fodder. Petioles twisted into ropes for drawing water from wells and its fibre used for paper making materials.

References : Agarwal (2003), Ambasta (1986), Chakraborty & Bhattacharjee (2003), Dash & Mishra (2000), Dastur (1964), Jain (1991), Raju (1998), Singh (1992), Singh & Pandey (1998), Singh & Kumar (2004), Talbot (1976b), Varghese (1996), Watt (1972).

Phormium tenax Forst.

Family: Liliaceae.

English Name:

Eng. : New Zealand Flax.

Distribution : S. India.

Part Used : Stem.

Utilization : Soft fibre used for making sail cloths, cordage and paper.

References : Agarwal (2003), Watt (1972).

Phragmites communis Trin.

Syn: *Arundo phragmites* Linn.
Family: Poaceae.

Regional Name:

Pub. : Dila.

Distribution : Plains of N.W. India upto 4,500 m.

Part Used : Plant.

Utilization : Eaten by cattle in Ladakh.

Reference : Duthie (1888).

Phragmites karka (Retz.) Trin *ex* Steud.

Syn: *Arundo corea* Herb. Rottle.; *A. donax* Herb. Ham. *ex* Wall.; *A. karka* Retz.; *A. roxburghii* Kunth Revis.; *A. roxburghiana* Kunth *ex* Steud.; *A. tibialis* Roxb.; *Oxyanthe japonica* Steud.; *Phragmites bifaria* Wight Herb.; *P. nepalensis* Nees *ex* Steud.; *P. roxburghii* Steud.; *Sericura japonica* Steud.; *Trichoon karka* Roth.
Family: Poaceae.

Distribution : Throughout India from Punjab to Myanmar. Ascending the Himalaya in Hot Valley upto 1,000 m.

Parts Used : Clums. Leaf.

Utilization : Clums used for making paper pulp. Leaf fibre good for making fish traps, brooms, mats, etc.

References : Das *et al.* (1996), Duthie (1888), Hooker (1897), Islam (2002).

Picrasma guassioides Bennet.

Syn: *Nima quassioides* Ham.; *Simba quassioides* Ham.
Family: Simarubaceae.

English, Hindi and Regional Names:

Eng. : Simaru bee;
Hindi : Kashshing;
Beng : Bhurungi;
Pub. : Bera, Bering, Birgo, Hala, Pesho, Tuthai.

Distribution : Common in Shady Forest; Sub-Tropical Himalaya from Jammu to Nepal.

Part Used : Shoots.

Utilization : Commonly used as fodder for sheep and goats wherever available.

Active Constituents : Leaves contain quassin.

References : Purohit & Samant (1995), Watt (1972).

Pilea glaberrima Blume.

Syn: *Pilea smilacifolia* Wedd.

Family: Urticaceae.

Hindi Name:

Hindi : Pahari- sunn.

Distribution : It grows in the E. Himalaya.

Part Used : Plant.

Utilization : Yields fibre for cordage.

References : Agarwal (2003), Ambasta (1986).

Pipturus incanus Wedd.

Syn: *Pipturus velutinus* Wedd.

Family: Urticaceae.

Hindi Name:

Hindi : Genti.

Distribution : Outermost Himalaya.

Part Used : Bark.

Utilization : Bark yields a fibre for sails and mats.

References : Agarwal (2003), Ambasta (1986).

Pistacia integerrima Stewart *ex* Brandis.

Syn: *Pistacia khinjak* Stew *ex* Brandis.; *Rhus integerrima*Wall.; *R. kakrasingee* Royle.

Family: Anacardiaceae.

Vern. Name : Kakadsingi.

English, Hindi, Sanskrit and Regional Names:

Eng. : Crab's Claw;
Hindi : Kakra;
Sans. : Chakra, Ghosha, Karkali, Vakra;
Beng : Kakrsingi;
Bom. : Kakrasingi;
Gar. : Kakar;
Guj. : Kakrasingi;
Kash. : Kakkar, Kakkeri;
Kum. : Kakra;
Mal. : Karkatasringhi;
Mar. : Kakrasingi;
Ori. : Kakadashringi;
Pub. : Kaklashne, Kakra,Tungy;
Tam. : Kakkatashingi;
Tel. : Kakarashingi.

Distribution : W. Himalaya (often cultivated): 600-2,000 m.

Part Used : Leaves.

Utilization : Leaves lopped for all classes of livestock during the summers, autumn and rainy season.

References : Agarwal (2003), Ambasta (1986), Dwivedi (1888), Purohit & Samant (1995), Raju (1998), Singh (1982), Sood & Thakur (2004), Watt (1972).

Pistia stratiotes Linn.

Syn: *A. cuneataengl*, *Kiambam kitsii* Rumph.; *Pistia crispata* Blume.; *P. stratiotes* Linn.

Family: Araceae.

English, Hindi, Sanskrit and Regional Names:

Eng. : Tropical Duck Weed, Water Lettuce;

Hindi	:	Jalkumbhi;
Sans.	:	Kumblika, Jalakumbhi, Vaariparni;
Beng	:	Takapana;
Bom.	:	Gondala, Prashni;
Kan.	:	Anthara tomara;
Mar.	:	Prasni;
Ori.	:	Borajanji;
Tam.	:	Kogitamari;
Tel.	:	Anthara Tamara.

Distribution : Bengal. The Coromandal Coast. Deccan. Myanmar. Sri Lanka.

Part Used : Leaves.

Utilization : Excellent fodder for big fish like crap, catla, rohu and merigal and also for pigs.

Active Constituents : Rich in vit. A, B.& C., proteins 1.4%; fats 0.35; carbohydrates 2,65; fibre 0.9%;Ca 0.20% and phosphorus 0.06%.

References : Das *et al.* (1996), Khare (2004), Raju (1998).

Pithecellobium dulce Benth.

Syn: *Inga dulcis* Willd.; *Mimosa dulcis* Roxb.

Family: Fabaceae.

English, Hindi and Regional Name:

Eng.	:	Madras thorn;
Hindi	:	Jangli jalebi. Vilayatibabul, Vilayatiimli;
Beng	:	Dekhani Babul;
Bom.	:	Chinch, Vilayatiyamli;
Kan.	:	Kottampuli, Seemae hunase;
Mal.	:	Korrukapuli;
Mar.	:	Vilayati chinch;
Ori.	:	Simakoina;
Tam.	:	Kodukkaapuli;
Tel.	:	Sima chinta.

Distribution : Throughout Hot & Dry Regions of India.

Parts Used : Leaves. Oil Cakes.

Utilization : Leaves and left over cake after extracting oil from seeds used as fodder.

Active Constituents : Leaves contain crude proteins 29%, 4.4%, crude fibre 17.5%, Ca.14% and P 0.35%.

References : Agarwal (2003), Ambasta (1986), Dwivedi (1993), Hooker (1879), Kirtikar & Basu (1984), Rastogi & Mehrotra (1993, 1995b), Singh (1982).

Plumbago auriculata Lam.

Syn: *Plumbago capensis* Thunb.

Family: Plumbaginaceae.

Part Used : Leaves.

Utilization : Foliage eaten by poultry and stock, particularly the sheep.

Active Constituents : Rich in apigenine, luteolin, á-amyrin, palmitic acid and â-sitosterol.

References : Ambasta (1986), Rastogi & Mehrotra (1995a).

Plumeria annua Linn.

Syn: *Plumeria nepalensis* Wall. *ex* Duthie.

Family: Poaceae.

English and Regional Names:

Eng. : Dwarf & Annul Meadow Grass;
Pub. : Chirua.

Part Used : Plant.

Utilization : Used as a very nutritious fodder.

Reference : Ambasta (1986).

Plumeria pratensis Linn.

Family: Poaceae.

English Names:

Eng. : Kentucky Blue grass, Smooth Stalk Meadow Grass.

Part Used : Plant.

Utilization : Excellent pasture and lawn grass rich in proteins when young and relished by all classes of livestock.

Active Constituents : Rich in proteins.

Reference : Ambasta (1986).

Poa annua L.

Syn: *Poa alpina* Herb.; *P. royleana* Nees *ex* Steud.; *P. supina* Schrad.

Family: Poaceae.

Regional Name:

Pub. : Chirua.

Distribution : Temperate & Sub-Alpine Himalaya. W. Tibet. Punjab Plains. Meghalaya. Nilgiris 1,500- 2,000 m.

Part Used : Grass.

Utilization : Plant used for fodder purposes.

References : Hooker (1897), Islam (2002), Watt (1986).

Poa khasiana Stapf.

Syn: *Poa himalayana* Nees *ex* Steud.

Family: Poaceae.

Distribution : Sikkim Himalaya, Khasia Hills.

Part Used : Leaves.

Utilization : A good fodder of alpine Himalaya.

References : Ambasta (1986), Hooker (1897).

Poa sterilis Bieb.

Family: Poaceae.

Part Used : Leaf Blades.

Utilization : A fodder grass found in Alpine Himalaya.

Reference : Ambasta (1986).

Poa trivialis Linn.

Family: Poaceae.

Distribution : Grows in Assam. S. India.

Part Used : Grass.

Utilization : Grass used as fodder for animals.

References : Agarwal (2003), Ambasta (1986), Watt (1972).

Pogostemon benghalenses (Burm.) Kuntz.

Syn: *Mentha fruticulosa* Roxb.; *M. secunda* Roxb.: *Origanum benghalenses* Burm. f.; *O. indicum* Roth.; *Pogostemon plectranthoides* Desf.

Family: Lamiaceae.

Vern. Names: Ghento, Kali Basunti.

Regional Name:

Ori. : Dumobadokkota.

Distribution : H. P. upto 1,300 m.

Part Used : Leaves.

Utilization : Leaves fed to cattle to enhance lactation.

References : Rastogi & Mehrotra (1991), Sur & Halder (2002).

Pollinia argentea Trin.

Syn: *Erianthus roxburghii* F. Muell.; *E. tristachyinus* Trin.; *Pollinia tristachya* Thw.

Family: Poaceae.

Part Used : Plant.

Utilization : Affords good fodder for cattle when young.

Reference : Duthie (1888).

Pollinia eriopoda Hance.

Syn: *Andropogon invalutus* Steud.; *A. notopogon* Nees.; *Spodiopogon angustifolius* Trin.; *S. laniger* Nees.

Family: Poaceae.

Part Used : Plant.

Utilization : Used as an excellent fodder for cattle and also a good material for cordage.

Reference : Duthie (1888).

Polyalthia suberosa (Roxb.)

Family: Annonaceae.

Part Used : Bark.

Utilization : Bark fibre valued for making ropes by 'Chenchus'.

Reference : Rao & Henry (1996).

Polygonum alpium All.

Syn: *Polygonum acidum* Pall.; *P. divaricatum* Vill.; *P. polymorphum* Ledeb.; *P. sibiricum* Linn.

Family: Polygonaceae.

Vern. Name: Alipap.

Hindi and Regional Names:

Hindi : Chita- hola;
Kash. : Tsok- Ladar.

Distribution : N.W. Himalaya.

Part Used : Plant.

Utilization : Plant used as fodder.

Active Constituents : Contains myricetin, quercetin and kaempferol.

References : Hooker (1890), Kaul (1997), Rastogi & Mehrotra (1990), Sood *et al.* (2001).

Polygonum aviculare Linn.

Syn: *Polygonum heterophyllum* Lindm.

Family: Polygonaceae.

English, Hindi, Sanskrit and Regional Names:

Eng. : The Knot grass;
Hindi : Ban natia, Huranj, Machoti;
Sans. : Nisomali;

Beng	:	Machutie;
Kash.	:	Drop;
Pub.	:	Banduka, Kesru.

Distribution	:	Kashmir to Kumaon Hills upto 4,000 m. W.Tibet.
Part Used	:	Leaves.
Utilization	:	Leaves used as fodder for sheep and goats.
Active Constituents	:	Rich in kaempferol, luteolin, quercetin, rhamnetin.
References	:	Agarwal (2003), Ambasta (1986), Hooker (1890), Rastogi & Mehrotra (1995b), Watt (1972).

Polygonum barbatum Linn.

Syn: *Persicaria barbatum* (Linn.) Hara.; *Polygonum fluviatile* Herb.; *P. horemanni* Meissn.; *P. marmoramae* Herb.; *P. rivulare* Koen.; *P. stagninum* Buch.-Ham. *ex* Mesissn.

Family: Polygonaceae.

Vern. Name: Rohini.

Hindi and Regional Names:

Hindi	:	Bekhor Jangli Machotti;
Beng	:	Bokh unjubaz;
Kan.	:	Konde malle;
Mal.	:	Vellutamodellamukku;
Mar.	:	Dhaktasheral;
Pub.	:	Narri;
Tam.	:	Atalari;
Tel.	:	Miru ganneru.

Distribution	:	Throughout the Warmer Regions of India.
Part Used	:	Aerials parts.
Utilization	:	Fodder for cattle.
References	:	Agarwal (2003), Ambasta (1986), Hooker (1890), Raju (1998), Sood & Thakur (2004), Watt (1972).

Polygonum chinense Meissn.

Syn: *Ampelygonum chinense* Lindl.; *Coccoloba crispata* Ham.; *Polygonum adenotrichum* Wall.; *Polygonum brachiatum* Poir.; *P. corymbosum* Willd.; *P. cymosum* Roxb.; *P. ovatum* Heyne.; *P. patens* Don.; *P. polycephalum* Wall.; *P. ovatum* Heyne.; *P. panduraeforme* Ham.

Family: Polygonaceae.

Vern. Name: Tarod.

English and Regional Names:

Eng.	:	Chinese Knot weed;
Kan.	:	Bilichini, Ganiglu;
Mar.	:	Paral.

Distribution	:	Sub-Tropical & Temperate Himalaya (Shimla to Bhutan) 1,000-2,000 m.
Part Used	:	Leaves.
Utilization	:	Leaves used as fodder.
References	:	Hooker (1890), Prajapati *et al.* (2003), Saklani & Jain (1994).

Pongamia pinnata Piere.

Syn: *Pongamia glabra* Vent.

Family: Fabaceae.

English, Hindi, Sanskrit and Regional Names:

Eng.	:	Indian Beech, Pongom oil tree;
Hindi	:	Karanj, Karanja;
Sans.	:	Karanj;
Ass.	:	Korach;
Beng	:	Dhara Karanja, Natakaranja;
Guj.	:	Karanj, Karanja;
Kan.	:	Honge Berru;
Mal.	:	Pungu, Punnu;
Mar.	:	Karanj;
Ori.	:	Koranjo;
Pub.	:	Sukhchien;
Tam.	:	Ponga, Pongam;
Tel.	:	Gannuga.

Distribution : Throughout India.

Parts Used : Leaves. Seeds. Bark.

Utilization : Leaves lopped for fodder. Bark yields a fibre for cordage. Seed cake used for poultry ration.

Active Constituents : Leaves contain 43.50% dry matter, 17.62% crude proteins, 62.02% neutral detergent fibre, 39.80% cell content, 40.60% acid detergent fibre, 20.20% hemicellulose, 14.60% lignin, 20.05 cellulose and 13.35% total minerals.

References : Agarwal (2003), Ambasta (1986), Dwivedi (1982), Kshirsagar (2003), Oommachan (1977), Singh (1982), Singh & Pandey (1998), Thakur *et al.* (1989), Yoganarasimhan *et al.* (1981).

Populus ciliata Wall. *ex* Royle

Syn: *Populus pyriformis* Royle.; *P. rotundifolia* Griff.
Family: Salicaceae.

Vern. Name: Phalsa.

English, Hindi and Regional Names:

Eng.	:	Himalayan Poplar;
Hindi	:	Piplas;
Gar.	:	Syan;
Kum.	:	Chan, Syan;
Pub.	:	Dudfrass, Falish, Kramali;
Sim.	:	Chelaun, Chelum.

Distribution : Throughout Himalaya (Kashmir to Bhutan): 2,000-3,000 m.

Part Used : Leaves.

Utilization : Considered nutritious fodder for goats.

References : Agarwal (2003), Ambasta (1986), Hooker (1890), Parker (1973), Purohit & Samant (1995), Sahni (1998), Sharma (2004), Singh (1982), Watt (1972).

Populus euphratica Olivier.

Syn: *Populus diversifolia* Schrank.;
Balsamiflora deltoides Griff.
Family: Salicaceae.

English, Hindi and Regional Names:

Eng. : Indian Poplar;
Hindi : Bahan;
Ladk. : Hotung, Hodung;
Pub. : Bahan, Bhan, Jungli, Safeda.

Distribution : W. Himalaya. Punjab Plains. W. Tibet.

Part Used : Leaves. Bark.

Utilization : Leaves valued as fodder for sheep, goats and camels. Bark browsed by animals in cold deserts of Ladakh.

References : Agarwal (2003), Ambasta (1986), Dwivedi (1993), Hooker (1890), Parker (1973), Sahni (1998), Singh (1982), Watt (1972).

Populus laurifolia Ledeh.

Syn: *Populus balsamifera* Hook. f. non Linn.
Family: Salicaceae.

Distribution : Himalaya, Lahaul- Spiti: 2,400- 4,000 m.

Part Used : Leaves.

Utilization : Leaves lopped for cattle fodder.

References : Ambasta (1986), Rastogi & Mehrotra (1995a).

Pouzolzia viminea Wedd.

Syn: *Boehmeria frondosa* Don.; *B. ovalis* Miq.;
B. sanguinea Hassk.; *B. viminea* Wall.; *Leptocnide borbonica* Blume.;
Urtica viminea Wall.;*U. chiple, punchata & parviflora* Ham.
Family: Urticaceae.

Hindi and Regional Names:

Hindi : Khojo;
Ass. : Misagi jallaphang.

Distribution : Tropical & Sub-Tropical Himalaya from the Sutlej Eastwards to Sikkim.

Part Used : Bark.

Utilization : Bark yields a fibre for fishing nets and cordage.

References : Agarwal (2003), Ambasta (1986), Hooker (1890), Watt (1972).

Prangos pabularia Lindl.

Family: Apiaceae.

English, Hindi and Regional Names:

Eng. : Stony Herb;
Hindi : Avipriva, Niyamak;
Bom. : Phattarsalam;
Kash. : Kurungas;
Pub. : Fiturasaliya.

Distribution : Kashmir & Baluchisthan: 2,000 - 4,000 m & H.P. upto 4,000 m.

Part Used : Leaves.

Utilization : Given as fodder for sheep, cattle and goats during winter in high altitude.

Active Constituents : Leaves contain proteins 10.4%, fats 3.55 and crude fibre 22.6%.

References : Agarwal (2003), Ambasta (1986), Dwivedi (1993), Hooker (1879), Kaul (1997), Rastogi & Mehrotra (1991), Sharma (2003), Singh *et al.* (1996), Watt (1972).

Premna latifolia Roxb.

Syn: *Premna mucronata* Roxb.

Family: Verbenaceae.

Vern. Name: Vikar.

Hindi and Regional Names:

Hindi : Baker, Bakarcha, Basota;
Kum. : Agniu;
Mal. : Knappa;

Ori.	:	Gondhono;
Pub.	:	Bankar, Ganhila, Gian;
Tam.	:	Erumaimunai, Munnai;
Tel.	:	Pedda - nelle- koora.

Distribution : Throughout India. (Lowers Hills of the Coromandel Coast. Bengal. Rajmahl to the S. Circars).

Part Used : Leaves.

Utilization : Valued as fodder.

Active Constituents : Furan isolated from the leaves.

References : Ambasta (1986), Dwivedi (1993), Hooker (1885), Raju (1998), Rastogi & Mehrotra (1993), Singh (1982), Singh & Kumar (2000), Watt (1972).

Prinsepia utillis Royle

Family: Rosaceae.

Vern. Names: Bhaeklu, Bhekal.

Hindi and Regional Names:

Hindi	:	Bhekal, Bekkra, Karanga;
Kum	:	Bhekla, Chirara, Dhatela;
Pub.	:	Behhul, Bhekal, Chamba, Karangaru, Phulwara.

Distribution : Temperate Himalaya from Hazara to Sikkim & Bhutan. Khasia Hills. Nilgiris.

Part Used : Leaves.

Utilization : Eaten by sheep and goats in winter.

References : Agarwal (2003), Dwivedi (1993), Hooker (1879), Purohit & Samant (1995), Watt (1972).

Prosopis chilensis Stuntz. (Pl. 7C)

Syn: *Prosopis juliflora* DC.

Family: Fabaceae.

English and Hindi Names:

Eng.	:	Mesquite;
Hindi	:	Vilayati Kikkar, Vilayati Babul, Vilayati Khejra.

Distribution : Cultivated in the Drier Parts of India.

Part Used : Foliage.

Utilization : Foliage fed to livestock both in fresh condition and as hay.

References : Agarwal (2003), Ambasta (1986), Dwivedi (1993), Rastogi & Mehrotra (1995b), Singh (1982).

Prosopis cineraria Druce.

Syn: *Mimosa cineraria* Linn.; *Prosopis spicigera* Linn.
Family: Fabaceae.

English, Hindi, Sanskrit and Regional Names:

Eng. : Algarobo;
Hindi : Jand;
Sans. : Shami;
Beng : Shami;
Guj. : Khijado, Kandi, Sami, Semru;
Mal. : Parampu, Tambu;
Mar. : Saunder, Shemri;
Ori. : Shami;
Pub. : Jand;
Raj. : Khejri;
Tam. : Jambu, Perumbay;
Tel. : Jammi Chettu.

Distribution : Throughout Alluvial Plains of India, Punjab, Haryana, Rajasthan, U.P., Gujarat, M. P. & S. India.

Parts Used : Pods. Foliage.

Utilization : Foliage and pods considered as fodder for camel and goats.

Active Constituents : Leaves contain crude proteins 32.53%, crude fibre 50.59% and patulitrin.

References : Agarwal (2003), Ambasta (1986), Dhiman (2003), Dwivedi (1993), Rastogi & Mehrotra (1990), Sahni (1998), Singh (1982), Singh & Pandey (1998).

Prunus cerasoides Don.

Syn: *Prunus puddum* Roxb.
Family: Rosaceae.

Vern. Name: Pajja, Paaja.

English, Hindi, Sanskrit and Regional Names:

Eng.	:	Birds Cherry, Himalayan Wild Cherry;
Hindi	:	Paddam;
Sans.	:	Padmaka, Padmavriksha, Salmagundi;
Guj.	:	Padmak, Padmakathi;
Kan.	:	Padmaka;
Mal.	:	Patimukam, Patimukham, Patumukam;
Mar.	:	Padmakashtha;
Pub.	:	Amalguchh, Pajja;
Tam.	:	Patumugam
Tel.	:	Padmakla.

Distribution : Distributed between 1,000- 2300 m.

Part Used : Leaves.

Utilization : Fed to all classes of livestock.

References : Agarwal (2003), Purohit & Samant (1995), Prajapati *et al.* (2003), Saklani & Jain (1994), Sharma (2004), Singh (1982).

Prunus cornuta Wall.

Syn: *Prunus padus* Lim.
Family: Rosaceae.

English, Hindi and Regional Names:

Eng.	:	Bird Cherry;
Hindi	:	Jamunoi;
Kash.	:	Jaman;
Kum.	:	Bombaksing, Jamun, Jamuna;
Pub.	:	Bart, Chule, Dudla, Jammun, Kalakat, Sun, Zam.

Distribution : Distributed in Assam to Punjab upto 3,600 m.

Part Used : Leaves.

Utilization : Leaves consumed as fodder by sheep and goats and other class of ruminants.

Active Constituents : All parts of plant contain glucoside amygdalin, glucose, β-sitosterol and quercetin.

References : Agarwal (2003), Dwivedi (1993), Kirtikar & Basu (1984), Purohit & Samant (1995), Rastogi & Mehrotra (1995a), Singh (1982).

Prunus persica Stokes.

Syn: *Amygdalus persica* Linn.; *A. collinus* Wall.; *Persica vulgaris* Miller.

Family: Rosaceae.

Vern. Name: Aadu.

English, Hindi and Regional Names:

Eng. : Nectarine, Peach;
Hindi : Aru, Khirola;
Gar. : Khirola;
Kum. : Aru;
Ori. : Pishu;
Pub. : Aru, Bembeimi, Chimnanu, Sunu;
Raj. : Aru.

Distribution : Kashmir to Kumaon Hills & in the Sub-Tropical Regions of India: 2,000-3,000 m.

Part Used : Leaves.

Utilization : Leaves fed to ruminants.

References : Agarwal (2003), Purohit & Samant (1995), Watt (1972).

Pterocarpus marsupium Roxb. (Pl. 7D)

Syn: *Forma bilocha* (G. Don) Prain.; *Pterocarpus bilobus* Roxb *ex* G. Don.; *P. marsupium* Roxb.

Family: Fabaceae.

English, Hindi, Sanskrit and Regional Names:

Eng. : Indian Kino Tree, Malabar Kino Tree;
Hindi : Bija, Bijasal;

Sans.	:	Asanaka, Bijasara, Pitsara;
Beng	:	Pitsal, Pitshal;
Bom.	:	Bibla, Bija, Honi, Huni, Piasal;
Guj.	:	Biyo, Hiradakhan;
Kan.	:	Bange, Honne;
Mal.	:	Venga;
Mar.	:	Dhorbenla;
Ori.	:	Byasa;
Pub.	:	Chandan Lal, Channan Lal;
Tam.	:	Vengai;
Tel.	:	Peddagi, Yegi.

Distribution : Throughout Deccan, Gujarat, M.P., Bihar & Orissa.

Part Used : Leaves.

Utilization : Leaves used as an excellent fodder.

References : Agarwal (2003), Ambasta (1986), Bhandary *et al.* (1996), Dwivedi (1993), Hooker (1879), Parrotta (2001), Singh (1982), Watt (1972).

Pterocarpus santalinus Linn. f.

Family: Fabaceae.

English, Hindi, Sanskrit and Regional Names:

Eng.	:	Red Sandal Wood, Red Sanders;
Hindi	:	Raktachadan;
Sans.	:	Raktachandana;
Beng	:	Lal Chandan, Ruki;
Bom.	:	Lal Chandan, Ratanjali;
Guj.	:	Ratanjali;
Kan.	:	Agaru, Honne, Kempugandha;
Mal.	:	Patrangam, Tilaparnni;
Mar.	:	Tambada Chandana;
Ori.	:	Raktachndan;
Pub.	:	Chandan Lal;
Tam.	:	Atti, Sivappu Chandanam;
Tel.	:	Raktachandan.

Distribution : S. India, W. Peninsula. Bombay. Bengal.

Part Used : Leaves.

Utilization : Leaves used as a fodder.

Active Constituents : Leaves comprise of lupeol, epilupeol, lupenone, β-amyrone and sitosterol.

References : Agarwal (2003), Ambasta (1986), Hooker (1879), Rastogi & Mehrotra (1991), Singh (1982), Watt (1972).

Pterocymbium tinctorium Merrill.

Syn: *Sterculia campanulata* Wall. *ex* Mast.
Family: Sterculiaceae.

Hindi Names:

Hindi : Pipita, Tshaw.

Distribution : W. Coast. Andaman & Nicobar Islands.

Part Used : Bark.

Utilization : Bark yields a fibre for ropes.

References : Agarwal (2003), Ambasta (1986).

Pterospermum xylocarpum (Gaertn.) Sant. & Wagh.

Syn: *Pterospermum heyneanum* Wall. *ex* W. & A.
Family: Sterculiaceae.

Part Used : Bark.

Utilization : Bark fibre used for making ropes by 'Jatapus', 'Khonds' and 'Savaras' tribals.

References : Rao & Henry (1996), Watt (1972).

Pueraria lobata (Willd.) Ohwi.

Syn: *Pueraria hirsuta* Kurz.; *P. thunbergiana* (Sieb. & Zucc.) Benth.
Family: Fabaceae.

English, Hindi and Sanskrit Names:

Eng. : Kudzu Hemp, Lobed kudzuvine;
Hindi : Jermei.
Sans. : Kudzu.

Distribution : Tropical Regions of India upto 2,000 m.

Parts Used : Leaves. Stalks.

Utilization : Leaves used as fodder. Stalks yield a fibre for cordage and fabrics for summer wear.

Active Constituents : Leaves contain robinin.

References : Agarwal (2003), Ambasta (1986), Purohit & Samant (1995), Rastogi & Mehrotra (1991), Saklani & Jain (1994).

Pueraria tuberosa DC.

Syn: *Hedysarum tuberosum* Roxb.

Family: Fabaceae.

English, Hindi, Sanskrit and Regional Names:

Eng. : Indian Kudzu;
Hindi : Bankumra, Bharda, Bilaikand, Sural, Tirra;
Sans. : Svadukanda, Vidari, Vidarika;
Beng : Shimia batraji;
Bom. : Dari;
Guj. : Khakarvel, Phagvelo, Vidarikand;
Kan. : Gumadigida;
Mal. : Palamutukku;
Mar. : Badra, Ggorbel, Pithana;
Ori. : Bhuiankakharu;
Pub. : Siali;
Raj. : Gorabel;
Tam. : Darigummadi, Niala -gumodi;
Tel. : Darigummadi.

Distribution : Distributed Throughout India: Kumaon, Hills of S. India. Orissa.

Part Used : Leaves.

Utilization : Leaves used as fodder for horses and cattle.

Active Constituents : Rich in crude proteins 23.8%, fats 0.7%, crude fibre 30.75, other carbohydrates 31.65, Ca 1.6% and P 0.2%.

References : Agarwal (2003), Ambasta (1986), Dwivedi (1993), Hooker (1879), Purohit & Samant (1995), Prajapati *et al.* (2003), Watt (1972).

Punica granatum Linn.

Syn: *Punica nana* Linn.
Family: Punicaceae.

Vern. Name: Daadu.

English, Hindi, Sanskrit and Regional Names:

Eng.	:	Pomegranate;
Hindi	:	Anar;
Sans.	:	Bijapura, Dadima, Granal;
Ass.	:	Dalim;
Beng	:	Dalimgachh;
Guj.	:	Dadam, Daramphal;
Kan.	:	Darim;
Mal.	:	Pumatalam, Uruyampalam;
Mar.	:	Dalimba;
Ori.	:	Dalim, Dalimba;
Pub.	:	Anar, Daan, Daru;
Tam.	:	Kalumal, Madasai;
Tel.	:	Dadimba, Dadimanu.

Distribution : Widely Cultivated for its Fruits Throughout India.

Part Used : Leaves.

Utilization : Leaves browsed by sheep and goats.

References : Agarwal (2003), Hooker (1879), Parrotta (2001), Purohit & Samant (1995), Rastogi & Mehrotra (1990).

Putranjiva roxburghii Wall.

Syn: *Nageia putranjiva* Roxb.; *Putranjiva sphaerocarpa & amblyocarpa* Muell. Arg.
Family: Euphorbiaceae.

Hindi and Regional Names:

Hindi	:	Joti, Juti, Patgia, Patji;
Beng	:	Jiaputa, Putranjiva;

Bom. : Jewan- Putr, Jiv- Putrak, Puta - Jan;
Mar. : Jewan putr, Puta jan;
Tam. : Irukolli, Karupalai.

Distribution : Cultivated Throughout Tropical India from the Lower Himalaya in Kumaon Eastwards & Southwards to Sri Lanka.

Part Used : Twigs.

Utilization : Twigs used as fodder.

References : Agarwal (2003), Dastur (1964), Dwivedi (1993), Oommachan (1977), Talbot (1976a), Watt (1972).

Pyrus foliolosa Wall.

Syn: *Pyrus ursina* Wall.; *Sorbus foliosa* Dcne.; *S. ursina* Wenzing in Linnaea.

Family: Rosaceae.

Regional Names:

Kum. : Huliya, Martili, Suliya;
Pub. : Huliya, Suliya.

Distribution : Temperate Himalaya (Kunawar to Sikkim): 3,000-4,000 m.

Part Used : Leaves.

Utilization : Considered to be good fodder for sheep and goats during July- August.

References : Purohit & Samant (1995), Watt (1972).

Pyrus pashia Buch.-Ham. *ex* Don.

Syn: *Pyrus nepalensis* Herb. Ham.; *P. variolosa* Wall.; *P. verruculosa* Bertol.; *Pyrus nepalensis* Herb. Ham.

Family: Rosaceae.

Vern. Name: Kaenth.

Hindi and Regional Names:

Hindi : Mehal;
Ass. : Chalthei, Soh- jhur, Soh - shur;

Kum.	:	Mehal, Mol;
Pub.	:	Batangi, Kaenth, Keitha, Kiat, Ku, Tang.

Distribution : Occurs in Himalaya, Assam, M. P., Khasia Hills, Manipur: 500- 2,500 m.

Part Used : Twigs.

Utilization : Twigs lopped for cattle fodder.

References : Agarwal (2003), Hooker (1879), Purohit & Samant (1995), Raju (1998), Rastogi & Mehrotra (1990), Sahni (1998), Singh (1982), Sood & Thakur (2004), Watt (1972).

Quercus dilatata Lindl. *ex* Royle

Syn: *Quercus floribunda* Lindl.; *Q. himalayana* Lindl.
Family: Fagaceae.

English, Hindi and Regional Names:

Eng.	:	Green Oak, Moru Oak;
Hindi	:	Chora, Moru;
Pub.	:	Chora, Kali ring, Moru.

Distribution : Temperate Himalaya from Kashmir to Nepal.

Parts Used : Leaves. Shoots.

Utilization : Leaves and shoots used for fodder for sheep and goats.

Active Constituents : Leaves contain 9.56% crude protein, 29.06% crude fibre, 51.75%, 1.6% Ca and 0.11% P.

References : Agarwal (2003), Ambasta (1986), Dwivedi (1993), Hooker (1890), Purohit & Samant (1995), Singh (1982), Singh (2000), Watt (1972).

Quercus glauca Thunb.

Syn: *Quercus annulata* Sm.; *Q. dentosa* Lindl.; *Q. laxiflora* Ham.; *Q. phullata* Ham.
Family: Fagaceae.

Vern. Name: Banni.

English, Hindi and Regional Names:

Eng.	:	Blue Japanase Oak, Green Oak;
Hindi	:	Pharonj;
Kum.	:	Phaniant;
Pub.	:	Banni, Bran, Banku, Imbri.

Distribution : Banks of streams, cool & shady ravines between 1,000-2,500 m. Khasia Hills upto 2,000 m.

Part Used : Leaves.

Utilization : Leaves used as fodder for cattle during summer and winter.

Active Constituents : Leaves contain total digestive nutrients 39.8%, digestive proteins 4.6% and digestive carbohydrates 33.8%.

References : Agarwal (2003), Ambasta (1986), Dwivedi (1993), Hooker (1890), Purohit & Samant (1995), Rastogi & Mehrotra (1991), Sharma (2004), Singh (1982), Watt (1972).

Quercus incana Roxb.

Syn: *Quercus dealbata* Wall.; *Q. lanata* Smith.; *Q. oblongata* DC.; *Q. leucotrichophora* A. Camus *ex* Bahadur.

Family: Fagaceae.

English, Hindi and Regional Names:

Eng.	:	Ban Oak, Grey Oak, Himalayan Oak
Hindi	:	Baj, Shindar;
Gar.	:	Phanat;
Kash.	:	Iri, Sila Supari;
Kum.	:	Ban, Bang;
Pub.	:	Ban, Dagunban, Rin, Ring, Vari.

Distribution : Kashmir, H.P., Nepal, Garhwal Regions 800- 2,000 m.

Parts Used : Leaves.

Utilization : Leaves used as fodder for cattle, sheep and goats.

Active Constituents : Leaves contain 10-11.42% crude protein, 31.34-32.1% crude fibre, quercetin and 3-ψ- disaccharide.

References : Agarwal (2003), Ambasta (1986), Dwivedi (1993), Hooker (1890), Rastogi & Mehrotra (1990), Sahni

(1998), Sharma (2003), Singh (1982, 2000), Singh & Kumar (2000), Watt (1972).

Quercus semecarpifolia Sm.

Syn: *Quercus cassura* Don.; *Q. obtusifolia* Don.
Family: Fagaceae.

Vern. Name: Kharsu.

English, Hindi and Regional Names:

Eng.	:	Brown oak;
Hindi	:	Kharssu;
Pub.	:	Bhanchar, Karshu, Karsu, Sanj.

Distribution : Temperate Himalaya. Manipur upto 3,600 m.

Part Used : Leaves.

Utilization : Leaves used as fodder for cattle, sheep and goats during winter.

Active Constituents : Leaves contain 4.45%, ash and 1.37% nitrogen.

References : Agarwal (2003), Hooker (1890), Nautiyal *et al.* (2003), Purohit & Samant (1995), Sahni (1998), Singh (1982, 2000), Watt (1972).

Randia spinosa Poir.

Syn: *Randia brandisii* Gamble.; *R. dumetorum* Poir.; *R. longispina* Wight & Arn.; *R. tomentosa* Wight & Arn., non Blume.; *Xeromphis spinosa* Keay.
Family: Rubiaceae.

English, Hindi, Sanskrit and Regional Names:

Eng.	:	Common Emetic Nut;
Hindi	:	Mainphal;
Sans.	:	Chhardanna, Nadana, Pindi, Shalayaka;
Guj.	:	Mindhal;
Kan.	:	Banegara, Mangri;
Mal.	:	Kara;
Mar.	:	Ghela, Mindhal;

Ori.	:	Patova;
Pub.	:	Arara;
Tam.	:	Mad Karai, Marukkallankay.

Distribution : Tropical & Sub-Tropical Parts of India including Sikkim & Peninsular India.

Part Used : Twigs.

Utilization : Plant browsed by goats and also lopped for fodder.

Active Constituents : Leaves contain 1.77% N and 2.51% Ca.

References : Agarwal (2003), Ambasta (1986), Chatterjee & Pakrashi (1991), Singh (1982).

Ranunculus diffusus DC.

Syn: *R. geranioides* Blume Bijdr.; *Ranunculus hydrocotyloides* Wall.; *R. mollis* Wall.; *R. nepaulensis* DC.; *R. obtectus* Wall.; *R. subpinnatus* W.& A.; *R. trilobatus* Don Prodr.; *R. vitifolius* Royle

Family: Ranunculaceae.

Distribution : Temperate Himalaya from the Chenab to Bhutan. 2,000-3,000 m & Higher Mountains of the W. Peninsula.

Part Used : Plant.

Utilization : Plant used as fodder.

Reference : Singh & Kumar (2000).

Raphia farinifera Hylander.

Syn: *Raphia pedunculata* Beauv.; *R. ruffia* Mart.

Family: Arecaceae.

English and Hindi Names:

Eng.	:	Raffia;
Hindi	:	Raffiatar.

Part Used : Leaves.

Utilization : Leaf fibre (Raffia fibre) used for binding and weaving mats, bags, ceremonial aprons, hats, curtains, etc.

References : Agarwal (2003), Ambasta (1986).

Raphia hookeri Mann & Wendl.

Family: Arecaceae.

English Name:

Eng. : Wine palm.

Part Used : Leaves.

Utilization : Yields a strong fibre, Pissava which is used for brooms, mats, baskets and cordage.

References : Agarwal (2003), Ambasta (1986).

Raphidophora glauca (Wall.) Sch.

Family: Araceae.

Distribution : Climbers in Tropical & Sub -Tropical Himalaya.

Part Used : Aerial Parts.

Utilization : Aerial parts used as fodder.

Reference : Saklani & Jain (1994).

Rhamnus pentapomica Parker.

Syn: *Rhamnus persica* M. Laws. in part, non Boiss.

Family: Rhamnaceae.

Hindi and Regional Names:

Hindi : Chila;

Gar. : Chhetula;

Kum. : Charyula;

Pub. : Kukai, Kander, Nikki, Wurak.

Distribution : N. W. Himalaya from Jhelum to Sarda 800-1,000 m.

Part Used : Leaves.

Utilization : Leaves browsed by sheep, goats and camels.

References : Ambasta (1986), Singh (1982).

Rhamnus purpurea Edgew.
Family: Rhamnaceae.

Regional Names:

Gar. : Bakauro;
Kum. : Payan;
Pub. : Bat Sinjal, Kri, Karn, Rari, Tadru.

Distribution : W. Himalaya from Marri to Kumaon.

Part Used : Twigs.

Utilization : Twigs lopped for fodder at times of scarcity.

References : Dwivedi (1993), Raju (1998), Watt (1972).

Rhamnus triqueter (Wall.) Brandis.
Syn: *R. triqueter* Wall.
Family: Rhamnaceae.

Vern. Name: Katheru.

Regional Names:

Gar. : Gaunth;
Pub. : Fagora, Gardhan, Gudlei.

Distribution : Kashmir to Nepal. Deccan Region.

Part Used : Twigs.

Utilization : Twigs used as cattle fodder.

References : Agarwal (2003), Ambasta (1986), Dwivedi (1993), Hooker (1872), Watt (1972).

Rhamnus virgatus Roxb.
Syn: *Rhamnus dahuricus & R. davurica* Pall.
Family: Rhamnaceae.

Hindi Name:

Hindi : Chadua.

Distribution : Throughout Himalaya. C. India & Nilgiris.

Part Used : Twigs.

Utilization : Fed to all ruminants, especially for sheep and goats.

References : Agarwal (2003), Dwivedi (1993), Purohit & Samant (1995), Singh (1982).

Rhizophora apiculata Bl.

Syn: *R. conjugata* auct. (non L.)
Family: Rhizophoraceae.

Distribution : Tidal Marshes of Coastal Regions of India, Andaman & Nicobar Islands.

Part Used : Leaves.

Utilization : Leaves used for fodder purposes.

Active Constituents : Triterpenoid isolated from the leaves.

References : Das *et al.* (1996), Rastogi & Mehrotra (1995a).

Rhodiola imbricate Edgew.

Family: Crassulaceae.

Vern. Name: Shalo.

English Name:

Eng. : Rose Root.

Distribution : Himalaya from Kashmir to Kumaon: 3,600- 5,100 m.

Part Used : Aerial parts.

Utilization : Grazed by yak and sheep.

References : Bennet *et al.* (1991), Singh *et al.* (1996).

Rhus cotinus Linn.

Family: Anacardiaceae.

Hindi Names:

Hindi : Tung, Tunga.

Distribution : N. W. Himalaya from Sarda River to East.

Part Used : Leaves.

Utilization : Leaves used as fodder.

References : Dwivedi (1993), Singh (1982).

Rhus mysurensis G. Don.

Family: Anacardiaceae.

Hindi and Regional Names:

Hindi : Dashar;

Kan. : Sabale;

Raj. : Dasani, Dasarni;

Tel. : Sitha.

Distribution : N. W. India, extending Southwards into Deccan Peninsula.

Part Used : Leaves.

Utilization : Considered as nutritive fodder for goats.

Active Constituents : Leaves comprise of quercetin, aempferol, myricetin and rhamnoides.

References : Hooker (1879), Rastogi & Mehrotra (1995b), Singh & Pandey (1998).

Rhus parviflora Roxb.

Family: Anacardiaceae.

Hindi Names:

Hindi : Ninas, Tumra.

Distribution : N.W. Himalaya from Satluj to Nepal: 600-1,500 m. Panchmari.

Part Used : Leaves.

Utilization : Leaves valued as fodder.

References : Dwivedi (1993), Singh (1982).

Rhus sinuta Thunb.

Syn: *R. mysurensis* Heyne.
Family: Anacardiaceae.

Regional Names:

Mar. : Amooni
Raj. : Assani, Dannonia.

Distribution : Rajasthan, Dry Areas of Punjab. Haryana.

Part Used : Leaves.

Utilization : Leaves used for fodder purposes.

References : Dwivedi (1993), Singh (1982).

Ribes glaciale Wall.

Syn: *Ribes acuminatum* Wall.
Family: Grossulariaceae.

Regional Name:

Pub. : Kukulia, Mangle.

Distribution : Temperate & Alpine Himalaya: Bhutan to Kashmir.

Part Used : Leaves.

Utilization : Leaves fed to sheep and goats during October-November.

References : Hooker (1879), Purohit & Samant (1995), Roy *et al.* (1998), Watt (1972).

Ribes rubrum Linn.

Syn: *Ribes himalayense* Dcne.
Family: Grossulariaceae.

English, Hindi and Regional Names:

Eng. : Red currant;
Hindi : Kinkolia;
Pub. : Ans, Dak, Kagh dak, Nabar, Nabre.

Distribution : Himalaya from Kashmir to Kumaon upto 3,500 m.

Part Used : Leaves.

Utilization : Leaves browsed by sheep and goats during September-October.

Active Constituents : Fresh leaves contain hydrocyanic acid.

References : Agarwal (2003), Hooker (1879), Purohit & Samant (1995), Watt (1972).

Robinia pseudo-acacia Linn.

Family: Fabaceae.

Vern. Name: Ravinia.

English and Regional Names:

Eng. : Black locust, False- Acacia, Robinia;
Kull. : Kikar.

Distribution : Throughout India, N. W. Himalaya 1,500-2,000 m.

Part Used : Leaves.

Utilization : Leaves used as fodder.

Active Constituents : Leaves contain crude proteins 25.5%, crude fibre 17.2%, mineral matter 7.55, Ca- 1.55 and P- 0.32 %.

References : Agarwal (2003), Ambasta (1986), Dwivedi (1993), Rastogi & Mehrotra (1993), Sharma (2004), Singh (1982, 2000), Singh & Kumar (2000), Watt (1972).

Roccella montagnei Bel.

Family: Roccellaceae.

Distribution : Karnataka & Coastal Regions of Orissa, Andhra Pradesh & Pondicherry.

Part Used : Leaves.

Utilization : Used as good fodder.

Active Constituents : Rich in β-carotene.

References : Ambasta (1986), Rastogi & Mehrotra (1990).

Rosa brunonii Lindl.

Syn: *Rosa moschata* Herrm.

Family: Rosaceae.

English, Hindi, Sanskrit and Regional Names:

Eng. : Himalayan Musk Rose;
Hindi : Karer, Kuji, Kunja;
Sans. : Sewati;
Pub. : Tarni.

Distribution : Kashmir to Bhutan: 600-2,000 m.

Part Used : Leaves.

Utilization : Used as a fodder plant for cattle.

References : Ambasta (1986), Dwivedi (1993), Sharma (2003), Singh & Kumar (2000).

Rosa glandulosa Bell.

Family: Rosaceae.

Distribution : Distributed between 3,000- 4,000 m.

Part Used : Leaves.

Utilization : Leaves browsed by sheep, goats and cattle.

Reference : Purohit & Samant (1995).

Rosa macrophylla Lindl.

Syn: *Rosa hoffmeisteri & guliemi waldemarii* Klotzsch.; *R . hookeriana* Bertol.

Family: Rosaceae.

Hindi and Regional Names:

Hindi : Ban- gulab;
Pub. : Akhiari, Beri, Shingari, Tunbi, Yal.

Distribution : Temperate Himalaya from the Indus to Sikkim.

Part Used : Leaves.

Utilization : Leaves browsed by sheep and goats throughout the year.

References : Hooker (1879), Purohit & Samant (1995), Watt (1972).

Rosa sericea Lindl.

Syn: *Rosa tetrapetala* Royle.; *R. wallichii* Trattin.
Family: Rosaceae.

English and Regional Names:

Eng. : Thorny rose;
Kum. : Chapala.

Distribution : Common in Temperate Himalaya from Kumaon to Sikkim: 2,100-4,200 m.

Part Used : Leaves.

Utilization : Browsed by sheep and goats throughout the year.

References : Dwivedi (1993), Hooker (1879), Purohit & Samant (1995), Sharma (2003), Watt (1972).

Rosa webbiana Wall. *ex* Royle

Syn: *Rosa pimpinellifolia* Herb.; *R. unguicularis* Bertal.
Family: Rosaceae.

Vern. Name: Shaybala.

Regional Names:

Pub. : Kugina, Sikanda.

Distribution : Drier Inner Himalaya from Kashmir to W. Tibet: 1,500-4,000 m.

Part Used : Leaves.

Utilization : Leaves browsed by sheep and goats throughout the year.

References : Hooker (1879), Purohit & Samant (1995), Watt (1972).

Rotala densiflora (Roth.) Koehne.

Family: Lythraceae.

Part Used : Plant.

Utilization : Plant used as fodder.

Reference : Islam (2002).

Rotala rotundifolia (Buch -Ham.) Koehne.

Family: Lythraceae.

Part Used : Leaves.

Utilization : Plant used as fodder.

Reference : Islam (2002).

Rubia tinctorum Linn.

Family: Rubiaceae.

English, Hindi and Regional Names:

Eng. : Alizari, European Madder;
Hindi : Bacho;
Pub. : Bacho.

Distribution : Distributed in Kashmir & Sindh.

Part Used : Plant.

Utilization : Yields green fodder for cattle (especially for camel).

References : Agarwal (2003), Ambasta (1986), Hooker (1882), Watt (1972).

Rubus ellipticus Smith.

Syn: *Rubus affinis* Madden.; *R. flavus* Ham.; *R. gowry- phul* Roxb.; *R. gowreephul* Roxb.; *R. hirsutus* Roxb.; *R. rotundifolius* Wall.; *R. sessilifolius* Miq.; *R. wallichianus* W. & A.

Family: Rosaceae.

Vern. Name: Aakhe, Kharyadi.

English, Hindi and Regional Names:

Eng. : Himalayan Yellow Raspberry;
Hindi : Hinsalu, Hisara, Hisol;
Ass. : Jote lupoka;
Gar. : Kashyem;
Kum. : Kashyem;
Pub. : Akhi, Ankri, Knachi.

Distribution : Temperate to Sub-Tropical Himalaya from Sirmur (2,000 m) to Sikkim (1,000-2,000 m). Bhutan.

Part Used : Leaves.

Utilization : Leaves browsed by sheep and goats during winter.

Active Constituents : Deoxysericoside isolated from the leaves.

References : Dwivedi (1993), Hooker (1879), Purohit & Samant (1995), Rastogi & Mehrotra (1995b), Watt (1972).

Rubus niveus Thunb.

Syn: *Rubus albexens* Roxb.;
R. gracilis Roxb.; *R. pedunculosus* D. Don.
Family: Rosaceae.

Vern. Name: Kala- Akha.

English, Hindi and Regional Names:

Eng. : Mysore raspberry;
Hindi : Gowriphal, Kali anchhi;
Kash. : Kandiare, Kharmuch;
Kum. : Kala Hinsal;
Pub. : Kalga.

Distribution : Temperate Himalaya from Kashmir to Bhutan.

Part Used : Leaves.

Utilization : Browsed by sheep and goats during rainy season.

References : Agarwal (2003), Hooker (1879), Parmar & Kaushal (1982), Purohit & Samant (1995), Watt (1972).

Rubus rosaefolius Smith.

Syn: *Rubus asper* Don.; *R. pinnatus* Willd.;
R . paniculatus Clarke in Journ.
Family: Rosaceae.

Distribution : Temperate Himalaya from Kumaon to Sikkim. Khasia Hills.

Part Used : Leaves.

Utilization : Relished by sheep, goats and cattle during rainy season.

References : Hooker (1879), Purohit & Samant (1995), Watt (1972).

Rumex hastatus D. Don.

Family: Polygonaceae.

Regional Names:

Kan. : Almora, Chalmor;
Pub. : Amla, Amlora, Ami, Khattimal.

Distribution : Common in Himalaya from Kashmir to Kumaon.

Part Used : Leaves.

Utilization : Leaves browsed by sheep and goats.

References : Hooker (1890), Purohit & Samant (1995), Raju (1998), Watt (1972).

Sabal palmetto Lodd. *ex* Roem. & Schult. (Pl. 8A)

Family: Arecaceae.

English and Hindi Names:

Eng. : Cabbage Palm, Sabal Palm;
Hindi : Ghobitar.

Distribution : Grown in Indian Gardens.

Part Used : Petioles.

Utilization : Palmetto fibre valued for different types of brushes and cordages.

References : Agarwal (2003), Ambasta (1986).

Saccharum officinarum Linn.

Syn: *Saccharum sinense* Roxb.

Family: Poaceae.

Vern. Name: Ganna.

English, Hindi, Sanskrit and Regional Names:

Eng. : Noble cane, Sugarcane;
Hindi : Ganna, Ikh, Kumad;
Sans. : Ikshu, Khanda, Sarkara;
Ass. : Pura- khuiyar;
Beng : Ak, Kushiar, Puri;

Bom.	:	Gol, Serdi;
Guj.	:	Naisakar, Sheradi, Uns;
Kan.	:	Basari mara, Khabbu;
Mal.	:	Karimbu;
Mar.	:	Aos, Kabho;
Ori.	:	Aku, Gudodaru;
Pub.	:	Ghanna, Khaud, Paunda;
Tam.	:	Poovam karumbu;
Tel.	:	Tellacheraku.

Distribution : Throughout India, Widely Cultivated upto 1,000 m.

Part Used : Leaf.

Utilization : Fibre used for making paper, ropes, rough cordage, and tying the logs into rafts.

References : Agarwal (2003), Ambasta (1986), Duthie (1888), Oommachan (1977), Watt (1972).

Saccharum spontaneum Linn.

Syn: *Imperata spontanea* Beauv.;
Saccharum canaliculatum Roxb.; *S. semidecumbens* Roxb.
Family: Poaceae.

English, Hindi, Sanskrit and Regional Names:

Eng.	:	Kans Grass;
Hindi	:	Kans, Kansi;
Sans.	:	Kasa, Khasa;
Guj.	:	Kans, Kanado;
Kum.	:	Jasha, Jhansh;
Mal.	:	Kusha, Nannana;
Mar.	:	Kagara;
Ori.	:	Chattiagaso, Inkora, Kaso, Khhodi;
Pub.	:	Kahi;
Raj.	:	Kash, Kashi;
Tam.	:	Koasangan, Nanal, Nanmugappul;
Tel.	:	Rellu.

Distribution : Throughout the Warmer Parts of India & Sri Lanka ascending to 2,000 m in the Himalaya.

Parts Used : Plant. Leaf. Stem.

Utilization : Used for feeding the buffaloes and cattle.

References : Dastur (1964), Dwivedi (1993), Duthie (1888), Hooker (1897), Islam (2002), Oommachan (1977), Parrotta (2001), Raju (1998), Varghese (1996), Watt (1972).

Sacciolepis interrupta Retz.

Syn: *Panicum interruptum* Willd.

Family: Poaceae.

Regional Names:

Beng : Nardula;
Bom. : Pakalia;
Kan. : Modike hullu;
Tam. : Tandan pillu;
Tel. : Wolam.

Distribution : Assam. Bengal. Bihar. Orissa. Maharashtra. Tamil Nadu upto 900 m.

Part Used : Plant.

Utilization : A good fodder grass relished by elephants.

References : Ambasta (1986), Bennet *et al.* (1991), Das *et al.* (1996).

Sageretia oppositifolia Brongn.

Syn: *Rhamnus filiformis* Wall.; *R. parvifolius* Roem.& Schult.; *R. reggynus* Don.; *Sageretia filiformis* G. Don.; *S. oppositifolia* (Wall.)3 Brogn.; *S. parviflora* (Roem. & Schult.) G. Don.; *Zizyphus oppositifolia* Wall.

Family: Rhamnaceae.

Vern. Name: Bakkar Gha.

Regional Names:

Kash. : Gidardak, Kanak;
Pub. : Drange, Girthan.

Distribution : Found in the N.W. Himalaya from Peshawar to Nepal. S. India from Konkan Southwards.

Part Used : Leaves.

Utilization : Leaves used for fodder purposes.

References : Sharma (2004), Watt (1972).

Salix acmophylla Boiss.

Syn: *Salix glauca & S. acmophylla* Anderss.; *S. ocandra* Del.

Family: Salicaceae.

Hindi and Regional Names:

Hindi : Bada, Bed;
Bom. : Budha;
Kum. : Godh- bhain;
Pub. : Bisu, Jalmala.

Distribution : Throughout Sub- Himalayan Tract from U.P. to Punjab.

Part Used : Leaves.

Utilization : Leaves lopped for fodder and fed to sheep and goats.

References : Agarwal (2003), Ambasta (1986), Purohit & Samant (1995), Singh (1982), Watt (1972).

Salix alba Linn.

Family: Salicaceae.

English, Hindi and Regional Names:

Eng. : European Willow, White Willow;
Hindi : Vivir;
Kash. : Vivir;
Pub. : Bis, Bhushan, Madnu.

Distribution : W. Himalaya upto 2,500 m from Kashmir to U. P.

Part Used : Twigs.

Utilization : Tender twigs lopped for fodder.

Active Constituents : Glucoside- alboside isolated from the leaves.

References : Agarwal (2003), Ambasta (1986), Rastogi & Mehrotra (1990), Singh (1982), Singh *et al.* (1996).

Salix daphnoides Vill.

Syn: *Salix acutifolia* Wall.
Family: Salicaceae.

English, Hindi and Regional Names:

Eng.	:	Violet Willow;
Hindi	:	Betsa;
Kash.	:	Yur;
Pub.	:	Bedi, Beli, Betsa, Bushan, Shun, Thail.

Distribution	:	Himalaya. Kumaon to Westwards.
Parts Used	:	Leaves. Branches. Bark.
Utilization	:	Branches, leaves and bark of the larger branches used as cattle fodder.
Active Constituents	:	Leaves contain apigenin, cinaroside, cosmosiin, luteolin, quercetin and rutin.
References	:	Agarwal (2003), Ambasta (1986), Dwivedi (1993), Rastogi & Mehrotra (1995a), Singh (1982), Watt (1972).

Salix tetrasperma Roxb.

Syn: *Salix disperma* Don.; *S. horsefieldiana* Miq.; *S. nilagirica* Miq.; *S. tetrasperma â- nilagirica* Anderss.
Family: Salicaceae.

Vern. Name: Buins.

English, Sanskrit and Regional Names:

Eng.	:	Indian Willows;
Sans.	:	Varuna;
Ass.	:	Bhi, Yeh;
Beng	:	Boishaki, Punijami;
Bom.	:	Bacha, Wallung;
Kan.	:	Baricho, Niravanji;
Mal.	:	Attupala, Vanji;
Mar.	:	Boch, Bitasa , Wallunj;
Ori.	:	Baisi, Barija;
Pub.	:	Bis, Bain. Beis, Bitsa;
Tam.	:	Vanji;
Tel.	:	Eti Palla.

Distribution : Throughout India from the Punjab Eastwards to Mishmi, Assam & Manipur.

Part Used : Leaves.

Utilization : Leaves used as cattle fodder.

Active Constituents : Leaves contain 2.71% Ca, 45.06% carbon and 2.07% nitrogen.

References : Agarwal (2003), Ambasta (1986), Dwivedi (1993), Hooker (1890), Rastogi & Mehrotra (1993), Sharma (2004), Singh (1882), Singh & Kumar (2000), Watt (1972).

Salix viminalis L.

Family: Salicaceae.

English, Hindi and Regional Names:

Eng. : Basket Willlow;
Hindi : Bibsu;
Pub. : Bibsu.

Distribution : Kashmir to Sikkim upto 3,000 m.

Part Used : Twigs.

Utilization : Twigs good for cordage.

References : Agarwal (2003), Ambasta (1986), Watt (1972).

Salix wallichiana Anders.

Family: Salicaceae.

Regional Name:

Pub. : Bwir.

Distribution : Temperate Himalaya (Kashmir to Bhutan). Punjab Plains.

Part Used : Leaves.

Utilization : Fed to sheep and goats.

References : Dwivedi (1993), Hooker (1890), Purohit & Samant (1995), Singh (1982), Watt (1972).

Salmalia insignis (Wall.) Schott. & Endl.

Family: Bombacaceae.

Distribution : Growing in Assam, Andamans, W. Coast to Kerala.

Part Used : Fruits.

Utilization : Fruits yield cotton silk fibre for stuffing.

Reference : Agarwal (2003).

Salsola baryosma Dandy.

Syn: *Salsola foetida* Delile *ex* Spreng.

Family: Chenopodiaceae.

English and Regional Names:

Eng. : Saltwort;
Pub. : Gora lane, Lana shora;
Tel. : Ellakura.

Distribution : Found in Punjab. Upper Gangetic Plains. Gujarat. Rajasthan.

Part Used : Twigs.

Utilization : Used as camel fodder.

Active Constituents : These are sitosterol (29.9), stigmasterol (36.9) and campesterol (5.2%).

References : Ambasta (1986), Bennet *et al.* (1991), Dwivedi (1993), Kalpash *et al.* (2004), Rastogi & Mehrotra (1995a).

Salvadora olooides Decne

Syn: *Salvadora indica* Royle.; *S. persica* T. And.; *S. stocksii* Wight.

Family: Salvadoraceae.

Hindi, Sanskrit and Regional Names:

Hindi : Bahapilu, Chootapilu, Jhal, Pilu;
Sans. : Dhani, Gudaphala, Lagupilu, Pilu, Pluka, Shali;
Bom. : Kakhan, Kankhina;
Pub. : Konkan, Mitijar, Pilava, Pilu;
Tam. : Kalawa, Karkol, Kohu, Ughat.

Distribution : Arid Regions of Punjab. W. India.

Part Used : Seed Cake.

Utilization : Seed cake fed to cattle to enhance lactation.

Active Constituents : It contains 12.4% moisture, 26.66% crude proteins, 22.0% carbohydrates and 1.68% P.

References : Agarwal (2003), Ambasta (1986), Dwivedi (1993), Hooker (1882), Kalpash *et al.* (2004), Singh (1982), Watt (1972).

Samanea saman Merrill

Syn: *Enterolobium saman* Prain.; *Pithecolobium (Pithecolobium) saman* Benth.
Family: Fabaceae.

Part Used : Leaves.

Utilization : Green and tender leaves relished by cattle.

References : Ambasta (1986), Oommachan (1977).

Sansevieria lyacinthoides (L.) Druce.

Family: Liliaceae.

English Name:

Eng. : Ceylon Bowstring Hemp.

Part Used : Leaf.

Utilization : Leaf fibre good for making commercial mats.

Reference : Agarwal (2003).

Sansevieria roxburghina Schult. f.

Syn: *Sanseveria zeylanica* Roxb.
Family: Liliaceae.

English, Hindi, Sanskrit and Regional Names:

Eng. : Indian bowstring hemp;
Hindi : Marul, Marva;
Sans. : Murva, Maurvi;
Beng : Gorachaka, Murba;

Ori.	:	Murga;
Tam.	:	Marul, Mottamanji;
Tel.	:	Niyanda sagal.

Distribution : Coromandal Coast of India.

Part Used : Leaves.

Utilization : Fibre used for making bow-string, cordage, mats and fine cloths.

References : Agarwal (2003), Ambasta (1986), Geetha *et al.* (1996), Jain (1991).

Sanseveria trifasciata Prain

Family: Liliaceae.

English and Hindi Names:

Eng.	:	African bowstrins hemp;
Hindi	:	African Sun.

Distribution : Cultivated in the Gardens & in the Eastern Coastal Belts.

Part Used : Plant.

Utilization : Fibre used for fishing lines, nets, bow-string, cordage, ropes, mats and fabrics, etc.

References : Agarwal (2003), Ambasta (1986).

Sapindus mukorossi Gaertn.

Syn: *Sapindus abruptus* Lour.; *S. detergents* Roxb.; *S. trifoliatus* Linn.

Family: Sapindaceae.

Vern. Name: Ritha.

English, Hindi, Sanskrit and Regional Names:

Eng.	:	Soap nut tree;
Hindi	:	Ritha;
Sans.	:	Aristha , Phenila;
Ass.	:	Haithaguti;
Beng	:	Aritha;

Bom.	:	Kanmar, Ritha;
Kum.	:	Ritha;
Mar.	:	Ita;
Ori.	:	Ita;
Pub.	:	Thali;
Tel.	:	Kunkudu.

Distribution : Cultivated Throughout N.W. India, Bengal, Kumaon, Assam & upto 1,600 m in the Himalaya.

Part Used : Leaves.

Utilization : Leaves valued as cattle fodder.

Active Constituents : Leaves contain 1.79% N & 1.99% CaO.

References : Hooker (1872), Kirtikar & Basu (1935), Raju (1998), Singh (1982), Watt (1972).

Sapium insigne Trimen.

Syn: *Carambium insigne* Kurz.; *Falconeria insignes* Royle; *F. wallichiana* Royle.
Family: Euphorbiaceae.

Vern. Name: Balodar.

Hindi and Regional Names:

Hindi	:	Kendwa;
Ass.	:	Mah kalo;
Bom.	:	Duella;
Gar.	:	Khinna, Kindra, Khinna;
Kan.	:	Kurda, Nanaka;
Mar.	:	Dudla, Hure;
Pub.	:	Bilodhar, Biloja, Dulda, Karalla;
Tam.	:	Garphashola,Tillai;
Tel.	:	Garpushula, Kadura.

Distribution : H.P. to Assam (Rocky Areas) & Peninsular India.

Part Used : Twigs.

Utilization : Twigs lopped for fodder.

References : Agarwal (2003), Ambasta (1986), Hooker (1890), Singh (1982).

Saraca asoca (Roxb.) Dewilde.

Syn: *Jonesia asoca* Roxb.; *J. pinnata* Willd.; *Saraca arborescens* Burm.; *S. indica* auct. non L.; *S. zollingeriana* Miq.

Family: Fabaceae.

Vern. Name: Asoka.

English, Hindi, Sanskrit and Regional Names:

Eng. : Asoka;
Hindi : Ashok;
Sans. : Karnapuraka;
Beng : Ashok;
Bom. : Ashok, Jasundi;
Guj. : Asok;
Kan. : Asasunkar;
Mal. : Ashok;
Mar. : Ashok;
Ori. : Ashok;
Pub. : Ashok;
Tam. : Asogam;
Tel. : Asoka chettu.

Distribution : C. & E. Himalaya. E. Bengal. S. India.

Part Used : Pods.

Utilization : Pods valued as a fodder to livestock.

Active Constituents : Contain gallic acid, leucocyanidin, sitosterol, flavonoides, kaempferol and apigenin.

References : Khare (2004), Parrotta (2001), Raju (1998), Rastogi & Mehrotra (1995b), Watt (1972).

Sarcochlamys pulcherrima Gaudich.

Syn: *Urtica pulcherrima* Roxb.

Family: Urticaceae.

English, Hindi and Regional Names:

Eng. : Dogal tree;

Hindi : Mesaki;
Ass. : Mesaki.

Distribution : Assam, Khasia Hills, Silhet and Cochar.

Part Used : Bark.

Utilization : Bark fibre used as a subsititute for 'ramie', for cordage and ropes.

References : Agarwal (2003), Ambasta (1986), Hooker (1890), Watt (1972).

Sarcosperma arboreum Hook. f.

Syn: *Sideroxylon arboreum* Ham.
Family: Sapotaceae.

Distribution : E. Himalaya (Sikkim). Khasia Hills.

Part Used : Leaves.

Utilization : Leaves lopped for fodder.

References : Agarwal (2003), Ambasta (1986), Singh (1982), Watt (1972).

Sarcostemma secamone (Linn.) Bennet.

Syn: *Oxystelma esculentum* (L. f.) R. Br. *ex* Schult *O. secamone* (L.) Karst.
Family: Asclepiadaceae.

Vern. Name : Dudhya.

English, Sanskrit and Regional Names:

Eng. : Rosy milk weed;
Sans. : Dughhiko, Dudhialata;
Tam. : Usipplai.

Distribution : Throughout the Plains, Lower Hills and Paddy Fields of India.

Part Used : Stem.

Utilization : Stem fibre used for making ropes and fishing nets.

References : Chatterjee & Pakrashi (1991), Khan & Khanum (2005), Sur & Halder (2002).

Saurauia napaulensis DC.

Syn: *Saurauria paniculata* Wall.; *Ternstroemia racemosa* Don.; *Zanthozylum serra* Turce.

Family: Saurauiaceae.

Vern. Name: Gogna.

Hindi and Regional Names:

Hindi	:	Goganda, Gogina, Pangaru;
Kum.	:	Gugna.

Distribution	:	W. & E. Himalaya & S. India.
Part Used	:	Leaves.
Utilization	:	Leaves lopped for fodder. Also fed in conjuction with wheat straw for positive balance of Ca in buffaloes.
Active Constituents	:	Leaves contain 12.24% crude proteins, 18.40% fibre, 2.71% Ca, 0.24% P, β-sitosterol, arborinie and vannilic acid.
References	:	Agarwal (2003), Ambasta (1986), Dwivedi (1993), Purohit & Samant (1995), Rastogi & Mehrotra (1995b), Saklani & Jain (1994), Sharma (2004), Singh (1982).

Saurauria roxburghii Wall.

Family: Saurauriaceae.

Regional Name:

Ass.	:	Bon Poshala.

Distribution	:	N. E. India. Khasia Hills. Kashmir. Gujarat.
Part Used	:	Leaves.
Utilization	:	Leaves used as fodder.
References	:	Roy *et al.* (1998), Saklani & Jain (1994), Singh (1982).

Saussurea costus (Falc.) Lipsch.

Syn: *Saussurea lappa* (Decne.) Sch. Bip.

Family: Asteraceae.

Vern. Name: Kuth.

Distribution : W. Himalaya all along in the Temperate conditions.

Part Used : Dried Stem.

Utilization : Dried stem given as fodder in Lahaul (H.P.)

Reference : Kaul (1997).

Scaphium scaphigerum (G. Don) Guibourt

Syn: *Scaphium wallichii* Schoyl & Endl.; *Sterculia scaphigera* Wall.

Family: Sterculiaceae.

Hindi Name:

Hindi : Jelly kat.

Distribution : Introduced in Indian Botanic Gardens, Howrah.

Part Used : Inner Bark.

Utilization : Inner bark yields a fibre cordage.

References : Agarwal (2003), Ambasta (1986).

Schefflera impressa (C. B. Clarke) Harms

Syn: *Heptapleurum impressum* C. B.Clarke.

Family: Araliaceae.

Distribution : Tropical & Sub-Tropical India. Himalaya (Kashmir to Bhutan & N. E. India): 1,800- 3,300 m.

Part Used : Leaves.

Utilization : Leaves afford a good fodder.

References : Ambasta (1986), Saklani & Jain (1994), Singh (1982).

Schizachyrium semiberbi Nees

Syn: *Andropogon semiberbis* Kunth.

Family: Poaceae.

Hindi Name:

Hindi : Nalle hullu.

Distribution : Tropical Himalaya to Assam. M.P. S. India.

Utilization : Good fodder for ruminants.

References : Agarwal (2003), Ambasta (1986).

Schleichera oleosa (Lour.) Oken.

Syn: *Schleichera trijuga* Willd. & Klein.

Family: Sapindaceae.

English, Hindi, Sanskrit and Regional Names:

Eng.	:	Ceylon Oak, Lac-tree, Macassar oil tree;
Hindi	:	Kusum;
Sans.	:	Kosamra;
Beng	:	Kosano, Kusum;
Guj.	:	Kusum, Kosumb;
Kan.	:	Chakota, Jendala, Kendalle;
Mal.	:	Puvam;
Mar.	:	Kusumb, Koon, Kohan, Peduman;
Ori.	:	Kusamo, Susam, Swad, Kussum;
Pub.	:	Sama, Jamoa, Gausam, Kussam;
Tam.	:	Puvathipuram, Pulachi, Karanchi, Kula, Pumaratha;
Tel.	:	Botanga, Busi, Kodali- pulusanarajati.

Distribution	:	Kashmir to Garo Hills. Bengal. Peninsular India upto 900 m.
Part Used	:	Young Shoot.
Utilization	:	Young shoots lopped for fodder.
Active Constituents	:	Leaves contain C_{26-35} alkanes, C_{26-34} alcohols, stigmasterol, b-sitosterol, campasterol, cholesterol, glucose, fructose, tartaric acid and oxalic acid.
References	:	Agarwal (2003), Ambasta (1986), Dwivedi (1993), Prajapati *et al.* (2003), Raju (1998), Rastogi & Mehrotra (1991), Singh (1982), Upadhye & Kumbhojkar (1998).

Scirpus grossus Linn. f.

Syn: *Scirpus kysoor* Roxb.

Family: Cyperaceae.

Sanskrit and Regional Names:

Hindi	:	Kasuru, Kesar;
Sans.	:	Kaseruka;
Beng	:	Kasuru, Kesar;
Guj.	:	Gundaro;

Ori.	:	Santara;
Pub.	:	Dila, Kaseru;
Tel.	:	Gundatungaddi.

Distribution : Practically Throughout Indian Swamps.

Part Used : Rhizomes.

Utilization : Green plant rhizomes used as fodder.

References : Agarwal (2003), Ambasta (1986), Das *et al.* (1996).

Scirpus tuberosus Desf.

Syn: *Scirpus maritimus* C. B. Clarke, non Linn.

Family: Cyperaceae.

English, Hindi and Regional Names:

Eng.	:	Sea club-Rush;
Hindi	:	Maruk;
Beng	:	Mahat-Kesar;
Pub.	:	Dila, Murak;
Tel.	:	Gurrapu sakatunga.

Distribution : Marshy Areas of India.

Part Used : Plant.

Utilization : Fresh plant useful as good forage.

References : Agarwal (2003), Ambasta (1986).

Sclerostachya fusca (Roxb.) A. Camus

Syn: *Saccharum fuscum* Roxb.

Family: Poaceae.

Hindi and Regional Names:

Hindi	:	Khilut, Retwa, Tilluk;
Beng	:	Khuree, Pata-Khuru;
Tel.	:	Kandu-Rellugaddi.

Distribution : Kashmir to U.P. Khasia Hills. Assam.

Part Used : Leaf-Sheaths.

Utilization : Leaf sheaths provide a fibre for cordage.

References : Agarwal (2003), Ambasta (1986).

Scoparia dulcis Linn.

Family: Scorphulariaceae.

English and Hindi Names:

Eng. : Sweet Broomweed, Tooth brush;

Hindi : Chini Buti.

Distribution : Commonly in Many Parts of India.

Parts Used : Aerial Parts. Leaves.

Utilization : Plant used as cattle fodder.

Active Constituents : These are dulcital (aerial parts) scutellasein, its 7-0-methyl ether and 7-0-b-D glucuronide (leaves).

References : Agarwal (2003), Ambasta (1986), Rastogi & Mehrotra (1990, 1991).

Sechium edule Sw.

Family: Cucurbitaceae.

Vern. Name: Lanku.

English, Hindi and Regional Names:

Eng. : Chaw-Chow, Chayote, Choco;
Hindi : Squash kadu;
Tam. : Seema kattirikkai.

Distribution : Sub-Tropical Regions of India. Nilgiris.

Part Used : Woody Stem.

Utilization : Woody stem yields a fibre.

References : Agarwal (2003), Ambasta (1986), Watt (1972).

Sehima nervasum (Rottl.) Stapf.

Syn: *Ischaemum laxum* (R. Br.) Hook. f .

Family: Poaceae.

Hindi and Regional Names:

Hindi : Sedwa;
Kan. : Chikka Sali hullu, Karikodda, Nalai hullu;
Mar. : Shara, Sheda;

M.P.	:	Sain, Sainar, Sen;
Raj.	:	Gandhi, Gundra, Sagen;
Tam.	:	Kuraitti, Nendra pul;
Tel.	:	Nandro gaddi, Telagadi;
U. P.	:	Sain, Sairan, Senda.

Distribution : Throughout India.

Part Used : Grass Blades.

Utilization : Considered as one of the best palatable and nutritious fodder grasses of hills. It can be made into hay and silage.

References : Agarwal (2003), Ambasta (1986).

Sehima sulcatum (Hack.) A. Camus

Syn: *Ischaemum sulcatum* Hack.
Family: Poaceae.

Part Used : Plant.

Utilization : Valued for green fodder, hay and silage; most nutritious at the flowering stage.

Reference : Ambasta (1986).

Selinum candollei DC.

Syn: *Selinum tenuifolium* Wall. *ex* DC.
Family: Apiaceae

Regional Names:

J. K.	:	Bhootakeshi;
Sim.	:	Khes havo.

Distribution : Kashmir (2,000-3,000 m). Nepal. Kumaon.

Part Used : Plant.

Utilization : Used as fodder for sheep.

References : Ambasta (1986), Hooker (1879).

Sequoia sempervirens Endl.

Syn: *Taxodium sempervirens* D. Don
Family: Taxodiaceae.

English and Hindi Names:

Eng. : Californian Red wood, Red wood;
Hindi : Rakta kat.

Distribution : Distributed Throughout S. India.

Part Used : Bark.

Utilization : Cleaned bark fibre, upto 9.9 mm long used as a blending agent with wool, cotton and the jute.

References : Agarwal (2003), Ambasta (1986).

Sesbania cannabina (Retz.) Pers.

Syn: *Sesbania aculeata* Pers.; *S. aculeata* Pers. var. *cannabina* Baker.
Family: Fabaceae.

English, Hindi, Sanskrit and Regional Names:

Eng. : Daincha, Prickly, Sesban;
Hindi : Dunchi;
Sans. : Jayanti, Itakata;
Ass. : Daincha;
Guj. : Sasilkad;
Kan. : Dhaincha;
Mal. : Kitannu;
Mar. : Ran Shevari;
Ori. : Dhaincha, Tentua;
Pub. : Jayanti, Jhijan, Jhinjan;
Tam. : Mullagathhi, Muleaembai, Neer-saembai;
Tel. : Errajiluga.

Distribution : Widely Cultivated Throughout India.

Parts Used : Leaves. Shoots.

Utilization : Cultivated for fibre, fishing nets, lines, ropes, rough fabrics and cordage.

References : Agarwal (2003), Ambasta (1986), Dwivedi (1993).

Sesbania grandiflora Pers.

Syn: *Agati grandiflora* Desv.; *Caronilla grandiflora* Willd.
Family: Fabaceae.

English, Hindi, Sanskrit and Regional Names:

Eng.	:	Agathi, Swamp Pea, Sesbean;
Hindi	:	Agasti, Bak, Bat, Hatiya;
Sans.	:	Agathi, Agasti, Anari, Buka;
Beng	:	Agathi, Agusta, Bak, Bajaphal;
Bom.	:	Augastha, Basna;
Guj.	:	Agatoio, Ayathio;
Kan.	:	Agas;
Mal.	:	Akathi, Athi;
Mar.	:	Madga, Agasta, Shevari;
Ori.	:	Buko, Ogosti;
Tam.	:	Agathi, Peragathi;
Tel.	:	Avici.

Distribution : N. & E. India including Andamans.

Parts Used : Bark. Leaves.

Utilization : Bark fibre valued for cordage. Leaves useful as fodder in Maharashtra and M.P.

References : Agarwal (2003), Ambasta (1986), Dwivedi (1993), Hooker (1879), Parrotta (2001), Raju (1998), Singh (1982), Talbot (1976b), Watt (1972).

Sesbania sesban Merrill

Syn: *Sesabnia argyptiaca* Pers.
Family: Fabaceae.

English, Hindi, Sanskrit and Regional Names:

Eng.	:	Common Sesban, Egyptian Rattle Pod;
Hindi	:	Jainti;
Sans.	:	Jayantika, Jayanti;
Ass.	:	Jintri, Jayantri;
Beng	:	Jainti, Jayante;
Guj.	:	Jayati, Raishingin;
Mal.	:	Sempa, Nellithalai, Ledangu;

Mar. : Shewarie, Jayat, Jarjan;
Ori. : Thaitimul, Surya-jantis, Jayontri;
Pub. : Jaint, Jait;
Tam. : Chithagathi, Karunchembai, Champai.

Distribution : Cultivated Throughout the Plains of India upto 1,200 m.

Parts Used : Young Leaves. Branches. Bark.

Utilization : Bark fibre used for rope making. Young leaves and branches lopped for fodder.

Active Constituents : Leaves rich in proteins.

References : Agarwal (2003), Ambasta (1986), Diwakar & Ansari (2003), Dwivedi (1993), Khanna (2002), Parrotta (2001), Raju (1998), Rastogi & Mehrotra (1993).

Sesbania speciosa Taub. *ex* Engl.

Family: Fabaceae.

Hindi and Regional Names:

Hindi : Seemaiagathi;
Tam. : Seemai, Agathi.

Distribution : Throughout S. India mainly in Kerala.

Part Used : Bark.

Utilization : Yields a fibre for coir and ropes.

References : Agarwal (2003), Ambasta (1986).

Setaria glauca Beauv.

Syn: *Panicum glaucum* Linn.; *Pennisetum glaucum* R. Br.; *Setaria aurea* Hoschst.; *S. auricoma* Link. *ex* Loud.; *S. chrysantha* Heynh.; *S. helvala* Roem & Sch.; *S. humifusa & rubicunda* Dum.; *S. pumila* Sch.

Family: Poaceae.

English, Hindi and Regional Names:

Eng. : Pigeon or Bottle grass, Yellow Foxtail Millet, Yellow Bristle Grass.

Hindi	:	Bandra;
Beng	:	Pinginatchi;
Guj.	:	Kollat, Kunchi, Zipiti ghass;
Kan.	:	Bilikorla hullu;
Mar.	:	Barati (cultivated); Bhadli (wild);
Pub.	:	Bandra, Dissi, Kotu, Ban Kangni;
Raj.	:	Kuta chutti, Soma;
Tel.	:	Nakka kora.

Distribution : Distributed Throughout India.

Part Used : Straw.

Utilization : As a nutritious good green fodder for cattle.

References : Agarwal (2003), Ambasta (1986), Duthie (1888), Khanna (2002), Hooker (1879), Watt (1972).

Setaria italica (Linn.) Beauv.; Hook. f.

Syn: *S. ariastica* Hort. *ex* Reichb.; *S. compacta* Schur; *S. erythrosperma, flayida, persica* & *vailacea* Hornen. *ex* Reichb.; *S. germanica* Beauv.; *S. globularis* Persl.; *S. macrochaeta* Sch.; *S. maritime* R.& S.; *S. melinis* Link *ex* Steud.; *S. panis* Jessen Deutschl.

Family: Poaceae.

English, Hindi and Regional Names:

Eng.	:	Italian or Fox tail Millet;
Hindi	:	Kalakangni, Koni, Kanghanuni;
Anda.	:	Tanahal;
Beng	:	Kara, Kangu;
Bom.	:	Kangani, Kara Kang, Kang, Vavani;
Guj.	:	Kang, Karang;
Kash.	:	Shali, Pinghi;
Mal.	:	Tena, Thina;
Mar.	:	Kangu, Sala, Chenna;
Ori.	:	Tangun, Kangu;
Pub.	:	Kangni, Chiurr, Khar, Khauni, Salau, Shak;
Tam.	:	Tenai;
Tel.	:	Korallu, Korra.

Distribution : Cultivated Throughout India, both in the Plains and on the Hills.

Part Used : Straw.

Utilization : Straw useful as a good fodder and also makes good hay.

References : Agarwal (2003), Ambasta (1986), Duthie (1888), Hooker (1897), Watt (1972).

Shorea robusta Gaertn. f.

Family: Dipterocarpaceae.

Hindi Name:

Hindi. : Sal.

Distribution : N. Parts of India excepting Higher Himalaya.

Part Used : Leaves.

Utilization : Leaves useful as fodder.

Active Constituents : Leaves contain 10.6% crude proteins, 27.43% fibre, 0.77% Ca and 0.12% P.

References : Dwivedi (1993), Singh (1982).

Sida acuta Burm. f.

Syn: *Sida acuta* Burm.; *S. carpinifolia* Mast. Non Linn.; *S. lanceolata* Roxb.; *S. scoparia* Lourex W. & A.; *S. stipulata* Cav.; *S. stauntoniana* DC.

Family: Malvaceae.

English, Hindi, Sanskrit and Regional Names:

Eng. : The Hornbeam-Leaved Sida;
Hindi : Bariara, Kerata, Kharenta;
Sans. : Bala, Pata;
Bom. : Pila barela- shikar, Sweetberela;
Guj. : Fungli-methi, Balajungli-methi, Baladungaraubal;
Kan. : Cheruparuva, Malatanni;
Mal. : Malatanni, Shiruparuva;
Mar. : Chikana, Pata, Tupkaria, Tukati.
Ori. : Ancharna, Siobola, Sunakhodika;
Pub. : Bariara, Karenta;
Tam. : Malaitangi, Mayir-manikham, Pazhampausi, Vattatirippi.

Distribution : Distributed Throughout Hotter Parts of India; Tropics generally.

Part Used : Stem.

Utilization : Fibre used as a substitute for jute for making brooms, bags, baskets, mats, wicks, etc.

References : Agarwal (2003), Ambasta (1986), Prasad *et al.* (2002), Watt (1972).

Sida cordata (Burm. f.) Borssum

Syn: *Melochia cordata* Burm. f.; *Sida humilis* Cav.; *S. humilis* var. *veronicaefolius* (Lam.) Masters.; *S. multicaulis* Cav. *S. veronicifolia* Lam.
Family: Malvaceae.

Hindi, Sanskrit and Regional Names:

Hindi : Bananiyar;
Sans. : Bhumibala, Nagabala;
Beng : Junka;
Kan. : Bekkenatalegida;
Tel. : Gayapurwaku.

Distribution : Hotter Parts of India upto 1,500 m.

Part Used : Stem Bark.

Utilization : Tribals in the E. of Aravalli use bast fibre extracted from the stem bark for making ropes and cordage.

References : Ambasta (1986), Parrotta (2001), Prajapati *et al.* (2003), Singh & Pandey (1998).

Sida codifolia Linn.

Syn: *S. althaeifolia* SW.; S. *herbacea* Micans.
Family: Malvaceae.

Vern. Names: Dredae, Dredain.

English, Hindi, Sanskrit and Regional Names:

Eng. : Country-Mallow;
Hindi : Kangyi;
Sans. : Badiyalaka, Bala, Brela, Jayanti, Kanaka, Krura, Nilaya, Sursana;
Beng : Bala, Brela;

Guj. : Bala baladona, Mahabala, Khapat;
Kan. : Hettuthi, Kisangi;
Mal. : Kurun Thott, Vellurum, Kathuram;
Mar. : Chikana, Khiranti;
Ori. : Badianaula, Bisvokopari;
Pub. : Kowar, Simak;
Tam. : Nilatutti, Paniar-tuthi;
Tel. : Tellagorra, Tellantisa.

Distribution : Throughout Tropical Regions of India upto 1,050 m.

Part Used : Stem Bark.

Utilization : Yields a fibre for brooms, mats, fabrics, ropes and cordages.

References : Agarwal (2003), Ambasta (1986), Blatter & Hallberg (1984), Hooker (1872), Parker (1973), Singh & Pandey (1998), Sood & Thakur (2004), Watt (1972).

Sida glutinosa Cav.

Syn: *Sida myoorensis* Wight & Arn.

Family: Malvaceae.

Part Used : Stem.

Utilization : Yields fibre used for ropes.

Reference : Ambasta (1986).

Sida ovata Forsk.

Family: Malvaceae.

Part Used : Stem.

Utilization : Stem bark yields a fibre of rather low quality. Tribals of 'Kota' district use these for making ropes and cordages.

Reference : Singh & Pandey (1998).

Sida rhombifolia Linn. Mast.

Syn: *S. canariensis* Willd; *S. compressa* Wall.
Family: Malvaceae.

English, Hindi, Sanskrit and Regional Names:

Eng.	:	Broomjute Sida;
Hindi	:	Sweet Brela, Sahadebi;
Sans.	:	Atibala;
Beng	:	Lalbrela;
Guj.	:	Khetraubat-atibala, Baladana Tenacham;
Kan.	:	Bolam gadale, Gobetade gida;
Mal.	:	Anakurun thalti;
Mar.	:	Sahadevi;
Ori.	:	Dholabadiamla;
Tam.	:	Chitra mutti, Athibala Chedi, Sirramutti;
Tel.	:	Athibala, Chettu, Muttava.

Distribution : Throughout India upto 2,000 m in the Himalaya.

Part Used : Stem.

Utilization : Stem exploited for cordage and considered as good substitute to jute.

References : Agarwal (2003), Ambasta (1972), Hooker (1872), Jain (1991), Singh & Kumar (2000), Singh & Pandey (1998).

Siegesbeckia orientalis L.

Syn: *S. brachiata* Roxb.
Family: Asteraceae.

English, Sanskrit and Regional Names:

Eng.	:	Common St. Paul's Wort;
Sans.	:	Lickura;
Guj.	:	Pilibadkadi;
Tam.	:	Katampam.

Distribution : Throughout Tropical and Sub-Tropical Regions of India.

Part Used : Plant.

Utilization : Plant used as fodder.

Active Constituents : Plant contains essential oil, salicylic acid and bitter glucoside.

References : Ambasta (1986), Caius (2003), Khan & Khanum (2005), Kirtikar & Basu (1984), Prajapati *et al.* (2003), Singh & Kumar (2000).

Smilax aspera L.

Syn: *S. fulgens* Wall.; *S. maculate* Roxb.; *S. nilgarensis* Steud.
Family: Liliaceae.

Vern. Name: Bragh Bel.

English Name:

Eng. : Rough Bird Weed;

Distribution : Found Throughout India.

Utilization : Used for cordages.

References : Chopra & Chopra (1969), Bhattacharjee (1998).

Smilax ovalifolia Roxb.

Family: Liliaceae.

Vern. Names: Bara, Datoon.

English, Hindi and Regional Names:

Eng. : Green braiar, Sarasparilla;
Hindi : Jangali aush bah;
Beng : Kumarika;
Kan. : Birrubeta;
Mal. : Kaltamara;
Mar. : Ghotrel;
Ori. : Mitri;
Tam. : Malaittamarai;
Tel. : Kondatamara.

Part Used : Stem.

Utilization : Strong stem fibre used for brushes.

References : Jain (1991), Raju (1998).

Smilax pavifolia Wall.

Family: Liliaceae

Distribution : Near Ravines and Shady Moist Forests between 1,000-1,500 m.

Parts Used : Leaves. Branches.

Utilization : Relished by sheep and goats and fed to cattle.

Reference : Purohit & Samant (1995).

Smithia sensitiva Ait.

Syn: *S. abyssinica* Hochst.

Family: Fabaceae.

Hindi and Regional Names:

Hindi : Oda birni;
Beng : Nulla Kashina;
Mar. : Kaola;
Pub. : Odabrini.

Distribution : All over India in the Himalaya ascending to 1,000 m.

Part Used : Leaves.

Utilization : Leaves useful as fodder and also an excellent source of hay.

Reference : Dwivedi (1993), Hooker (1879), Roy *et al.* (1998), Saklani & Jain (1994), Watt (1972).

Solanum indica L.

Syn: *S. anguivi* Bojer.; *S. canescens* Blume.; *S. cuneatum* Moinch.; *S. heynii* Roem Sch.; *S. himaense* Dunal.; *S. junghuhnii* Miq.; *S. pinnatifidum* & *agrest* Roth.; *S. violaceum* Facq.

Family: Solanaceae.

English, Hindi, Sanskrit and Regional Names:

Eng. : Poison berry;
Hindi : Barhanta;
Sans. : Vrihati;
Ass. : Tid bhagnri;
Beng : Baykura;

Guj.	:	Mhotiringni, Motaringni, Ubhibharingni;
Mal.	:	Cheruchenda;
Mar.	:	Ringani;
Ori.	:	Bonobryhoti;
Pub.	:	Katang-kari;
Tam.	:	Mulli;
Tel.	:	Tellamulaka.

Distribution : Throughout Tropical India ascending to 1,600 m.

Part Used : Leaves.

Utilization : Leaves used as fodder.

References : Parrotta (2001), Raju (1998), Watt (1972).

Sorghum bicolour (Linn.) Moench (Race Bicolor)

Syn: *Andropogon bicolour* Kunth. Non Roxb.; *A. cernuus* Roxb.; *A. laxus* Linn.; *Halcus bicolour* Linn.; *Sorghum vulgare* Pers.

Family: Poaceae.

English, Hindi, Sanskrit and Regional Names:

Eng.	:	Indian Millet;
Hindi	:	Juvar;
Sans.	:	Dirghamala;
Beng	:	Kala Mucha;
Bom.	:	Sundia;
Guj.	:	Jowar;
Kan.	:	Jalah;
Mar.	:	Juari;
Tam.	:	Jowar;
Tel.	:	Bonda Janu, Tella.

Distribution : Cultivated by the Inhabitants of Manipur and other Mountainous Districts Adjoining Bengal.

Part Used : Twigs.

Utilization : Fodder relished by cattle and it is said to manifest a power in them to survive for more than one year. Extracted fibre used for making brushes and brooms.

References : Agarwal (2003), Ambasta (1986), Dhiman (2003), Duthie (1888), Harborne & Baxter (2001), Varghese (1996), Watt (1972).

Sorghum dochna (Forsk.) Snowden

Syn: *Andropogon sorghum* Brot. var. *saccharatus* Koern.; *Sorghum sacchoratum* Pers. (Race Bicolor).

Family: Poaceae.

Distribution : Cultivated in Plains of India.

Part Used : Plant.

Utilization : Mostly grown for fodder.

References : Agarwal (2003), Ambasta (1986).

Sorghum milliforme (Hack.) Snowden

Syn: *Andropogon sorghum* Brot. var. *miliformis* (Hack) Hook. f.

Family: Poaceae.

Part Used : Twigs.

Utilization : Used as fodder.

References : Agarwal (2003), Ambasta (1986).

Sorghum purpurea-sericeum (Hochst. *ex* A. Rich.) Aschers. & Schweinf.

Syn: *Andropogon purpurea-sericeus* Hochst *ex* A. Rich.

Family: Poaceae.

Part Used : Twigs.

Utilization : Relished by cattle in green as well as dry form; also good for silage.

References : Agarwal (2003), Ambasta (1986).

Sorghum verticelliforum (Steud.) Stapf.

Family: Poaceae.

Part Used : Twigs.

Utilization : Used as fodder.

Reference : Ambasta (1986).

Soymida febrifuga A. Juss.

Syn: *S. soymida* Duncan.; *S. rubra* Rottl.;
Swietenia febrifuga Roxb.
Family: Meliaceae.

English, Hindi, Sanskrit and Regional Names:

Eng.	:	Indian Red wood, Bastard Cedar;
Hindi	:	Rakat, Ruhun, Ruhunna, Rohan;
Sans.	:	Patranga, Rohuna;
Beng	:	Rohan, Rohina;
Bom.	:	Rohan, Rohing
Guj.	:	Rohina;
Kan.	:	Some, Suani;
Mar.	:	Ruhin, Ruhan;
Ori.	:	Karwai;
Tam.	:	Shem, Wond;
Tel.	:	Sonidsa manu, Sumi.

Distribution : Throughout India Excepting W. Region in Maharashtra.

Part Used : Bark.

Utilization : Strong red bark fibre used for making ropes in Chota Nagpur.

References : Agarwal (2003), Ambasta (1986), Singh (1982), Watt (1972).

Sparrmannia africana Linn. f.

Family: Tiliaceae.

English and Hindi Name:

Eng.	:	African Hemp, Stockrase;
Hindi	:	Afiricisuun.

Distribution : Cultivated in Sub-Tropical Regions.

Part Used : Bark.

Utilization : Bark fibre used for cordage.

References : Agarwal (2003), Ambasta (1986).

Spartium junceum Linn.
Family: Fabaceae.

English Names:

Eng. : Jo Jhau vilayati, Spanish or Weaver's Broom.

Distribution : Grown in H.P. and Punjab.

Part Used : Stem.

Utilization : Fibre employed in the manufacture of mats, ropes, paper, rubberized belts for conveyors in mines and also for filling mattresses and pillows.

References : Agarwal (2003), Ambasta (1986).

Spathodea campanulata Beaov.
Family: Bignoniaceae.

English, Hindi and Regional Names:

Eng. : African Tulip Tree;

Hindi : Rugtoora;

Kan. : Kujjekaye;

Tam. : Patadi;

Tel. : Patade, Patatiya.

Distribution : Cultivated in Coffee States for shade.

Part Used : Plant.

Utilization : Used for fibre purposes.

References : Agarwal (2003), Parrotta (2001).

Spermadictyon suaveolens Roxb.
Syn: *Hamiltonia suaveolens* Roxb.
Family: Rubiaceae.

Part Used : Leaves.

Utilization : Leaves useful as fodder.

Reference : Singh & Kumar (2000).

Sphaeranthus africanus Linn.

Syn: *S. africanus* Wall.; *S. africanus* and *microcephalus* Willd.; *S. indicus* Kurtz in Journ.; *S. microcephalus* DC.

Family: Asteraceae.

English, Hindi and Regional Names:

Eng. : Sverta hapusa;
Hindi : Safed Hapusa;
Mal. : Velutha adakkamaniyan.

Distribution : Coast of Bengal, Kerala and Maharashtra.

Part Used : Herb.

Utilization : Used as cattle fodder.

References : Agarwal (2003), Ambasta (1986), Hooker (1879).

Spiraea conescens D. Don.

Syn: *Spiraea cuniefolia* Wall.

Family: Rosaceae.

Vern. Name: Chakhu.

Distribution : Temperate Himalaya from Kashmir to Kumaon: 1,500- 3,000 m.

Part Used : Leaves.

Utilization : Leaves grazed by all ruminants during May-June.

Active Constituents : Plant contains hydrocyanic acid, saponin and ethanolic compound.

References : Hooker (1879), Nautiyal *et al.*(2003), Purohit and Samant (1995).

Spiraea lindleyana Wall.

Family: Rosaceae.

Distribution : Distributed between 2,000-3,500 m.

Part Used : Leaves.

Utilization : Browsed by sheep and goats during the winter.

Active Constituents : Leaves have hydrocyanic acid.

Reference : Purohit & Samant (1995).

Sporobolus diander Beauv.

Syn: *Agrostis contracta* Herb. Heyne *ex* Wall.; *A. diandra* Retz.; *S. indicus* R. Br.; *S. tenacissimus* Beauv.; *Panium seundum* Herb. Heyne *ex* Wall.; *Vilfa diandra* Wight.; *V. erosa* Trin.; *V. tenacissima* Trin.

Family: Poaceae.

Hindi and Regional Names:

Hindi : Tandlan;
Beng : Bena joni;
Delhi. : Doob;
Kan. : Navilu dendi hullu;
Pub. : Khir, Nonak.

Distribution : Dry Regions Throughout India. Plains. Moderate elevations on the Hills.

Utilization : Fodder for cattle and horses in Punjab.

References : Agarwal (2003), Ambasta (1986), Duthie (1888), Hooker (1897), Watt (1972).

Sporobulus helvolus Dur. & Sohina

Syn: *Sporobolus glaucifolius* Hochst. *ex* Steud.

Family: Poaceae.

Parts Used : Leaves. Branches.

Utilization : A good fodder and hay for cattle, camels and horses.

Reference : Ambasta (1986).

Sporobolus maderaspatanus Bor.

Syn: *Sporobolus orientalis* Kunth.

Family: Poaceae .

Hindi and Regional Names:

Hindi : Usar ki ghas;
Pub. : Kheo, Tandua.

Part Used : Leaves.

Utilization : A good fodder and hay for cattle and horses.

Reference : Ambasta (1986).

Sporobolus pallidus Nees.

Syn: *Velfa pallida* Nees.
Family: Poaceae.

Distribution : Common in the Punjab & Rajasthan on Sandy Grounds.

Part Used : Leaves.

Utilization : Used as fodder.

Reference : Duthie (1888).

Stemona tuberosa Lour. (Pl. 8B)

Family: Stemonaceae.

Regional Name:

Tel. : Jedagadda.

Distribution : Coastal Regions of A.P., Tamil Nadu & Arunachal.

Part Used : Twigs.

Utilization : Used for making ropes.

Reference : Agarwal (2003).

Stephania glabra Miers.

Syn: *Cissampelos glabra* Roxb.; *S. delavayi* Auct non Diels.; *S. rotunda* Hook. f. & Thombs. Non Lour.
Family: Menispermaceae.

Vern. Name: Bish-Khapar.

Regional Name:

Pub. : Gindaru.

Distribution : Himalaya. Assam. W. Ghats. Tamil Nadu upto 2,000 m.

Part Used : Stem.

Utilization : Stem fibre useful for fishing lines.

References : Agarwal (2003), Saklani & Jain (1994), Sood *et al.* (2005).

Sterculia colorata Roxb.

Family: Sterculiaceae.

Hindi and Regional Names:

Hindi : Bodula, Samari, Walena;
Mar. : Hhoswey;
Tel. : Karaka.

Distribution : Sub-Himalayan Tract, ascending to 1,200 m.

Part Used : Leaves.

Utilization : Leaves used as fodder.

Reference : Singh (1982).

Sterculia guttata Roxb.

Syn: *S. alata* Wall.; *S. cuneata* Heyne ms.
Family: Sterculiaceae.

Regional Names:

Mal. : Kithondi;
Mar. : Kuhar, Goldava;
Tam. : Kavalam, Thondi.

Distribution : S. India. W. Peninsula. Koncan. Nilgiris. Malabar. E. Peninsula. Andaman Islands.

Part Used : Bark.

Utilization : Bark fibre said to be good for making coarse paper, cordage and rough fabrics.

References : Agarwal (2003), Ambasta (1986), Hooker (1872), Watt (1972).

Sterculia indica Merrill

Syn: *Sterculia coccinea* Roxb. non Jack.
Family: Sterculiaceae.

Regional Name:

Ass. : Nakchepeta.

Part Used : Bark.

Utilization : Bark yields a strong fibre for cordage.

References : Ambasta (1986), Singh (1982).

Sterculia urens Roxb.

Syn: *Cavallium urens* Scholt & Endl.

Family: Sterculiaceae.

Vern. Name: Karaya.

English, Hindi, Sanskrit and Regional Names:

Eng.	:	Arjuna, Balika;
Hindi	:	Gular, Gullu, Kulu;
Sans.	:	Arjuna;
Ass.	:	Hatchanda, Odla;
Bom.	:	Kovalee, Kandol, Korai;
Guj.	:	Karai, Eagdol;
Kan.	:	Kempudale;
Mal.	:	Thondi, Tonti;
Mar.	:	Karal, Kandol;
Ori.	:	Kavili;
Pub.	:	Kulu;
Tel.	:	Kavile, Tabasi, Tansi, Tabsu.

Distribution : Practically Throughout N.W. India, Assam, Bihar, W. Peninsula & Sri Lanka.

Part Used : Bark.

Utilization : Bark yields a fibre for cordage, rope-making and coarse cloth.

References : Agarwal (2003), Ambasta (1986), Hooker (1872), Jain (1968, 1991), Khan & Khanum (2005), Oommachan (1977), Sahni (1998), Singh & Prakash (1996), Talbot (1976a), Watt (1972).

Steruilia villosa Roxb.

Family: Stericuliaceae.

Vern. Name: Leandrein.

English, Hindi and Regional Names:

Eng.	:	Udal;
Hindi	:	Udar;
Ass.	:	Udol;
Bom.	:	Anni-nar;
Guj.	:	Gulbodla, Gulkandar;
Kan.	:	Chauri, Savaya;
Mal.	:	Vakka;
Mar.	:	Sardal, Sarda, Kuthada;
Ori.	:	Kodalo;
Pub.	:	Gulbodla, Gulkandar;
Tam.	:	Muratham, Orni Vakenar;
Tel.	:	Erra pullike, Kummari.

Distribution : Throughout Greater Part of N.W. India. Andamans. Bengal. Malabar. Tropical Himalaya from Kumaon Eastwards.

Parts Used : Bark. Leaves.

Utilization : Bark yields a strong fibre for cordage, rough bags, broad flakes, nets and paper industry. Leaf fodder is reported to be of poor quality.

Active Constituents : Leaf contains 1.84% N and 2.70% Ca.

References : Agarwal (2003), Ambasta (1986), Hooker (1872), Jain (1991), Saklani & Jain (1994), Singh (1982), Talbot (1976a), Watt (1972).

Stereospermum personatum (Hassk.) D. Chatterjee

Syn: *Stereospermum chelonoides* Auct.; A. D.; *Stereaspermum tetragonum* DC.

Family: Bignoniaceae.

English, Hindi and Regional Names:

Eng.	:	Trumpet-flower, Yellow Snaketree;
Hindi	:	Padar, Parrel;
Ass.	:	Parolli;
Beng	:	Dharmur;
Bom.	:	Padal, Padri, Paral, Toutuka;
Guj.	:	Paddi;

Kan.	:	Geri, Pudika;
Mal.	:	Pathiri,Piimbathiri;
Mar.	:	Kissel, Koosga, Padhri, Padvale,Tuatuka;
Ori.	:	Pamphunia;
Tam.	:	Ambuvagini, Padri,Pampadri, Pathiri, Vellippadri.

Distribution : Throughout Moist Parts of India from the Tarai of Oudh & Assam to Sri Lanka and Myanmar.

Part Used : Leaves.

Utilization : Leaves lopped for fodder.

Active Constituents : Leaves contain 1.81% N, 43.4% C, 1.67% Cu and 54.10% Mn.

References : Agarwal (2003), Ambasta (1986), Singh (1982), Watt (1972).

Streblus asper Louor.

Syn: *Epicarpurus asper* Steud.; *E. orientalis* Bl.; *Trophis aspera* Retz.

Family: Moraceae.

English, Hindi, Sanskrit and Regional Names:

Eng.	:	Siamese Rough-Bush;
Hindi	:	Siora;
Sans.	:	Shathotaka;
Beng	:	Sheora;
Bom.	:	Karvati, Karera, Kaeaoli, Rusa;
Kan.	:	Mitli;
Mal.	:	Paruka, Parava, Tintapparuya;
Mar.	:	Karera,Kharaoli, Kharota, Paraya, Poi;
Pub.	:	Jindi, Sihora, Dahya;
Tam.	:	Barivenkachetu, Barinki, Takki, Kakkabedi, Sitanite.
Tel.	:	Baridiki.

Distribution : H.P. to Assam. Tripura. Andaman in Dry Areas upto 600 m.

Part Used : Bark.

Utilization : Bark yields a fibre for paper making. Leaves lopped for fodder for cattle and elephants.

Active Constituents : Leaves rich in b-sitosterol, a-amyrin and upeol.

References : Agarwal (2003), Ambasta (1986), Dwivedi (1993), Prajapati *et al.* (2003), Rastogi & Mehrotra (1993), Singh (1982), Watt (1972).

Strobilanthes artopurpureus Nees.

Family: Acanthaceae.

Distribution : Occurs between 2,000- 3,000 m. N. E. India & N. Bengal.

Part Used : Mature Leaves.

Utilization : Considered nutritive for sheep and goats, but poisonous when young.

References : Purohit & Samant (1995), Singh & Kumar (2000).

Strobilanthus glutinosis Nees.

Syn: *Ruellia glutinosa* Wall.; *R. jacquemontiama* Nees.

Family: Acanthaceae.

Vern. Name: Neela Phool.

Ditribution : Temperate Himalaya: 1,000- 2,000 m.

Part Used : Leaves.

Utilization : Leaves used as an excellent fodder for sheep and goats.

Reference : Sharma (2004).

Strobilanthes wallichii Nees.

Family: Acanthaceae.

Vern. Name: Janila.

Distribution : Occurs between 2,000- 3,000 m.

Part Used : Leaves.

Utilization : Mature leaves considered to be very nutritive fodder.

Reference : Purohit & Samant (1995).

Suaeda monoica Forsk. *ex* J. F. Gmel.

Family: Chenopodiaceae.

Regional Names:

Ori. : Nunia;
Tam. : Karu- Vumari;
Tel. : Koyyalakoora.

Distribution : Southwards Gujarat to Maharashtra.

Part Used : Aerial Parts.

Utilization : A good fodder for sheep, goats and camels.

References : Ambasta (1986), Das *et al.* (1996), Dwivedi (1993).

Syagrus coronata Becc.

Syn: *Cocos coronata* Mart.
Family: Arecaceae.

Part Used : Leaves.

Utilization : Leaves used for making sacks, mats and ropes.

Reference : Ambasta (1986).

Symphytum asperum Lepech.

Syn: *Symphytam asperrimum* Don. *ex* Sims.
Family: Boraginaceae.

English Names:

Eng. : Prickly comfrey, Rough comfrey.

Parts Used : Leaves. Branches.

Utilization : Used as fodder.

References : Ambasta (1986), Watt (1972).

Symplocos paniculata Miq.

Syn: *Prunus panuculatus* Thunb.;
Symplocos crataegoides Buch.-Ham. *ex* D. Don.
Family: Symplocaceae.

Vern. Names: Lodar, Lodh, Lodhra.

English, Hindi, Sanskrit and Regional Names:

Eng.	:	Sappire Berry, Sweet leaf;
Hindi	:	Ludh, Mota Bhomlati;
Sans.	:	Patti, Shabara, Valka, Valkalodhra;
Bom.	:	Lodh;
Kum.	:	Lodh;
Mal.	:	Pachoti;
Pub.	:	Lodhar;
Tam.	:	Valli Lothi;
Tel.	:	Lodduga.

Distribution : Himalaya. Assam. Khasia Hills: 600- 2,400 m.

Part Used : Leaves.

Utilization : Fodder for sheep and goats.

References : Agarwal (2003), Ambasta (1986), Dwivedi (1993), Prajapati *et al.* (2003), Singh (1982), Sood *et al.* (2005), Thakur *et al.* (1989).

Symplocos ramosissima Wall.

Family: Symplocaceae.

Distribution : Temperate Himalaya from Garhwal to Bhutan. Khasia Hills.

Part Used : Leaves.

Utilization : Leaves fed to cattle almost throughout the year.

References : Hooker (1882), Purohit & Samant (1995), Watt (1972).

Syzygium cuminii (Linn.) Skeels.

Syn: *Eugenia cuminii* Druce.; *E. jambolana* Lam.; *Myrtus cumini* Linn.

Family: Myrtaceae.

Vern. Name: Jaamun.

English, Hindi, Sanskrit and Regional Names:

Eng.	:	Black plum, Jaman, Java plum;
Hindi	:	Jamun, Jam;
Sans.	:	Jambu, Jambula, Mahaskanda;

Ass.	:	Jamu;
Beng	:	Jam, Kalajam;
Guj.	:	Jamli, Jambu;
Mal.	:	Naval, Perinnaral;
Mar.	:	Jaman, Jambul;
Ori.	:	Jamo, Jamakoli;
Pub.	:	Jammu;
Tam.	:	Naga, Naval, Neredam, Sambal;
Tel.	:	Nasudu, Neredu, Nairari.

Distribution : Widely Cultivated Throughout India upto 1,800 m.

Part Used : Leaves.

Utilization : Leaves form palatable fodder for sheep, goats and cattle.

Active Constituents : Leaves contain dihydromyricetin.

References : Agarwal (2003), Ambasta (1986), Das *et al.* (2000), Dwivedi (1993), Parrotta (2001), Rastogi & Mehrotra (1995a), Singh (1986), Sood *et al.* (2005), Sood & Thakur (2004).

Tacca leontoprtaloides (L.) Kunth.

Syn: *Tacca pinnatifida* Forst. & Forst. f.

Family: Taccaceae.

English, Hindi, Sanskrit and Regional Names:

Eng.	:	Figi arrow root;
Hindi	:	Diva;
Sans.	:	Surna;
Bom.	:	Diva;
Mal.	:	Kattuchena-Kizhangu;
Mar.	:	Karachunai;
Tam.	:	Karachuni;
Tel.	:	Kanda.

Distribution : Grown in M.P., Bihar and Deccan.

Part Used : Stem.

Utilization : Stem used for making mats.

Reference : Agarwal (2003).

Tamarindus indica L.

Syn: *Tamarindus officinalis* Hook.; *T. occidentalis* Gaertn.
Family: Fabaceae.

English, Hindi, Sanskrit and Regional Names:

Eng.	:	Tamarindus Tree;
Hindi	:	Amli, Imli;
Sans.	:	Amlika;
Ass.	:	Tetuli;
Beng	:	Ambli;
Bom.	:	Ambli, Chinch, Chincha;
Guj.	:	Aamali;
Kan.	:	Huli;
Mal.	:	Amlam;
Mar.	:	Chinch;
Ori.	:	Tentuli;
Pub.	:	Imbli;
Tam.	:	Puli;
Tel.	:	Chintachettu.

Distribution : Throughout India.

Part Used : Leaves.

Utilization : Leaves used as fodder for sheep and goats.

Active Constituents : Tender leaves contain moisture 70.7%, proteins 5.8%, fat 2.1%, fibre, 1.9%, carbohydrates 18.2%, minerals 1.5%, Ca 10, Mg 71, P 140, Cu 2.09, Cl 94, thiamine 0.24, riboflavin 0.17, niacin 4-1 and vit - C 3.0 mg/ 100g.

References : Dwivedi (1993), Hooker (1879), Parrotta (2001), Prajapati *et al.* (2003), Raju (1998), Rastogi & Mehrotra (1990), Singh (1982), Watt (1972).

Tamarix dioica (L.) Roxb.

Syn: *Tamarix articulata* Wall.; *T. gallica* Wight.
Family: Tamaricaceae.

Hindi, Sanskrit and Regional Names:

Hindi	:	Fau, Jaru;

Sans. : Pisula;
Beng : Lal Jahu;
Pub. : Lei, Lie, Panj;
Tam. : Attumari, Nirumarri;
Tel. : Palivela, Palligi.

Distribution : Punjab to Assam. W. Peninsula near sea Coasts.

Parts Used : Stem. Leaves.

Utilization : Leaves used as fodder; and stem for making baskets, mats, house walls, etc.

References : Hooker (1872), Islam (2002), Singh (1982), Watt (1972).

Tamarix troupii Hole.

Family: Tamaricaceae.

Hindi and Regional Names:

Hindi : Jahu;
Pub. : Pilchi.

Distribution : Gujarat. Rajasthan. Punjab. U. P.

Part Used : Leaves.

Utilization : Leaves used as fodder.

Reference : Singh (1982).

Tanacetum longifolium Wall. *ex* DC.

Family: Asteraceae.

English Name:

Eng. : Long Leaved Tansy.

Distribution : Himalaya: Kashmir to Nepal upto 3,000 m.

Part Used : Plant.

Utilization : Plant used as fodder for sheep and goats.

Reference : Singh & Kumar (2000).

Taxus baccata Linn.

Syn: *Taxus baccata* Hook.; *T. baccata* L. sub sps. *wallichiana* (Zucc.) Pilger.; *T. contorta* Griffth.; *T. orientialis* Bertoloni.; *T. nucefera* Wall.; *T. nepalensis* Jack.; *T. vieginata* Wall.; *T. wallichiana* L.
Family: Taxaceae.

Vern. Name : Rakhaal.

English, Hindi, Sanskrit and Regional Names:

Eng. : Himalayan Yew, The Yew;
Hindi : Thuno, Zirnebbirmi;
Sans. : Manduparni;
Beng : Bhirmik;
Bom. : Barmi;
Kash. : Chatung, Postil, Sungal;
Pub. : Barmi, Chogu, Rikhal.

Distribution : Temperate Zone of Himalaya (2,200 -3,000 m) in the Punjab & Khasia Hills.

Part Used : Leaves.

Utilization : Leaves used in times of scarcity as poor quality fodder.

References : Hooker (1890), Purohit & Samant (1995), Singh (1982), Watt (1972).

Tecomella undulata (Sm.) Seem.

Syn: *Bignonia undulata* Sm.; *Tectoma undulata* G. Don.
Family: Bignoniaceae.

English, Hindi, Sanskrit and Regional Names:

Eng. : Rohida tree;
Hindi : Rugtrora;
Sans. : Rohi, Rohina, Rohita;
Bom. : Lahuri, Lohero, Rakh treora, Roira;
Pub. : Lahura, Laur, Rohira.

Distribution : Punjab. Rajputana. Gujarat.

Part Used : Leaves.

Utilization : Leaves browsed by cattle.

References : Ambasta (1986), Chatterjee & Pakrashi (1991), Singh (1982).

Terminalia arjuna (Roxb. *ex* DC.)W. & A.

Syn: *Pentaptera angustifolia* Roxb.; *P. glabra* Roxb.; *Terminalia berryi* W. & A.; *T. glabra* W. & A.; *T. ovalifolia* Rottl.

Family: Combretaceae.

Vern. Name: Arjun.

English, Hindi, Sanskrit and Regional Names:

Eng.	:	Arjuna Myrobaln, White murdah;
Hindi	:	Arjun, Kowa;
Sans.	:	Arjuna, Sach, Sarya;
Ass.	:	Orjun;
Beng	:	Arjun, Kahu;
Bom.	:	Jamla;
Gar.	:	Arjun, Kauha;
Guj.	:	Sadado;
Kan.	:	Maddi;
Mal.	:	Vellamarutu;
Mar.	:	Sanmatta, Sadaru;
Ori.	:	Arjuno;
Pub.	:	Arjan, Jumla;
Tam.	:	Vellamatta;
Tel.	:	Kakubhamu, Tittumaddi.

Distribution : Sub-Himalyan Tract. Deccan. Chota Nagpur. Mayanmar. Sri Lanka.

Parts Used : Bark. Leaves.

Utilization : Bark used as fibre and leaves as fodder.

Active Constituents : Leaves contain 10.10% crude proteins, 5.75% starch, 17-2.86% Ca, 0.48% Mn and 1.36-2.05% N.

References : Agarwal (2003), Dastur (1964), Dey *et al.* (1973), Drury (1978), Dwivedi (1993), Khan & Khanum (2005), Parrotta (2001), Singh (1982), Sood *et al.* (2005), Watt (1972).

Terminalia belerica Roxb.

Family: Combretaceae.

Hindi and Regional Names:

Hindi	:	Bahera;
Kan.	:	Giting, Tare;
Tam.	:	Tani;
Tel.	:	Tandi.

Distribution : Scattered in the Sub -Himalayan Tract.

Part Used : Leaves.

Utilization : Leaves used as fodder.

Active Constituents : Leaves contain 8.63% crude proteins, 16.73-18.59% crude fibre, 7.64-11.72% ash, 2.08-3.6% Ca and 0.16-0.27% P.

References : Dwivedi (1993), Singh (1982).

Terminalia chebula Retz.

Syn: *Embryogonia arborea* Teys & Binn.; *Myrobalanus chebula* Gaertn.; *Terminalia aruta* Ham.; *T. reticulata* Roth.

Family: Combretaceae.

Vern. Name: Harad.

English, Hindi, Sanskrit and Regional Names:

Eng.	:	Chebulic myrobalan;
Hindi	:	Harra;
Sans.	:	Aralu, Arlu,Harra,Haritaki;
Ass.	:	Hulikha;
Beng	:	Haritaki;
Bom.	:	Hirda;
Gar.	:	Harr;
Guj.	:	Gardo;
Mal.	:	Kadduka;
Mar.	:	Hirda;
Ori.	:	Haridra;
Pub.	:	Har, harar;
Tam.	:	Kadukkai;
Tel.	:	Karaka.

Distribution : Throughout the Hills in the N. India from Kumaon to Bengal & Southwards to the Deccan 300-1,000 m.

Part Used : Leaves.

Utilization : Leaves used as fodder and fed to cattle.

Active Constituents : Leaves contain tannin, terflavins B, C & D, punicalin and punicalagin.

References : Agarwal (2003), Dwivedi (1993), Hooker (1879), Parrotta (2001), Purohit & Samant (1995), Rastogi & Mehrotra (1995a), Singh (1982), Watt (1972).

Terminalia tomentosa (DC) Wt. & A.

Syn: *Terminalia alata* Heyne *ex* Roth.; *T. crenulata* W. & A.
Family: Combretaceae.

Vern. Names: Alsan, Dhamni.

English, Hindi, Sanskrit and Regional Names:

Eng. : Black murdah;
Hindi : Asluna;
Sans. : Krishnatvaka, Nissaraphalako, Shyamaso- Raka;
Ass. : Amari;
Beng : Ashan, Pishel;
Bom. : Aim, Marthi;
Guj. : Hadri, Sadada, Dajadia;
Mal. : Karimaruti, Temparu;
Mar. : Madat, Sadada;
Ori. : Kolashalo, Pondosahajo;
Pub. : Aisan, Sein;
Tam. : Aruchannum;
Tel. : Innamaddi, Nalamaddi.

Distribution : N. India (Kumaon to Bengal) & Southwards to the Deccan at 1,000 m.

Part Used : Leaves.

Utilization : Leaves valued as good fodder.

Active Constituents : These contain crude proteins 10.92%, crude fibre 12.41%, total minerals 7.21%, reducing sugars 2.39%, total sugars 7.95% and starch 15.2%.

References : Ambasta (1986), Singh (1982), Watt (1972).

Thalictrum foliolosum DC.

Family: Ranunculaceae.

Vern. Names : Barmote, Mamira.

English, Hindi, Sanskrit and Regional Names:

- **Eng.** : Gold thread;
- **Hindi** : Pinjari, Shuprak;
- **Sans.** : Mamira, Pitmula, Pitranga;
- **Bom.** : Mamiran;
- **Kum.** : Pengla, Pila jari;
- **Pub.** : Gurbiani, Pashmarum

Distribution : Temperate Himalaya (1,000-2,500 m), Khasia Hills (1,000-2,000 m).

Part Used : Twigs.

Utilization : Twigs used for fodder purposes.

References : Prajapati *et al.* (2003), Singh (1982), Singh & Kumar (2000), Watt (1972).

Thamnocalamus falconeri Hook. f. *ex* Munro.

Syn: *Arundinaria falconeri* Benth. & Hook. f.

Family: Poaceae.

Distribution : Distributed between 2,000-3,000 m.

Part Used : Leaves.

Utilization : Leaves fed to all classes of herbivorus domestic animals.

Reference : Purohit & Samant (1995).

Thamnocalamus spathiflora (Trin) Munro

Syn: *Arundinaria spathiflora* Trin.

Family: Poaceae.

English, Hindi and Regional Names:

- **Eng.** : Spathe Bamoo;

Hindi	:	Ringal;
Kum.	:	Parikh.

Distribution : E. & W. Himalaya.

Parts Used : Leaves. Stem.

Utilization : Leaves fed to all ruminants in winter. Stem yields a fibre for ropes, mats, baskets, fishing nets and paper.

References : Agarwal (2003), Ambasta (1996), Purohit & Samant (1995), Raju (1998), Watt (1972).

Themeda arundinacea (Roxb.) Ridley

Syn: *Anthistiria gigantia* Hack.; *A. ciliata* L.; A. *arundinacea* Hack.; *Themeda quadrivalis* (L.)

Family: Poaceae.

English, Hindi and Regional Names:

Eng.	:	Kangaroo ghass, Ulla grass;
Hindi	:	Kapur Grass;
Gar.	:	Kapur grass;
Kum.	:	Kangua;
Pub.	:	Azkhum, Bharna.

Distribution : Himalaya & Plains of India.

Part used : Plant.

Utilization : Plant used as fodder and also yields a fibre for ropes and cordage.

References : Agarwal (2003), Ambasta (1986), Varghese (1996).

Themeda cymbaria Hack.

Syn: *Anthistiria cymbaria* Roxb.

Family: Poaceae.

Regional Names:

Guj.	:	Ful ghass, Rhatdu;
Kan.	:	Balai hullu;
Tam.	:	Noshia palai pullu.

Part Used : Twigs.

Utilization : Affords a good fodder and also can be made into hay.

Reference : Ambasta (1986).

Thespesia lampas etc.

Syn: *Hibiscus lampas* Cav.; *H. gangeticus & tetralocularis* Roxb.; *Paritium gangeticum* Don.

Family: Malvaceae.

Hindi, Sanskrit and Regional Names:

Hindi : Bankapasi, Jangli bhendi;
Sans. : Karpasi, Vanakarpasah;
Beng : Bankapas;
Guj. : Jangli paraspiplo, Paruspiplo;
Mal. : Draba, Kattuparatti;
Mar. : Jangli bhenda;
Ori. : Bonokopa;
Tel. : Kondapatti, Pattinga.

Distribution : Tropical Jungles of India from Kumaon Eastwards.

Part Used : Young Twigs.

Utilization : Extracted fibre used for binding loads of wood.

References : Parrotta (2001), Prasad *et al.* (2002), Varghese (1996), Watt (1972).

Thespesia populnea Soland. *ex* Correa

Syn: *Hibiscus populneus* L.; *H. populneoides* Roxb.; *Malvasiscus populneus* Gaertn.

Family: Malvaceae.

English, Hindi, Sanskrit and Regional Names:

Eng. : False rose wood, Indian tulip tree, Porita tree;
Hindi : Parsipu, Paras- papal;
Sans. : Gardha-bhanda;
Beng : Dumbla, Parsipal, Palaopipal;
Bom. : Bhendi, Parsipu;
Guj. : Paarsa peepala;
Mal. : Poovarasu;
Mar. : Bhendi-ki- jahr, Parsacha-jhada;

Ori.	:	Gunjausto, Porosopippoli;
Pub.	:	Poraspipal;
Tam.	:	Poovarasam kallal.

Distribution : Andaman Islands. Tropical Shores of Bengal. Peninsula. Sri Lanka.

Parts Used : Bark. Leaves.

Utilization : Leaves lopped for fodder. Bark fibre good for cordage, fishing nets, coffee bags, caulking boats and tying bundles of wood, etc.

References : Agarwal (2003), Ambasta (1986), Hooker (1872), Singh (1982), Watt (1972).

Thuja plicata D. Don.

Syn: *Thuja gigantean* Nutt.
Family: Pinaceae.

English Name:

Eng.	:	Giant arbour- vitae, Western Red Cedar.

Part Used : Wood.

Utilization : Wood used for paper pulp and yields a fibre for baskets, mats, hats, blankets, cordage and thatching.

References : Ambasta (1986), Jain (1991).

Thysanolaena maxima (Roxb.) Kuntze

Syn: *Agrostis maxima* Roxb.; *Thysanolaena acarifera* Arn. & Nees.; *T. agarostis* Nees.
Family: Poaceae.

English, Hindi and Regional Names:

Eng.	:	Broom Grass;
Hindi	:	Phuljanta;
Beng	:	Phuljanta;
Bom.	:	Barucha;
Kum.	:	Kuchi;
Ori.	:	Phulbadhuni.

Distribution : Throughout the Hills upto 2,000 m.

Parts Used : Leaves. Stem.

Utilization : Leaves and stem tips exploited for fodder.

Active Constituents : Leaves contain 9.5% crude proteins, 30.4% crude fibre, Ca 0.23% and P 0.11%.

References : Agarwal (2003), Ambasta (1986), Dash & Mishra (2000), Purohit & Samant (1995), Raju (1998), Samati (2004).

Tilia europaea Linn.

Syn: *Tilia vulgaris* L.
Family: Tiliaceae.

English Name:

Eng. : Common Lime.

Distribution : Found in H.P., Shimla.

Parts Used : Bark. Leaves.

Utilization : Leaves used as fodder for horses and cattle. Bark yields a fibre used for cordage.

Active Constituents : Leaves contain 16.2% proteins, 2.9% fat, 13.2% crude fibre and 11.4% ash.

References : Agarwal (2003), Ambasta (1986), Singh (1982).

Tinospora cordifolia (Willd.) Miers *ex* Hook. f. Thoms.

Syn: *Cocculus convolvulaceae* DC.; *C. cordifolia* DC.; *C. verrucocus* Wall.; *Menispermum cordifolia* Willd.; *Tinospora palmnervis* Miers.
Family: Menispermaceae.

Vern. Name: Gulja.

English, Hindi, Sanskrit and Regional Names:

Eng. : Gulancha, Heart leaved moonseed;
Hindi : Amrita, Gulancha;
Sans. : Amrita, Bishaka, Guluchi;
Ass. : Amarlata;
Beng : Galancha;

Bom.	:	Garol, Giroli;
Gar.	:	Gurch;
Guj.	:	Galo, Gulo, Gulvel;
Kash.	:	Amrita, Gilo;
Mal.	:	Amrytu, Chittamritam;
Mar.	:	Giroli, gulval, Gulvel;
Ori.	:	Galuchi;
Pub.	:	Batindu, Garun, Gilo;
Tam.	:	Amudom, Chindil;
Tel.	:	Tippateege.

Distribution : Throughout Tropical Parts of India from Kumaon to Assam & Bihar.

Part Used : Leaves.

Utilization : Leaves used as fodder

Active Constituents : Active constituents are cordifol, heptacosanal, octasanoe, proteins, Ca, P and β-sitosterol.

References : Agarwal (2003), Ambasta (1986), Kothari & Rao (2000), Jain (1991), Rastogi & Mehrotra (1991), Saklani & Jain (1994), Thakur *et al.* (1989), Varghese (1996).

Trachelospermum lucidum (Don) K. Schum.

Syn: *Tracheospermum fragrans* Hook. f.

Family: Apocynaceae.

Vern. Name : Naraun.

Hindi and Regional Names:

Hindi	:	Dudhi;
Ass.	:	Akhahilata;
Gar.	:	Dudh bel;
Kan.	:	Sohkyrmoti- krah.

Distribution : Throughout E. & W. Himalaya in Shady Moist Places & Rocks between 300-2,000 m.

Part Used : Leaves.

Utilization : Leaves fed to cattle throughout the year.

References : Agarwal (2003), Purohit & Samant (1995), Sharma (2004).

Trachycarpus fortunei (Hook.) Wendl.

Syn: *Trachycarpus excelsus* Wendl.
Family: Arecaceae.

English Names:

Eng. : Fortune's chusan palm, Windl mill palm.

Part Used : Plant.

Utilization : Yields a fibre for cordage and scrubbing brushes.

Reference : Ambasta (1986).

Trema cannabina Lour.

Syn: *Celtis amboinensis* Willd.; *Trema amboinensis* Blume, non auct.; *T. orientalis* auct, non Blume.; *T. virgata* Blume.
Family: Ulmaceae.

English, Hindi, Sanskrit and Regional Names:

Eng. : Charcoal tree, Indian Neetle Tre;
Hindi : Bakri pathi, Chickan;
Sans. : Jivanti;
Ass. : Padham Jola;
Beng : Jibon, Junpong;
Guj. : Gol;
Mal. : Ama, Malantotalli, Ratthi;
Mar. : gol, Kapashi, Kargol;
Ori. : Khaskhas, Jivani;
Tam. : Ambaratthi, Chenkolam.

Distribution : Throughout Moist Regions of India. Foot Hills of Nepal & Sikkim Himalaya.

Parts Used : Bark. Leaves.

Utilization : Leaves lopped for fodder and fed throughout the year. Bark fibre valued for fishing nets, ropes, twines, coarse cloth, tying rafters of native houses and for binding loads for making the coarse amphak cloth.

References : Agarwal (2003), Ambasta (1986), Hooker (1890), Jain (1991), Purohit & Samant (1995), Saklani & Jain (1994), Singh (1982), Watt (1972).

Trema politoria Planch.

Syn: *Celtis politoria* Wall.; *Sponia politoria* Planch.

Family: Ulmaceae.

English and Regional Names:

Eng. : Banharria, Khagshi, Khardol, Kuri;

Beng : Tila;

Ori. : Kharkas.

Distribution : Sub-Tropical Himalaya from the salt ranges Eastwards to Oudh & Sikkim; Southwards to Mount Aboo in Malabar & C. India.

Parts Used : Bark. Leaves.

Utilization : Leaves used as a good quality fodder throughout the year and bark fibre valued for ropes.

References : Ambasta (1986), Dwivedi (1993), Hooker (1890), Purohit & Samant (1995), Singh (1982).

Trifolium alexandrinum Linn.

Family: Fabaceae.

English Names:

Eng. : Barseem, Egyptian Clover.

Distribution : Cultivated Throughout India upto 1,500 m.

Part Used : Plant.

Utilization : Used as green fodder, silage and hay; valued because of good palatability.

Active Constituents : Rich in proteins, Ca, P, kaempferol and velutin.

References : Ambasta (1986), Khanna (2002).

Trifolium dubium Sibth.

Syn: *Trifolium minus* Sm.
Family: Fabaceae.

English Name:

Eng. : Yellow Sucking Clover.

Part Used : Plant.

Utilization : Used as fodder.

Reference : Ambasta (1986).

Trifolium pratense L.

Family: Fabaceae.

English and Hindi Names:

Eng. : Red Clover, Cow Grass;
Hindi : Tripatra.

Distribution : Kashmir to Garhwal upto 2,500 m. Nilgiris.

Part Used : Plant.

Utilization : Used as fodder to increase lactation in cows.

References : Agarwal (2003), Ambasta (1986), Chopra *et al.* (1969), Chandrasekar & Srivastava (2003), Hooker (1879), Rastogi & Mehrotra (1990, 1995b), Watt (1972).

Trifolium repens Linn.

Syn: *Trifolium venulosuni* Royle Mss.
Family: Fabaceae.

English, Hindi and Regional Names:

Eng. : Dutch clover, Shamrock, White Clover;
Hindi : Safed clover;
Pub. : Shaftul, Shotul.

Distribution : Himalaya, Nilgiris & Alpine Himalaya.

Part Used : Leaves.

Utilization : Used as an excellent hay, silage and forage to increase the milk.

Active Constituents : Rich in crude proteins, Ca and P.

References : Agarwal (2003), Ambasta (1986), Deshmukh *et al.* (2000), Hooker (1879), Singh (1982), Singh & Kumar (2000), Watt (1972).

Trifolium resupinatum Linn.

Family: Fabaceae.

Vern. Name : Perseen.

English Name:

Eng. : Persian clover.

Distribution : Sub- Tropical Regions.

Part Used : Whole Plant.

Utilization : Plant used as green fodder for livestock.

Active Constituents : Rich in proteins and P.

References : Ambasta (1986), Sood & Thakur (2004).

Trigonella polycerata Linn.

Syn: *Trigonella brahuica* Boiss.; *T. genminiflora* Bunge Rel.; *T. monantha* C. A.; *T. orthoceras* Kar.& Kir.; *T. pinnatifida* Cav.

Family: Fabaceae.

Vern. Name : Tongzil.

Regional Names:

Pub. : Chini, Khanda rore, Sainji.

Distribution : W. Himalaya. Upper Gangetic Plains ascending to 2,000 m.

Part Used : Plant.

Utilization : Used as fodder.

References : Dwivedi (1993), Hooker (1879), Sood *et al.* (2001).

Trisetum aestivum L.

Syn: *Trisetum sativum* Lam.; *T. vulgare* Vill.
Family: Poaceae.

English, Hindi, Sanskrit and Regional Names:

Eng.	:	Wheat;
Hindi	:	Giun;
Sans.	:	Arupa, Bahudugdha, Godumba, Rasala, Samam;
Bom.	:	Gahu, Ghawan, Gium;
Guj.	:	Gawan, Ghavum, Govum;
Kon.	:	Gahum;
Mal.	:	Gendum, Kotanpam;
Mar.	:	Gahum, Gahung;
Pub.	:	Do, Dro, Gehum, Kanak, Nis, Rozzatt, To, Zud;
Tam.	:	Godumai, Godumbaiyarisi;
Tel.	:	Godumulu.

Distribution : Cultivated Throughout India.

Parts Used : Straw. Green Plant.

Utilization : Both green plant and straw considered important fodder.

Active Constituents : Rich in lucenin-3, vicenin-2, isoswertisin-4-ψ-glucoside and tricin.

References : Kirtikar & Basu (1984), Rastogi & Mehrotra (1991), Singh *et al.* (1996).

Trisetum flavescense (Linn.) Beauv.

Syn: *Avena flavescens* Linn. non Hook. f.;
A. sikkimensis Hook. f.
Family: Fabaceae.

English Names:

Eng.	:	Yellow oat grass, Yellow trisetum.

Part Used : Plant.

Utilization : Valued for fodder and hay and considered nutritious.

Reference : Ambasta (1986).

Triumfetta pentendra A. Rich.

Syn: *Triumfetta neglecta* Wight & Arn.

Family: Tiliaceae.

Distribution : Throughout Tropical India.

Part Used : Stem.

Utilization : Stem yields soft fibre which matches jute in quality.

References : Ambasta (1986), Jain (1991), Singh & Kumar (2000).

Triumfetta pilosa Roth.

Syn: *Triumfetta glandulosa* Heyne Herb Wall.; *T. oblongata* Link.; *T. pilosa* var., *T. tomentosa* Wall.; *T. polycarpa* Wall.

Family: Tiliaceae.

Vern. Name: Kura.

Distribution : Throughout the Tropical areas of India from the Himalaya to Travencore & Sri Lanka

Part Used : Bark. Leaves.

Utilization : Fed to larger ruminants in times of scarcity. Bark yields a fibre of commerce.

References : Dastur (1978), Hooker (1872), Purohit & Samant (1995), Singh & Kumar (2000), Watt (1972).

Triumfetta rhomboidea Jacq.

Syn: *Triumfetta angulata* Lam.; *T. bartramia* Linn.

Family: Tiliaceae.

Vern. Name: Beedli.

English, Hindi, Sanskrit and Regional Names:

Eng. : Burbush, Burweed;
Hindi : Chitki;
Sans. : Jhinjharita;
Ass. : Agra;
Beng : Banokra;
Ori. : Bojoramuti, Jotojuti;
Tam. : Ottupullu, Puramutti;
U. P. : Bhora.

Distribution : Throughout Tropical Regions of India.

Part Used : Inner- bark.

Utilization : Yields a soft fibre which is used as substitute for jute for making fishing nets, ropes and fastening.

References : Agarwal (2003), Ambasta (1986), Hooker (1872), Jain (1991), Oommachan (1977), Parker (1973), Saklani & Jain (1994), Singh & Pandey (1998), Sood & Thakur (2004), Watt (1972).

Turpinia nepalensis Wall. *ex* Wight & Arn.

Syn: *Turpinia pomifera* Hiern in part, non DC.

Family: Staphyleaceae.

Regional Names:

Kan. : Nila;

K. Hi. : Dieng- ja- arthri;

Mal. : Attuneermulla, Pambavetti, Samtha;

Tam. : Kanali.

Part Used : Leaves.

Utilization : Leaves lopped for fodder.

References : Ambasta (1986), Dwivedi (1993), Singh (1982).

Tylophora indica (Burm. f.) Merrill.

Syn: *Tylophora asthmatica* Wight & Arn.

Family: Asclepiadaceae.

English, Hindi, Sanskrit and Regional Names:

Eng. : Emetic Swallow Wort, Indian or Country- Ipecacuanha;

Hindi : Jangli Pikvam;

Sans. : Antrapachaka, Arkaparini, Lataksiri;

Beng : Antomul, Ananthamul;

Guj. : Daminivel;

Kan. : Adumuttaha, Nepala;

Mal. : Villipalla;

Mar. : Khodiki Raasna, Pitkari;

Ori. : Mendi, Mulini;

Tam. : Nach- Churuppam, Naja- Murich- Chaan;
Tel. : Kakapalla, Kaakapalla.

Distribution : Assam. W. Bengal. Orissa. Peninsular India.

Part Used : Plant.

Utilization : Yields a strong fibre for manufacturing of extra-fine fabrics.

References : Agarwal (2003), Ambasta (1986), Jain (1991), Parrotta (2001), Prajapati *et al.* (2003), Rastogi & Mehrotra (1990).

Typha elephantina Roxb. non Grah Nee Schimp. *ex* Rohrb.

Syn: *Typha angustifolia* Watt. non Linn.
Family: Typhaceae.

English, Hindi, Sanskrit and Regional Names:

Eng. : Elephant Grass, Indian Reed Mace;
Hindi : Mothitrina, Bora;
Sans. : Gundra;
Beng : Hogla;
Guj. : Ghabajarin;
Kan. : Apu;
Mar. : Eraka, Rambdna;
Ori. : Hogla;
Pub. : Boj, Bori, Dib, Gond, Lukh, Patira;
Tam. : Anaikkorai, Anaipull, Chanba.

Distribution : Commonly in W. Bengal, Orissa, Kashmir, Punjab, Tamil Nadu, Gujarat & Assam.

Parts Used : Leaves. Stem.

Utilization : Leaves and stem used for making ropes and stuffing. Stem pulp suitable for marine ropes, fishing nets, carpets and coarse textile owing to its high fibre strength.

References : Ambasta (1986), Islam (1998), Raju (1998), Samanta & Das (2003).

Typha latifolia Linn.

Family: Typhaceae.

English, Hindi and Regional Names:

Eng.	:	Elephant Grass;
Hindi	:	Mothithirina;
Beng	:	Bora;
Bom.	:	Fungl;
Gar.	:	Kanda Tella;
Guj.	:	Ghabajarin;
Kan.	:	Apu;
Kash.	:	Pitz, yira;
Mar.	:	Eraka;
Ori.	:	Hogla;
Pub.	:	Patira;
Tam.	:	Anaikorrai;
Tel.	:	Enugu Jammu.

Distribution : Common in Punjab, Deccan & Kutch.

Part Used : Leaves.

Utilization : Leaves and stem used for making mats, stuffing material, coarse textiles, marine ropes and fishing nets.

References : Raju (1998), Watt (1972).

Ulex europaeus Linn.

Family: Fabaceae.

English and Hindi Names:

Eng.	:	Furze, Gorse, Whin;
Hindi	:	Gorse.

Distribution : Naturalized on High Altitude of India.

Part Used : Foliage.

Utilization : Browsed by sheep and goats and also useful as a substitute for hay.

Active Constituents : Proteins, Ca, Na and β- carotene.

References : Agarwal (2003), Ambasta (1986), Dwivedi (1993).

Ulmus laevigata Royle (Pl. 8C)

Syn: *Ulmus erosa* Wall.; *U. pedunculata* Herb.; *U. villosa* Brandis *ex* Gamble Man.; *U. wallichiana* Planch.

Family: Ulmaceae.

Hindi and Regional Names:

Hindi : Chambar, Maya, Mored, Moral;
Kash. : Brari, Breri;
Kum. : Emroi, Imroi, Mai;
Pub. : Brera, Chipal, Kai, Kain, Manu, Marrun

Distribution : Kashmir to U.P. upto 3,000 m.

Parts Used : Young Branches. Leaves. Bark.

Utilization : Young branches and leaves used as fodder. Bark contained a strong fibre useful for cordage, string, sandals and ropes.

References : Agarwal (2003), Ambasta (1986), Dwivedi (1993), Hooker (1890), Parker (1973), Purohit & Samant (1995), Sahni (1998), Singh (1982), Watt (1972).

Urena lobata Linn. Mast in part.

Syn: *U. cana* Wall.; *U. palmata* Roxb.

Family: Malvaceae.

Vern. Name: Badi- dredai.

English, Hindi, Sanskrit and Regional Names:

Eng. : Congo jute;
Hindi : Bachita, Kunjuya, Lopetua, Pithia;
Sans. : Atibala;
Beng : Ban orhra;
Kan. : Otte;
Mal. : Uram, Uran;
Mar. : Van Bhendi;
Ori. : Bilokapasiva;
Tam. : Ottatti, Ottututti;
Tel. : Peddabenda.

Distribution : Hotter Parts of India; Frequent in Waste Places, Bamboo & Mango Clumps of Bengal.

Part Used : Stem Bark.

Utilization : Yield a fibre, Aramina or Congo jute which is ideal for Hessian, sacks, ropes, cordage, carpets, linoleum, vakveteen, artificial silk, soil cloth, hammock, netting and fishing-lines. Waste fibre used for absorbing grease oil in machine industry and substitute for flax.

References : Agarwal (2003), Ambasta (1986), Dastur (1964), Hosagoudar & Henry (1996), Hooker (1872) Jain (1991), Oommachan (1977), Parker (1973), Prasad *et al.* (2002), Saklani & Jain (1994), Singh (1982), Singh & Kumar (2000), Sood & Thakur (2004), Varghese (1996), Watt (1972).

Urochola panicoides Beauv.

Syn: *Panicum javanicum* Poir.

Family: Poaceae.

Hindi and Regional Names:

Hindi : Kuri, Kuriya;
Beng : Barajal ganti, Jalghanti;
Guj. : Kuri;
Kan. : Kaadubilli- samal hullu;
Pub. : Chatta, Jhun, Kowin.

Distribution : Cultivated on the Plains of W. India.

Part Used : Aerial Parts.

Utilization : Used as a good fodder for horses and cattle.

References : Agarwal (2003), Ambasta (1986).

Urtica dioica Linn.

Family: Urticaceae.

Vern. Name: Choti Kogsi.

English and Hindi Names:

Eng. : Stinging Nettle;
Hindi : Bhichoo Buti.

Distribution : Kashmir to Kumaon Hills upto 3,500 m.

Parts Used : Tender Shoots. Stem.

Utilization : Tender shoots and dried plants useful as nutritive fodder and enhance lactation. Stem fibre said to rival the best hemp in strength and used to make cords, cloths, bags, sacks and fishing - nets.

Active Constituents : Leaves rich in proteins, mineral contents, chlorophyllin and an acidic carbohydrate protein polymer.

References : Agarwal (2003), Ambasta (1986), Hooker (1890), Monandhar (2000), Rastogi & Mehrotra (1991), Singh & Kumar (2000), Sood & Thakur (2004), Watt (1972).

Urtica parviflora Roxb.

Syn: *Urtica ardens* Link.; *U. virulenta* Wall.

Family: Urticaceae.

Regional Names:

Beng : Paharah- bichuti.

Distribution : Temperate Himalaya, Kashmir to Mishmi, (Nilgiri Hills) Ootacamunda: Very Common on Wastelands & in Forest Bordering Streams.

Parts Used : Leaves. Stems.

Utilization : Leaves used as fodder. Stem fibre used for making ropes and mats.

Active Constituents : Leaves contain acetyle-choline (318.8 Hg/g) and 5-hydroxyl- tryptanine (0.25Mg /g) and histamine.

References : Ambasta (1986), Dastur (1964), Dwivedi (1993), Hooker (1890), Purohit & Samant (1995), Rastogi & Mehrotra (1990), Singh (2000), Watt (1972).

Ventilago denticulata Willd.

Syn: *Ventilago calyculata* Tul.

Family: Rhamnaceae.

Hindi and Regional Names:

Hindi : Pitti, Raidhani;

Bom. : Kayel, Karkandichayeh;

Kan. : Gapsandiballi, Harugasuratichekka;
Kum. : Kala lag;
Mar. : Sakalyel;
Ori. : Pittoli;

Distribution : Throughout India from the Himalaya Southwards.

Part Used : Bark.

Utilization : Bark yields a fibre for cordage.

References : Agarwal (2003), Ambasta (1986), Jain (1991), Parrotta (2001), Talbot (1976b), Varghese (1996).

Viburnum continifolium Don.

Syn: *Viburnum polycarpum* Wall.
Family: Caprifoliaceae.

Vern. Name: Jangli dhakh.

Regional Names:

Pub. : Guch, Jawa, Khatip, Rajab, Richhuklu.

Distribution : Temperate Himalaya: (Kashmir to Kamaon) 2,000-4,000 m.

Part Used : Plant.

Utilization : Dried plant fed to animals for enhancing lactation.

References : Dwivedi (1993), Hooker (1882), Purohit & Samant (1995), Watt (1972).

Vicia hirsuta S. F. Gray.

Syn: *Ervum filiforme* Roxb.; *E. hirsutum* Linn.; *E. lens* Wall.
Family: Fabaceae.

Vern. Name: Choti Kaer.

English, Hindi and Regional Names:

Eng. : Hairy tare, Tiny Vetch;
Hindi : Jhunjhuni ankari, Gigla, Munmuna;
Beng : Musur chana;
Pub. : Akri.

Distribution : Widely Cultivated in Temperate & Sub-Tropical Regions.

Part Used : Leaves.

Utilization : Leaves used as fodder.

Active Constituents : Leaves yield 2-3% apiin and quercitrin.

References : Agarwal (2003), Ambasta (1986), Chopra *et al.* (1969), Khanna (2002), Sood & Thakur (2004), Watt (1972).

Vicia sativa Linn. (Pl. 8D)

Syn: *Vicia bobartii* Forst.; *V. pallida* Facquem.
Family: Fabaceae.

Vern. Names: Bada Kaer, Roothi.

English, Hindi and Regional Names:

Eng. : Common Vetch, Spring vetch;
Hindi : Akra, Ankara;
Beng : Ankari;
Ori. : Choni, Rothi.

Distribution : Throughout India.

Part Used : Whole Plant.

Utilization : Used as fodder.

Active Constituents : These are proteins, γ-L-glutamyl-β-alanine and γ-L-glutamyl-γ- aminobutyric acid (pods).

References : Ambasta (1986), Chopra *et al.*(1969), Dwivedi (1993), Khanna (2002), Rastogi & Mehrotra (1990), Singh & Kumar (2000), Sood & Thakur (2004), Watt (1972).

Vigna acontifolia (Jacq.) Marechal

Syn: *Phaseolus acontifolius* Jacq.
Family: Fabaceae.

English, Hindi, Sanskrit and Regional Names:

Eng. : Aconite Bean, Moth Bean;
Hindi : Basunta, Makushthaka, Vasunta;
Sans. : Bassunta, Makushthaka, Vasunta;

Beng	:	Kheri;
Guj.	:	Math, Muth;
Kan.	:	Madike;
Mar.	:	Math, Matki;
Tam.	:	Tulukkapayir.

Distribution : Cultivated Throughout India.

Part Used : Plant.

Utilization : Plant used as an excellent fodder for livestock.

Active Constituents : Rich in proteins.

References : Agarwal (2003), Ambasta (1986), Varghese (1996).

Vigna trilobata (Linn.)Verdcourt. (Pl. 9A)

Syn: *Dolichos trilobus* Linn.; *D. trilobatus* Linn.; *Phaseolus triloatus* Linn. Scherb.; *P. trilobus* Sensu Ait. & Auc.

Family: Fabaceae.

Hindi, Sanskrit and Regional Names:

Hindi	:	Mugam, Rakhalalai, Tringuli;
Sans.	:	Mudgaparni;
Beng	:	Mugan;
Mal.	:	Ceruvidukal;
Tam.	:	Panipayar.

Distribution : Throughout India upto 2,000 m.

Part Used : Plant.

Utilization : Plant used as green fodder, relished by cattle.

Reference : Ambasta (1986).

Vitis lanata Roxb.

Syn: *Cissus vitiginea* Roxb.; *V. cordifolia* Roth.; *V. heyneana* DC.; *V. indica* Hb. Ham.; *V. pentagona* Hb. Ham.; *V. rugosa* Wall.

Family: Vitaceae.

Regional Names:

Kum.	:	Pahar-phuta, Purain.

Distribution : Common in Forest, especially in the Hills of E. Bengal.

Part Used : Leaves.

Utilization : Relished by sheep and other animals in winter.

References : Hooker (1872), Purohit & Samant (1995), Watt (1972).

Vossia cuspidata Griff.

Family: Poaceae.

Part Used : Green Leaves.

Utilization : Green leaves used as fodder.

Reference : Ambasta (1986).

Wedlandia grandis Cowan.

Family: Rubiaceae.

Part Used : Leaves.

Utilization : Leaves used as fodder.

Reference : Saklani & Jain (1994).

Wendlandia exserata DC.

Syn: *Rondelatia cinerea* Wall.; *R . exerta* Roxb.; *R. heynei* Roem. & Sch.; *R. thyrsiflora & orissensis* Roth.; *Wendlandia cinera* Wall.; *W. heynei* (R. & S.) & Sant & Merch.

Family: Rubiaceae.

Hindi and Regional Names:

Hindi : Birsa, Tilai, Tilki;

Mar. : Padgira;

Pub. : Chilkiya, Pansira, Pudhara.

Distribution : Tropical Himalaya from Chenab Eastwards to Nepal & Sikkim, Orissa, C. India, Konkan.

Part Used : Leaves.

Utilization : Leaves browsed and also fed throughout the year.

References : Agarwal (2003), Dwivedi (1993), Hooker (1882), Purohit & Samant (1995), Singh (1982), Sood & Thakur (2004), Watt (1972).

Wikstroemia indica Mey.

Syn: *Wiksroemia viridiflora* Meissn.
Family: Thymelaeaceae.

English and Hindi Names:

Eng. : Small Leaf Salago;
Hindi : Chamlia.

Distribution : Gardens of Assam & Madras.

Part Used : Bark.

Utilization : Fibre used for ropes and high grade paper.

References : Agarwal (2003), Ambasta (1986), Hooker (1890).

Wissadula lescheaultiana Mast.

Syn: *W. rostrata* Planch.; *W. lescheultiana* Mast.
Family: Malvaceae.

Part Used : Bark.

Utilization : Bark fibre soft, silky and well adapted for spinning.

Reference : Ambasta (1986).

Withania sominifera Dunal.

Syn: *Physalis arborescens & tomentosa* Thunb.;
P. flexuosa Linn.; *P. sominifera* Link.
Family: Solanaceae.

English, Hindi, Sanskrit and Regional Names:

Eng. : Aswagandha, Winter cherry;
Hindi : Asgandh, Punir;
Sans. : Aswaganda;
Beng : Ashuagandh;
Bom. : Asgandh;
Guj. : Asan;
Kan. : Aswagandhi;
Mal. : Ammukirram;
Mar. : Askanda, Tilli;
Ori. : Asua gandha;
Pub. : Asgandh;

Raj.	:	Chirpotan;
Tam.	:	Amukurra;
Tel.	:	Pannerugudda.

Distribution : Throughout Drier, Sub-Tropical India; frequent in W. India.

Part Used : Tender Shoots.

Utilization : Used as fodder.

References : Dey *et al.* (1973), Hooker (1885), Raju (1998), Watt (1972).

Woodfordia fruticosa (Linn.) Kurz.

Syn: *Lythrum fructicosum* Linn.; *Woodfordia floribunda* Salisb.

Family: Lythraceae.

Vern. Name: Dhaval.

English, Hindi, Sanskrit and Regional Names:

Eng.	:	Fire Flame bush, Shiranjitea;
Hindi	:	Arishtas, Dhai;
Sans.	:	Agnijvala, Dhataki, Dhaatri;
Beng	:	Dawa, Dhadki, Dhan;
Guj.	:	Dhavadina, Dhavani;
Kash.	:	Thai, Thawi;
Mal.	:	Tatri;
Mar.	:	Dhayati, Dhavati;
Ori.	:	Dhatubiphula, Dgodo, Horibadi, Patalabeluri;
Pub.	:	Davi, Phul dhava;
Tam.	:	Kattati, Kattathi, Vakllakai;
Tel.	:	Gaji- godari.

Distribution : Throughout India as far Southwards as Karnataka & C. India upto 1,500 m.

Part Used : Leaves.

Utilization : Leaves fed to all ruminants.

Active Constituents : Contains ellagic acid, myricetin-3- galactoside and oleanolic acid.

References : Agarwal (2003), Purohit & Samant (1995), Rastogi & Mehrotra (1991, 1995), Sood *et al.* (1993).

Wrightia tomentosa Roem & Schult.

Syn: *Periploca arborea* Dennst.; *Wrightia arborea* (Dennst.) Mabb.; *W. coraia* Wall.; *W. hamiltoniana* Wall.; *W. mollissima* Wall.; *W. wallichii* A. DC.

Family: Apocynaceae.

Hindi and Regional Names:

Hindi : Dudhi;
Ass. : Atkuri;
Beng : Dudhkoraiya;
Bom. : Daria, Dudhi;
Guj. : Dudhlo;
Kan. : Bilikudigidda;
Kum. : Darula;
Mal. : Mailampala;
Mar. : Kala Inderjan;
Ori. : Pala Kurwan;
Pub. : Dudhi Keor, Kilawa;
Tam. : Pala;
Tel. : Pedda pala.

Distribution : Tropical India from the Indus Eastwards & Southwards to Sri Lanka.

Parts Used : Leaves. Floss.

Utilization : Tree yields leaf fodder of medium quality and lopped in M.P. Floss from the fruits used for stuffing.

References : Dwivedi (1993), Hooker (1982), Raju (1998), Singh (1982), Watt (1972).

Xylia xylocarpa (Roxb.) Taub.

Family: Fabaceae.

English, Hindi, Sanskrit and Regional Names:

Eng. : Irul;
Hindi : Jambu, Suria;

Sans. : Kanakakuli, Scimsapa;
Bom. : Jamba, Suria;
Kan. : Jambe, Shilpe;
Mal. : Irim pullam, Irumul, Irumulla;
Mar. : Jamba, Suria;
Ori. : Boja, Kongora;
Tam. : Aruvapalam , Iruly.

Distribution : C. & S. India & Orissa upto 600 m.

Part Used : Leaves.

Utilization : Leaves valued for fodder.

References : Bhandary *et al.* (1996), Bennet *et al.* (1991), Dwivedi (1993), Rastogi & Mehrotra (1991), Singh (1982).

Yucca aloiffolia Linn.

Family: Agavaceae.

English and Hindi Names:

Eng. : Beer Grass;
Hindi : Bhalu Ghas.

Distribution : Gardens of the Tropical Regions of India.

Part Used : Leaves.

Utilization : Fibre suitable for ropes, cordage and string.

References : Agarwal (2003), Ambasta (1986).

Yucca filamentosa L. (Pl. 9B)

Family: Agavaceae.

Hindi Name:

Hindi : Dhori ghas.

Distribution : Tropical Gardens of India.

Part Used : Leaves.

Utilization : Fibre suitable for ropes, cords, twine and coarse textiles.

References : Agarwal (2003), Ambasta (1986).

Yucca gloriosa Linn.

Syn: *Yucca recurvifolia* Salisb.
Family: Agavaceae.

English and Hindi Names:

Eng.	:	Adam's Needle;
Hindi	:	Arranot Grass.

Distribution : Tropical Gardens of India; Naturalized on the Himalaya & Tamil Nadu.

Part Used : Leaves.

Utilization : Leaf fibre used for mats, carpets and paper making.

References : Agarwal (2003), Ambasta (1986), Watt (1972).

Zanthoxylum acanthopodium DC.

Syn: *Zanthoxylum alatum* Wall.; *Z. armatum* DC.; *Z. hostile* Wall.; *Z. planispinum* Sieb & Zucc.
Family: Rutaceae.

Vern. Names: Tirmir, Turbar.

English, Hindi, Sanskrit and Regional Names:

Eng.	:	Toothache plant;
Hindi	:	Timur, Timoor;
Sans.	:	Mahamuni, Tejovati, Tumburu;
Beng	:	Gaira;
Guj.	:	Tejbal, Tejbala;
Kan.	:	Jimmi, Tejpatri;
Mal.	:	Thumboonal, Thumbooni;
Mar.	:	Tejabel;
Ori.	:	Tundapoda;
Pub.	:	Timmal, Tejbal;
Tam.	:	Thejyovathi;
Tel.	:	Tejovathi.

Distribution : Tropical Himalaya (Kumaon to Sikkim) ascending to 2,000 m.

Part Used : Leaves.

Utilization : Leaves browsed by sheep, goats and cattle.

Active Constituents : Leaves contain essential oils.

References : Agarwal (2003), Hooker (1872), Khare (2004), Kirtikar & Basu (1935), Purohit & Samant (1995), Singh (1982), Singh (1999), Singh (1982), Watt (1972).

Zea mays Linn.

Syn: *Zey alba, americana & vulgaris* Mill.; *Z. segetalis* Salisb.
Family: Poaceae.

English, Hindi, Sanskrit and Regional Names:

Eng. : Indian corn, Maize;
Hindi : Bhutta, Makka;
Sans. : Makayah, Yavanala;
Beng : Bhutta, Fanar, Junar;
Bom. : Buta, Makai;
Gar. : Junala, Mungari;
Guj. : Makkai;
Kum. : Bhutta Mukni;
Mal. : Cholam, Makka;
Mar. : Makka;
Ori. : Buta, Maka;
Pub. : Mak, Makki, Makkeii;
Tam. : Makka solam;
Tel. : Makka zonnalu, Makka jonna.

Distribution : Cultivated Throughout India.

Parts Used : Leaves. Stem.

Utilization : Leaves and stems used as fodder for cows and other animals.

Active Constituents : These contain proteins and glucoside.

References : Agarwal (2003), Dastur (1978), Drury (1964), Duthie (1888), Khan & Khanum (2005), Hooker (1897), Parrotta (2001), Rastogi & Mehrotra (1995a), Samati (2004), Watt (1972).

Zizyphus mauritiana Lam.

Syn: *Zizyphus jujuba* Lam.; non Mill.
Family: Rhamnaceae.

English, Hindi, Sanskrit and Regional Names:

Eng. : Common Jujube, Indian Jujube;
Hindi : Ber, Baer;
Sans. : Ajapriya Badara, Karkandu;
Beng : Ber, Boroi;
Guj. : Bor, Bordi;
Kan. : Elanji, Yalachi;
Mal. : Elentha;
Mar. : Bera, Bor;
Ori. : Bodokoli, Barkoli;
Pub. : Ber, Beri, Unab;
Tam. : Ellandai, Yellande.

Distribution : Cultivated Throughout India.

Part Used : Leaves.

Utilization : Tree yields fodder. Leaves fed to silkworms.

Active Constituents : Leaves contain crude proteins 15.32%, crude fibre 15.76%, total minerals 6.66%, reduced sugars 1.87%, total sugars 7.57% and starch 16.84%.

References : Agarwal (2003), Ambasta (1986), Dwivedi (1993), Prajapati *et al.* (2003), Rastogi & Mehrotra (1990, 1993), Sahni (1998), Singh (1982, 2002, 2003), Singh & Pandey (1998), Talbot (1976a).

Zizyphus nummularia (Brum. f.) Wight & Arn.

Syn: *Rhamnus nummularia* Burm. f.; *Zizyphus rotundifolia* Lam.
Family: Rhamnaceae.

Vern. Name: Ber.

English, Hindi, Sanskrit and Regional Names:

Eng. : Wild jujube;
Hindi : Jhahrberi, Jhadiaber;
Sans. : Bhukamtaka, Sukkhshamphala;

Bom. : Chanibor;
Guj. : Chanyobor;
Mar. : Gangar, Janglar;
Tam. : Korgodi;
Tel. : Neelareegu.

Distribution : Punjab, Rajasthan, Peninsular India & W. Ghats, Gujarat. Konkan.

Part Used : Leaves.

Utilization : Leaves used as fodder. They are also dried and stored for future use.

Active Constituents : Leaves contain zizynummin.

References : Agarwal (2003), Ambasta (1986), Dwivedi (1993), Hooker (1872), Oommachan (1977), Parker (1973), Rastogi & Mehrotra (1993), Singh (1982, 2002, 2003), Singh & Pandey (1998), Watt (1972).

Zizyphus rugosa Lam.

Family: Rhamnaceae.

Hindi and Regional Names:

Hindi : Suran, Churan;
Mal. : Todali;
Mar. : Churan, Turan.

Distribution : Sub -Himalayan Tract from the Ganges Eastwards upto 600 m, Bihar, C. India & W. India.

Part Used : Leaves.

Utilization : Leaves lopped for fodder.

Reference : Singh (1982).

Zizyphus xylopyrus Willd.

Syn: *Zizyphus xylocarpus* Willd.
Family: Rhamnaceae.

Vern. Name: Kath Ber.

Hindi and Regional Names:

Hindi	:	Kat- Ber, Kakor, Ghont;
Bom.	:	Ghat, Suti;
Kan.	:	Chalte, Mullukare;
Ori.	:	Got, Gotoboro, Kan tabohul;
Tam.	:	Kottei;
Tel.	:	Ghotte.

Distribution : Bihar, W. Peninsula from the Konkan Southwards & Sri Lanka.

Part Used : Leaves.

Utilization : Leaves useful as fodder for cattle and goats.

Active Constituents : Leaves contain 6.78% ash, 1.62% Ca, 46.78% C and 1.98% N.

References : Agarwal (2003), Ambasta (1986), Oommachan (1977), Singh (1982) Singh & Pandey (1998), Watt (1972).

□□□

5

Epilogue

Altogether 811 plants spp belonging to 507 genera under 132 families have been delineated as fibre and fodder resources of India. Of these, 3 spp (3 genera: *Abies, Thuja, Taxus*) belong to gymnosperms; 624 spp (403 genera) to dicots and 184 spp (101 genera: *Agave, Agropyron, Agrostis, Aloe, Alopecurus, Alpina, Apluda, Arundinaria, Arundinellia, Avena, Bambusa, Blyxa, Bothriochola, Brachiaria, Bromus, Canna, Carex, Cenchrus, Chimnobambusa, Chloris, Chrysopogon, Coix, Commelina, Cynodon, Cynotis, Cyperus, Dactylis, Dendrocalamus, Digitaria, Dinebra, Dracaena, Eichhornia, Eichinochola, Eleocharis, Ensete, Eragostiela, Eragrostis, Eremopogon, Erianthus, Eriochola, Euchlaena, Eulalia, Eulaliopsis, Floscopa, Furcarea, Glyceria, Hedychium, Hordeum, Hymenachne, Ichanthus, Imperata, Iris, Iseilema, Juncus, Kobresia, Koeleria, Kyllinga, Lasiurus, Leersia, Lolium, Mariscus, Microcos, Microstegium, Muhelenbergia, Muntinga, Musa, Nannorrhops, Ophiurus, Oryza, Panicum, Paspalidium, Paspalum, Pennisetum, Perotis, Phararis, Phormium, Phragmites, Plumeria, Poa, Pollinia, Raphidophora, Saccharum, Sacciolepis, Sansevieria, Schizachyrium, Scirpus, Sclerostachya, Sehima, Setaria, Smilax, Sorghum, Sporobolus, Thamnocalamus, Themeda, Thysanolaena, Trisetum, Typha, Urochola, Vossia, Yucca, Zea*) to monocots (Fig. 2; Appendix I).

Habit-wise analysis indicates (Table I; Fig.1) that 483 spp having scattered or distribution in entire India predominate in usage; followed by 92 spp in North East and Eastern Himalaya, 85 spp in North West and Western Himalaya, 56 spp in Entire Himalaya, 48 spp in Central Highlands and Plains of India and 47 spp in Peninsular India (Ghats, Coastal Plains, Deccan Plateau). Evidently, all the tribals and rural populace of India make plenty use of fibre and fodder resources available to them in their surroundings.

From the family-wise analysis (Table II; Fig. 3; Appendix II) it is inferred that Fabaceae with 58 genera and 110 spp predominate in usage amongst dicots. Other families are: Malvaceae (14 genera, 39 spp), Moraceae (7 genera, 32 spp), Arecaceae (13 genera, 21 spp), Asteraceae (21 genera, 21 spp), Rosaceae (8 genera, 18 spp), Tiliaceae (9 genera, 18 spp), Urticaceae (11 genera, 18 spp), Sterculiaceae (10 genera, 16 spp), Asclepiadaceae (11 genera, 13 spp), Euphorbiaceae (9 Genera, 12 spp), Apocynaceae (10 Genera, 11 spp), Rubiaceae (10 genera, 11 spp), Rhamnaceae (5 genera, 11 spp), Lauraceae (5 genera, 10 spp), Acanthaceae (6 genera, 9 spp), Salicaceae (2 genera, 9 spp), Scrophulariaceae (7 genera, 8 spp), Anacardiaceae (4 genera, 7 spp), Combretaceae (3 genera, 7 spp), Polygonaceae (4 genera, 7 spp), Thymelaeaceae (6 genera, 7 spp), Ulmaceae (3 genera, 7 spp), Brassicaceae (4 genera, 6 spp), Chenopodiaceae (5 genera, 6 spp), Fagaceae (3 genera, 6 spp), Sapotaceae (6 genera, 6 spp), Boraginaceae (3 genera, 6 spp), Aceraceae (1 genus, 5 spp), Lythraceae (4 genera, 5 spp), Amarantaceae (5 genera, 5 spp), Annonaceae (4 genera, 5 spp), Berberidaceae (1 Genus, 5 spp), Bignoniaceae (5 genera, 5 spp), Ranunculaceae (4 genera, 5 spp), Vitaceae (3 genera, 5 spp), Apiaceae (4 genera, 4 spp), Betulaceae (2 genera, 4 spp), Bombacaceae (4 genera, 4 spp), Convolvulaceae (2 genera, 4 spp), Cucurbitaceae (4 genera, 4 spp), Lamiaceae (4 genera, 4 spp), Menispermaceae (4 genera, 4 spp), Verbenaceae (3 genera, 4 spp), Sapindaceae (3 Genera, 4 spp), Capparidaceae (2 genera, 3 spp), Rutaceae (3 genera, 3 spp), Solanaceae (3 genera, 3 spp), Tamaricaceae (2 genera, 3 spp), Meliaceae (3 genera, 3 spp), etc. For monocots, the predominant families are Poaceae (72 genera, 129 spp), Cyperaceae (7 genera, 17 spp), Liliaceae (5 genera, 9 spp), Agavaceae (2 genera, 8 spp), Amaryllidaceae (1 Genus, 3 spp), Commelinaceae (3 genera, 3 spp) and Musaceae (2 genera, 3 spp).

Table I. Distribution Pattern of Fibre and Fodder Resources of India

Distribution	No. of Species
Entire India or Scattered Distribution	483
N.W. & W. Himalaya	85
N. E. & E. Himalaya	92
Peninsular India (Ghats, Coastal Plains, Deccan Plateau)	47
Central Highlands & Plains of India	48
Entire Himalaya	56

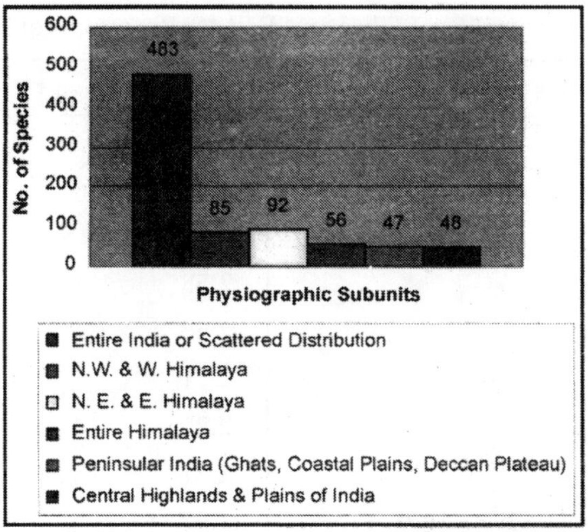

Fig. 1. Histogram Showing Distribution Pattern of Fibre and Fodder Resources of India

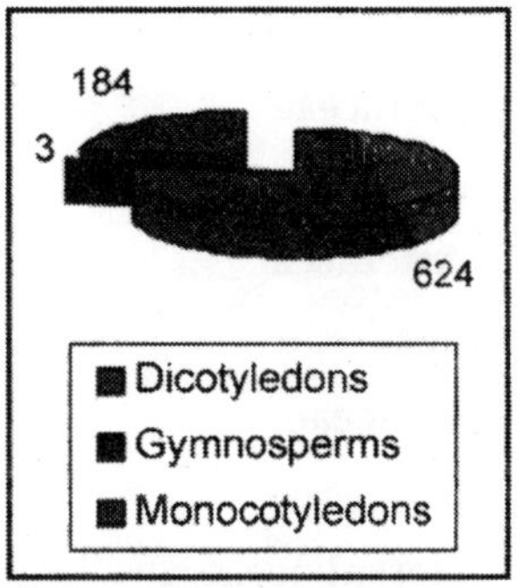

Fig. 2. Total Number of Fibre and Fodder Species Under Various Divisions of Plant Kingdom

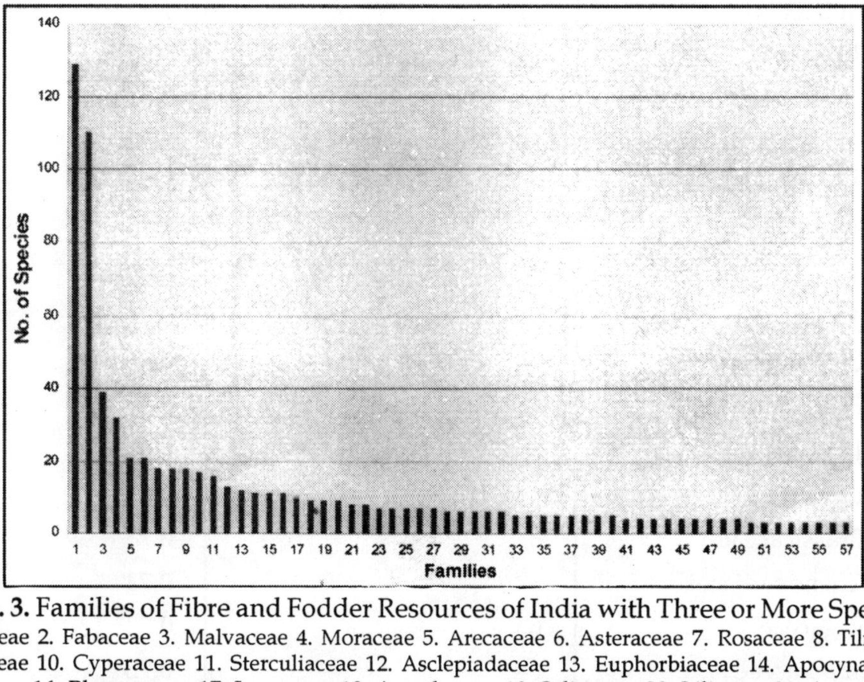

Fig. 3. Families of Fibre and Fodder Resources of India with Three or More Species

1. Poaceae 2. Fabaceae 3. Malvaceae 4. Moraceae 5. Arecaceae 6. Asteraceae 7. Rosaceae 8. Tiliaceae 9. Urticaceae 10. Cyperaceae 11. Sterculiaceae 12. Asclepiadaceae 13. Euphorbiaceae 14. Apocynaceae 15. Rubiaceae 16. Rhamnaceae 17. Lauraceae 18. Acanthaceae 19. Salicaceae 20. Liliaceae 21. Agavaceae 22. Scrophulariaceae 23. Anacardiaceae 24. Combretaceae 25. Polygonaceae 26. Thymelaceae 27. Ulmaceae 28. Brassicaceae 29. Chenopodiaceae 30. Fagaceae 31. Borangiaceae 32. Sapotaceae 33. Aceraceae 34. Amarantaceae 35. Annonaceae 36. Berberidaceae 37. Bignoniaceae 38. Lythraceae 39. Ranunculaceae 40. Vitaceae 41. Apiaceae 42. Betulaceae 43. Bombaceaeae 44. Convolvulaceae 45. Cucurbitaceae 46. Lamiaceae 47. Menispermaceae 48. Sapindaceae 49. Verbenaceae 50. Amaryllidaceae 51. Cappariadaceae 52. Commelinaceae 53. Musaceae 54. Rutaceae 55. Solanaceae 56. Tamaricaceae 57. Meliaceae.

Table II. Predominant Dicotyledonous Families of Fibre and Fodder Resources of India

Family	Species	% in the Taxa
Fabaceae	110	13.56
Malvaceae	39	4.80
Moraceae	32	3.94
Arecaceae	21	2.58
Asteraceae	21	2.58
Rosaceae	18	2.21
Tiliaceae	18	2.21
Urticaceae	18	2.21
Sterculiaceae	16	1.97
Asclepiadaceae	13	1.60
Euphorbiaceae	12	1.47
Apocynaceae	11	1.35
Rubiaceae	11	1.35
Rhamnaceae	11	1.35
Lauraceae	10	1.23
Acanthaceae	9	1.10
Salicaceae	9	1.10
Scrophulariaceae	8	0.98
Anacardiaceae	7	0.86
Combretaceae	7	0.86
Polygonaceae	7	0.86
Thymelaeaceae	7	0.86
Ulmaceae	7	0.86
Brassicaceae	6	0.73
Chenopodiaceae	6	0.73
Fagaceae	6	0.73
Sapotaceae	6	0.73
Boraginaceae	6	0.73
Aceraceae	5	0.61

Contd...

Table II. Contd...

Lythraceae	5	0.61
Amarantaceae	5	0.61
Annonaceae	5	0.61
Berberidaceae	5	0.61
Bignoniaceae	5	0.61
Ranunculaceae	5	0.61
Vitaceae	5	0.61
Apiaceae	4	0.49
Betulaceae	4	0.49
Bombacaceae	4	0.49
Convolvulaceae	4	0.49
Cucurbitaceae	4	0.49
Lamiaceae	4	0.49
Menispermaceae	4	0.49
Verbenaceae	4	0.49
Sapindaceae	4	0.49
Capparidaceae	3	0.36
Rutaceae	3	0.36
Solanaceae	3	0.36
Tamaricaceae	3	0.36
Meliaceae	3	0.36

Table III. Predominant Monocotyledonous Families of Fibre and Fodder Resources of India

Family	No. of Species	% in the Taxa
Poaceae	129	15.90
Cyperaceae	17	2.09
Liliaceae	9	1.10
Agavaceae	8	0.98
Amaryllinaceae	3	0.36
Commelinaceae	3	0.36
Musaceae	3	0.36

Table IV. Nutritional Components of Some Fodder Plants of India*

Name of Plant	Carbohydrates	Proteins	Fats	Ash	Fibres	Starch	Minerals	Vitamins
Amaranthus viridus	-	77.8%	-	-	-	-	-	-
Anogeissus pendula	-	7.60%	-	8.10%	19.0%	-	P, 0.10%; Ca, 3.50%	-
Artemisia sieversiana	-	15.5%	5.12%	-	-	-	-	-
Artocarpus heterophyllus	-	11-14%	-	-	18-22%	-	Ca, 0.52-2.15; P, 11-12	-
Avena sativa	-	87%	-	-	-	-	-	-
Bauhinia malabarica	-	-	-	8.83	-	-	N, 2.18; C, 45.6	-
B. retusa	-	-	-	7.84	-	-	Ca, 7.73; C, 46.20; N, 2.03	-
B. variegata	72.36	15.6	-	-	-	14.3	-	C
Boehmeria rugolosa	-	-	-	24.96	-	-	-	-
Bombax ceiba	11.95	-	-	1.09	1.38	-	-	-
Brassica oleracea	-	-	-	-	-	-	I, 1.4	-
Bridelia retusa	-	-	-	10.75	-	-	Ca, 1.30; C, 44; N, 2-3	-
Buchanania lanzan	-	21.6	-	11.10	-	-	C, 2.42; N, 1.6	-
Castanopsis tribuloides	-	-	-	3.6	-	-	-	-
Celtis australis	-	14	-	-	-	-	I, Cu	-
Corda dichotoma	-	12-15.13	-	-	16-25	-	Ca, 4.42; P, 0.30	-
Coriandrium sativum	-	-	-	-	-	-	-	A
Cynodon dactylon	-	10.47	-	-	-	-	-	-
Dactylis glomerata	50.0	-	-	-	-	-	-	-
Diospyros melanoxylon	-	25.28	-	-	-	-	-	-
Dolichos lablab	-	5	-	-	-	-	Ca, I	C
Eichnochola stagnina	10	-	-	-	-	-	-	-
Embelica officinalis	-	-	-	4.48	-	-	Ca, N	-
Eragrostiella bifaria	43	2.50	1.77	-	-	-	-	A
Eragrostis plumose	-	9.91	-	-	33.56	-	-	-
Eremopogon foveolatus	-	2.27	1.32	7.42	37.36	-	-	-
Eulalia cumingii	-	6.90	-	8.49	38.67	-	-	-
Ficus benghalensis	-	9.63	-	-	26.84	-	-	-
F. glomerata	-	9.63	-	-	-	-	-	-

Contd...

F. religiosa	-	13.99	-	15.06	22.36	-	P, 0.52; Cao, 4.64	-
Flacourtia indica	-	-	-	-	-	-	Ca, N, 2.42	-
Galinsoga parviflora	34.04	10.93	0.76	15.82	38.44	-	-	-
Garuga pinnata	-	-	-	-	-	-	C, 1.94; Ca, 44.15	-
Glyceria tonglensis	-	9.69	0.43	-	0.21	75.06	-	-
Grewia oppostifolia	-	10.1	-	14.2	-	-	-	-
Helicteres isora	-	13-25	-	-	-	-	CaO	-
Hordeum vulgare	69.3	11.5	-	-	3.9	-	I, 3.7; CaO	B
Ipomoea aquatica	-	-	-	-	-	-	Min. 2%	B, C, E
I. digitata	6.3	-	-	-	-	-	-	-
I. eriocarpa	35.12	15.26	-	-	20-70	-	Min. matter, 8.88	-
Iseilema laxum	-	5.67	-	-	28.03	-	N, 51.78; CaO, 1.30	-
Juglans regia	-	16	-	-	-	-	N, 2.75; K, 6.85; Ca, 6.1; Mg, 13; I, 2.8	B, C
Koeleria cristata	41.68	10.96	-	-	24.73	-	CaO, 0.87	-
Kydia calycina	41.68	11.09	-	-	14-23.71	-	Ca, 3.06; P, 0.31	-
Kyllinga brevifolia	-	8.47	0.94	-	29.86	-	-	-
Lepidium sativum	-	-	-	-	-	-	-	A
Leucaena glauca	-	24.2	-	-	13.55	-	-	-
Lolium multiflorum	-	-	-	-	-	-	-	B
Lycopersicon esculentum	-	-	-	-	-	-	Ca, 20; Fe, 2.4	-
Manilkara hexandra	-	23.14	-	-	54.94	-	Ca, 1.49; P, 0.35	A,C
Mikania cordata	-	-	-	-	-	-	-	A, C
Mitragyana parvifolia	-	7.73	-	-	19.97	-	Ca, 2.27; P, 0.24	-
Morus alba	-	71.20	-	-	54.13	-	-	C
M. serrata	-	11.48	-	-	34.02	-	Min., 1.76	-
Muhelenbergia huegelii	44.8	12.8	2.46	-	29.7	-	-	-
Opuntia ficusindica	-	-	-	-	-	-	-	C
Panicum maximum	-	-	-	-	-	-	-	B, C
P. miliaceum	-	18	-	-	-	-	-	-
Parkinsonia aculeata	-	7.5	-	-	29.0	-	Ca, 4.16; P, 0.17	-
Pennisetum purpureum	-	1.0	-	-	7.4	-	Min. matter, 3.1, Ca, 0.12, P, 0.07	A, B, C

Contd...

Pistia stratoides	2.65	1.4	0.35	-	0.9	-	Ca, 0.20; P, 0.06	-
Pithcellobium dulce	-	29	-	-	17.5	-	Ca, 14; P, 0.35	-
Pongamia pinnata	20.20	17.62	-	-	-	-	Min. 13.35	-
Prangos pabularia	-	10.4	3.55	-	22.6	-	-	-
Prosopis cineraria	-	32.53	-	-	50.29	-	-	-
Pueraria tuberosa	31.65	23.8	0.7	-	30.75	-	Ca, 1.6; P, 0.2	-
Quercus incana	-	10-11.42	-	-	31-32.1	-	-	-
Q. semecarpinifolia	-	-	-	4.45	-	-	N, 1.37	-
Randia spinosa	-	-	-	-	-	-	N, 1.77; Ca, 2.51	-
Robinia pseudocasia	-	25.5	-	-	17.2	-	Ca, 1.55; P, 0.32; Min. matter, 7.55	-
Salix tetrasperma	-	-	-	-	-	-	Ca, 2.71; N, 2.07	-
Salvadora olooides	22.0	26.66	-	-	-	-	P, 1.68	-
Sapindus mukorossi	-	-	-	-	-	-	N, 1.79; CaO, 1.99	-
Saurauia nepaulensia	-	12.24	-	-	18.40	-	P, 0.24; Ca, 2.71	-
Shorea robusta	-	10.6	-	-	27.43	-	Ca, 0.77; P, 0.12	-
Stereospermum personatum	-	-	-	-	-	-	N, 1.81; C, 43.4; Cu, 1.67; Mn, 54.0	-
Tamarindus indica	18.2	5.8	2.1	-	1.9	70.7	Ca, 10; Cl, 94; Mg, 71; P, 140; Cu, 2.09	-
Terminalia arjuna	-	10.10	-	-	-	5.75	Ca, 2.86; Mn, 0.48; N, 1.36-2.05	-
T. belerica	-	8.63	-	11.72	16-18.59	-	Ca, 3.06; P, 0.27	-
T. tomentosa	-	10.92	-	-	12.41	15.2	Min. matter, 7.21	-
Thysanolaena maxima	-	9.5	-	-	30.4	-	Ca, 0.23; P, 0.11	-
Tilia europaea	-	16.2	2.9	11.4	13.2	-	-	-
Trifolium repens	-	-	-	-	-	-	Ca, P	-
Zizyphus mauritiana	-	15.32	-	-	15.26	16.84	Min. matter, 6.66	-
Z. xylocarpus	-	-	-	6.78	-	-	Ca, 1.62; N, 1.98; C, 46.78	-

* Data Compiled from: The Wealth of India; Raw Materials. CSIR, New Delhi; Fodder Trees and Shrubs of Central Himalaya. Gyanodaya Prakashan, Nanital.

(-) value not available

For Fabaceae, the overall percentage in the distribution of taxa is 13.56 whereas for Poaceae amongst monocots, it is 15.90 (Tables II, III). It is apparent that the members of Poaceae far exceed than the predominant Fabaceae amongst dicotyledons in being used for fibre and fodder resources by the ethnic people of India.

As many as 224 spp are employed exclusively fibre purposes in India (Appendix III). Of these, some of the predominantly employed genera are: *Hibiscus* (7 spp), *Abutilon* (6 spp), *Sida* (6 spp), *Abelmoschus* (5 spp), *Agave* (5 spp), *Grewia* (4 spp), *Sterculia* (4 spp), *Cissus* (3 spp), *Crotolaria* (3 spp), *Furcarea* (3 spp), *Gossypium* (3 spp), *Sansevieria* (3 spp), *Yucca* (3 spp), *Annona* (2 spp), *Boehmeria* (2 spp), *Calamus* (2 spp), *Calotropis* (2 spp), *Caryota* (2 spp), *Corypha* (2 spp), *Daphne* (2 spp), *Livistonia* (2 spp), *Marsdenia* (2 spp), *Melochia* (2 spp), *Oreocnida* (2 spp), *Raphia* (2 spp), *Saccharum* (2 spp), *Sesbania* (2 spp), *Smilax* (2 spp), *Triumfetta* (2 spp), *Typha* (2 spp), etc. (Appendix IV). Over 519 spp are being used exclusively for fodder (Appendix V), and the most frequently employed genera for this purpose are: *Ficus* (11 spp), *Panicum* (9 spp), *Cyperus* (8 spp), *Pennisetum* (6 spp), *Acer* (5 spp), *Berberis* (5 spp), *Rosa* (5 spp), *Salix* (5 spp), *Albizia* (4 spp), *Astragalus* (4 spp), *Brachiaria* (4 spp), *Celtis* (4 spp), *Dalbergia* (4 spp), *Digitaria* (4 spp), *Indigofera* (4 spp), *Morus* (4 spp), *Litsea* (4 spp), *Mucuna* (4 spp), *Poa* (4 spp), *Polygonum* (4 spp), *Quercus* (4 spp), *Rhus* (4 spp), *Sorghum* (4 spp), *Sporobolus* (4 spp), *Zizyphus* (4 spp), *Acacia* (3 spp), *Agrostis* (3 spp), *Chrysopogon* (3 spp), *Desmodium* (3 spp), *Flemingia* (3 spp), *Ipomoea* (3 spp), *Oryza* (3 spp), *Paspalum* (3 spp), *Persea* (3 spp), *Populus* (3 spp), *Prunus* (3 spp), *Rubus* (3 spp), *Strobolanthes* (3 spp), *Terminalia* (3 spp), etc. (Appendix VI). Nutritional analysis provided in Table IV also point towards their high fodder value. Besides, about 68 spp are being used for fibre and fodder purposes in India (Appendix VII). Some of the important genera under this category are: *Bauhinia* (4 spp), *Ficus* (4 spp), *Acacia* (3 spp), *Artocarpus* (2 spp), *Bambusa* (2 spp), *Debregeasia* (2 spp), *Phoenix* (2 spp), *Trema* (2 spp), etc. (Appendix VIII).

Hopefully, the generated base-line data on traditional wisdom regarding fibre and fodder utilization would essentially be utilized by various government agencies and policy-makers in formulating policies for uplifting socio-economic life of the tribals and rural populace of India. Moreover, it would be of tremendous significance in the modern management practices of fibre and fodder resources; and also can serve as a genetic resource for breeding new varieties.

□□□

6

Literature Cited

Agarwal, V. S. 1986. *Economic Plants of India.* Bishen Singh Mahendra Pal Singh, Dehradun & Kailash Prakashan, Calcutta.

Agarwal, V. S. 2003. *Directory of Indian Economic Plants.* Bishen Singh Mahendra Pal Singh, Dehradun.

Ainslie, W. 1926. *Materia Medica.* Neeraj Publishing House, Delhi.

Alcorn, J. 1984. *Huaestec Mayan Ethnobotany* (Austin University Press: Texas, USA)

Ambasta, S. P. 1986. *The Useful Plants of India.* Publication & Information Directorate, CSIR. New Delhi.

Anonymous, (1948-1976). *The Wealth of India, Raw Materials.* Vol. 1 - 9. Govt. of India, New Delhi.

Anonymous, 1984. *Ethnobiology in India- A Status Report.* Ministry of Environment and Forest, Govt. of India.

Arora, R. K. 1971. Job's tears (*Coix lacryma-jobi*) – minor food and fodder crop of northeastern India. *Econ. Bot. 31*(3): 358-366.

Arora, R. K. 1990. Natural food plants of the tribals in northeast India: 137-152. In Jain, S. K. (*ed.*) *Contribution to Ethnobotany of India.* Scientific Publishers, Jodhpur (India).

Arora, R. K. 1995. Ethnobotanical studies on plant genetic resources- national efforts and concerns. *Ethnobotany 7* : 125-136.

Arya, K. R., Pande, P. C. & Prakesh, V. 1999. Ethnobotanical studies on tribal areas of Almora District-11. *Ethnobotany 11* (1 & 2): 100-104.

Atkinson, E. J. 1982. Economic Botany of the Himalayan Region. Cosmos Pub., New Delhi.

Awasthi, A. K. & Goel, A. K. 2000. Ethnobotanical studies on the tribes Onge from Little Andaman Island: 569-573. In Maheshwari, J. K. (*ed.*) *Ethnobotany and Medicinal Plants of Indian Subcontinent.* Scientific Publishers, Jodhpur, (India).

Bajpayee, K. K. & Dixit, G. 1996. Ethnobotanical studies on food stuffs of tribals of Tarai Region, Uttar Pradesh. *J. Econ. Taxon. Bot. Add. Series 12*: 128-132. Scientific Publishers, Jodhpur (India).

Basnet, D. B. 1999. *Edgeworthia gardneri* Meissn- A source of raw material of eco-friendly traditional hand-made paper of Eastern Himalaya. *Ethnobotany 11*: 115-118.

Bennet, S. S. R. 1986. *Name Changes in Flowering Plants of India and Adjacent Regions*. Triseas Publishers. Dehra Dun, India.

Bennet, S. S. R., Biswas, S., Chandra, V., Singh, P. J., Chandra, S. & Dayal, R. 1991. *Food From Forests*. Forest Resarch Institute. ICFRE, Dehradun.

Bhandary, M.J., Chandrashekar, K.R. & Kaveriappa, K.M. 1996. Ethnobotany of gowlis of Uttara Kannada Dist. Karnataka. *J. Econ. Taxon. Bot. Add. Series. 12*: 244-249 Scientific Publishers, Jodhpur (India).

Bhargava, N. 1981. Plants in folk life and folklore in Andaman and Nicobar Islands: 329-344. In Jain, S. K. (*ed.*) *Glimpses of Indian Ethnobotany*. Oxford, IBH Publishing Co., . New Delhi.

Bhargava, B., Katiyar, U. P. & Saxena, R. P. 1977. A note on the nutritive value of rainy (*Mallotus phillippinensis*) tree leaves as sole feed for sheep. *Indian J. Anim. Sci. 47*(9): 594-595.

Bhattacharjee, S. K. 1998. *Handbook of Medicinal Plants*. (Third Revised Edition) Pointer Publishers, Jaipur (Raj.) India.

Blatter, E. & Hallberg, F. 1984. *The Flora of the Indian Desert. (Jodhpur And Jaisalmer)*. Scientific Publishers, Jodhpur (India).

Binu, S., Nayar, T. S. & Pushpangadan, P. 1992. An outline of ethnobotanical research in India. *J. Econ. Tax. Bot. Addl. Series 10*: 405-428.

Bor, N. R. 1960. *The Grasses of Burma, Ceylon, India and Pakistan (Excluding Bambuseae)*. International Book Distributers, Dehradun.

Brandis, D. 1971. *Indian Trees*. Bishen Singh Mahendra Pal Singh. Dehradun.

Burkill, I. H. 1935. *Dictionary of Economic Products of India*.

Caius, Rev. Fr. Jean Ferdinand. 2003. *The Medicinal and Poisonous Plants of India*. Scientific Publishers, Jodhpur (India).

Chakraberty, C. 1923. *Comparative Hindu Materia Medica*. Neeraj Publishing House, Delhi.

Chakrabarty, T. & Balakrishan, N. P. 2003. Ethnobotany of the Andaman and Nicobar Islands, India- A Review: 869-889. In Singh, V. & Jain, A. P. (*eds.*) *Ethnobotany and Medicinal Plants of India and Nepal*. Vol. 2. Scientific Publishers, Jodhpur (India).

Chakraborty, M. K. & Bhattacharjee, A. 2003. Plants used for thatching purpose by the tribes of Purulia districts, West Bengal, India: 571-572. In Singh, V. & Jain, A. P. (*eds.*) *Ethnobotany and Medicinal Plants of India and Nepal*. Vol. 2. Scientific Publishers, Jodhpur (India).

Chandrasekar, K. & Srivastava, S. K. 2003. Ethnomedicinal studies in Pin Valley National Park, Lahaul- Spiti, Himachal Pradesh. *Ethnobotany 15* (1 & 2): 44-47.

Chatterjee, A. & Pakrashi, S. C. (*eds.*) 1997. *The Treatise on Indian Medicinal Plants.* Vol.1-Vol. 5. National Institute of Science Communication (CSIR), New Delhi.

Chaurasia, O. P. Singh, B. & Sareen, S. K. 2000. Ethnomedicinal plants of arctic desert Ladakh used in veterinary practices: 161-166. In Maheshwari, J. K. (*ed.*) *Ethnobotany and Medicinal Plants of Indian Subcontinent.* Scientific Publishers, Jodhpur (India).

Chet Ram & Ray, S. C. 1943. Tree leaves as cattle fodder II. Digestibility coefficient and nutritive value of some tree species. *Indian J. Vet. Sci. 13*: 191-195

Chopra, R. N., Chopra, I. C. & Verma, B. S. 1969. *Supplement to Glossary of Indian Medicinal Plants.* Publications & Information Directorate, New Delhi.

Chopra, R. N., Nayar, S. L. & Chopra, I. C. 1956. *Glossary of Indian Medicinal Plants.* CSIR, New Delhi.

Dagar, H. S. 2003. Ethnobotanical evaluation of some endemic plants in Andaman and Nicobar Islands: 134-147. In Khan, S. S. *Vistas in Ethnobotany.* Vol. 1. Classic Printers & Publishers. Bhopal, India.

Dagar, H. S. & Dagar, J. C. 1996. Ethnobotanical studies of the Nicobarese of Chowra island of Nicobar group of islands, *J. Econ. Taxon. Add. Series 12*: 381-388. Scientific Publishers, Jodhpur (India).

Dagar, S. S. & Dagar, J. C. 2000. Plant folk medicines for gynaecological, urino-genital and other related problems among arborigines of Andaman & Nicobar islands: 561-567. In Maheshwari, J.K. (*ed.*) *Ethnobotany and Medicinal Plants of Indian Subcontinent.* Scientific Publishers, Jodhpur (India).

Dam, D. P. & Hajara, P. K. 1881. Observation on ethnobotany of the Monpas of Kameng District, Arunchal Pradesh: 107-114. In Jain, S. K. (*ed.*) *Glimpses of Indian Ethnobotany.* Oxford, IBH Publishing Co., New Delhi.

Das, D. & Agarwal, V.S., 1991. *Fruits Drug Plants of India.* Kalyani Publishers. New Delhi.

Das, H. S., Panda, P. C. & Patnaik, S. N. 1996. Traditional use of wetlands plants of Eastern Orissa. *J. Econ. Taxon. Add. Series 12*: 306-313. Scientific Publishers, Jodhpur (India).

Dash, S. S. & Mishra, M. K. 2000. Taxonomic survey and systematic census of economic plants of Narayana Patna Hills of Koraput District, Orissa: 473-498. In Maheshwari, J. K. (*ed.*) *Ethnobotany and Medicinal Plants of Indian Subcontinent.* Scientific Publishers, Jodhpur (India).

Dastur, J. F. 1964. *Useful Plants of India and Pakistan.* Taraporevala Sons & Co. Private Ltd. Dadabhai Naro Ji Road, Bombay.

Datt, B. 1996. Ethnobotanical resources of Chhatarpur district (Madhya Pradesh): 400-402. In Jain, S. K. (*ed.*) *Ethnobiology in Human Welfare.* Deep Publication, New Delhi.

Dey, K. L., Bahadur, R. & C. I. E.1973. *The Indigenous Drugs of India: Short Descriptive Notices of the Principal Medicinal Products Met With in British India.* Pama Primlane. The Chronica Botanica. New Delhi, India.

Deshmukh, V. R., Muratkar, G. D. & Roth, S. P. 2000. Preliminary observation on the medicinal and economically important leguminous plant species from Amaravati Tehsil: 283-289. In Maheshwari, J. K. (*ed.*) *Ethnobotany and Medicinal Plants of Indian Subcontinent.* Scientific Publishers, Jodhpur (India).

Dhiman, A. K. 2003. *Sacred Plants and Their Medicinal Uses.* Daya Publishing House. Delhi.

Diwaker, P. G. & Ansari, A. A. 2003. Economic importance of plant resources of Buldhana district of Maharashtra: 592-600. In Singh, V. & Jain, A. P. (*ed.*) *Ethnobotany and Medicinal Plants of India and Nepal.* Vol. 2. Scientific Publishers, Jodhpur (India).

Drury, C. H. 1978. *The Useful Plants of India. With Notices of Their Chief Value in Commerce Medicine and the Arts.* International Book Distributors. Dehradun (India).

Duthie, J. E. 1888. *The Fodder Grasses of Northern India.* Scientific Publishers, Jodhpur (India).

Dutt, U. C. 1877. *Materia Medica of the Hindus With a Glossary of Indian Plants.* Sri Satguru Publications. Delhi.

Dutta, N. P. and Biswas, N. R. 1951. Trace elements in some Indian Grasses. *Indian J. Agric. Sci. 21*: 93-96.

Dwivedi, A. P. 1993. *Forest - The Non- Wood Resources.* International Book Distributors. Dehradun (India).

Ford, R. I. (*ed.*) 1978. *The Nature and Status of Ethnobotany.* Anthropological Paper No. 67. Museum of Anthrop. Unin. of Michigan, Ann. Arbor.

Gaur, R. D., Semwal, J. K. & Negi, K. S. 1981. Fodder plants of Garhwal hills and their impact on the quality and quantity of milk. *Himalayan Studies and Regional Development* 5-6: 87-89.

Geetha, S., Lakshmi, G. & Ranjithakani, P. 1996.Ethnobotanical review: wild fibre yielding plants of Kolli hills, Tamil Nadu. *J. Econ. Taxon. Bot. Addl. Series 12*: 250-252.

Gupta, P. C., Singh, D. C. Sangwan & Pradhan, K. 1975. Chemical composition and *in vitro* nutrient digestibility of some of the tree leaves. *Indian Forester 101*(11): 674-680.

Hajra, P. K. 1987. Plants in magico-religious beliefs and in Sanskrit literature: 98-103. In Jain, S. K. (*ed.*) *A Manual of Ethnobotany.* Scientific Publishers, Jodhpur (India).

Hajra, P. K. & Baishya, A. K. 1997. Ethnobotanical notes on the Miris (Mishing) of Assam plains: 161-168. In Jain, S. K. (*ed.*) *Contribution to Indian Ethnobotany.* Scientific Publishers, Jodhpur, (India).

Harborne, J. B. & Baxter, H. 2001. *Chemical Dictionary of Economic Plants.* John Wiley & Sons Ltd. Chichester. Toronto.

Heiser, C. B. 1995. Edgar Anderson botanist and curator of useful plants. *Ann. Miss. Bot. Gard. 82* (1): 54-60.

Hooker, J. D. 1982-1897. *The Flora of British India.* Vol. 1-VIII. Bishen Singh Mahendra Pal Singh. Dehradun, India.

Hosagoudar, V. B. & Henry, A. N. 1996. Ethnobotany of Tamil Nadu, Southern India. *J. Econ. Taxon. Bot. Add. Series* 12: (272-283). Scientific Publishers, Jodhpur (India).

Islam, M. 1984. Certain fibre yielding plants in North Eastern Region. *Indian J. Econ. Tax. Bot. 5* (4): 767-783.

Islam, M. 2002. Wetland biodiversity with reference to economic plants of Assam and its neighbouring areas: 169-204. In Trivedi, P.C. (*ed.*) *Ethnobotany*. Aavishkar Publishers, Distributors. Jaipur (Raj.) India.

Islam, M. & Hasin, F. 2003. Ethnobotany of certain asteraceous plants of Assam and its neighbouring areas: 442-446. In: Singh, V. & Jain, A. P. (*ed.*) *Ethnobotany and Medicinal Plants of India and Nepal.* Vol. 1. Scientific Publishers, Jodhpur (India).

Janaki Ammal, E. K. 1956. Introduction to the subsistence economy of India: 324-335. In William, L. T. (*ed.*) *Men's Role in Changing Face of the Earth.* Univ. of Chicago Press, Chicago.

Jain, S. K. (*ed.*). 1968. *Medicinal Plants.* National Book Trust, India.

Jain, S. K. 1986. Ethnobotany. *Interdiscipl. Sci. Rev.* II. 285-292

Jain, S. K. 1991. *Dictionary of Indian Folk- Medicine and Ethnobotany.* Deep Publishers, New Delhi.

Jain, S. K. 2002. *Bibliography of Indian Ethnobotany.* Scientific Publishers, Jodhpur (India).

Jain, S. K. & Srivastava, S. 2001. Indian ethnobotanical literature in last two decades- a graphic review and future directions. *Ethnobotany 13*: 1-8.

Jain, S. K., Mudgal, V. Banerjee, D. K., Guha, A. & Das, D. 1984. *Bibliography of Ethnobotany.* Bot. Surv. India, Calcutta :1-137.

Jayal, M. M. & Kehar, N. D. 1962. A study on the nutritive value of mulberry (*Morus indica*) tree leaves. *Indian J. dairy Sci. 15*(1): 21-27.

Joshi, D. C. & Talatatra, S. K. 1960. The chemical composition and nutritive value of the Himalayan tree fodder Bhimal (*Grewia oppostifolia*) Roxb. *Indian J. Dairy Sci. 13*(2): 68-76.

Kalpash, I., Rao, V. R., Parabia, F. M., Kothari, I. L. & Parabia, M. M. 2004. Ethnobotany of little Rann of Kachh. Gujarat. *Ethnobotany 16* (1 & 2): 40-43.

Kapur, S. K. 1996. Traditionally important medicinal plants of Bhaderwah Hills- Jammu province-3. J. *Econ. Bot. Additional Series 12*: 62-69. Scientific Publishers, Jodhpur (India).

Karnic, C. R. 1994. *Pharmacopoieal Standards of Herbal Plants.* Vol.1. Sri Satguru Publications. (Indological & Oriental Publishers.) A Division of Indian Book Centre. Delhi, India.

Kaul, M. K. 1997. *Medicinal Plants of Kashmir and Ladakh - Temperate and Cold Arid Himalaya.* Indus Publishing Company, New Delhi.

Kaul, O. N. & Mehta, V. 1996. Management of Bhabbar grass under joint forest management: 266-670. In Jain, S. K. (*ed.*) *Ethnobotany in Human Welfare.* Deep Publications, New Delhi.

Kehar, N. D. & Goswami, M. N. D. 1951. *Bauhinia variegata* leaves as cattle feed. *Sci. & Cult. 16*(10): 476-477.

Khajuria, H. N. & Chauhan, S. K. 2002. Biodiversity conservation and sustainable development. *Indian J. Ecol. 29*(2): 117-133.

Khan, Irfan Ali & Khanum, A. 2005. *Role of Biotechnology in Medicinal and Aromatic Plants.* Ukaaz publications, Hyderabad. Andhra Pradesh.

Khanna, K. K. 2002. Ethnobotany of five districts of terai, Uttar Pradesh: 128-149. In Trivedi, P. C. (*ed.*) *Ethnobotany.* Avishkar Publishers, Distributors. Jaipur, (Raj.) India.

Khare, P. C. (*ed.*) 2004. *Encyclopedia of Indian Medicinal Plants. Regional Western Therapy, Ayurvedic and other Traditional Usage, Botany.* Springer-Verlag, Berlin Heidelberg.

Kirtikar, K. R. & Basu, B. D. 1935. *Indian Medicinal Plants.* Vol. 1-4. Lalit Mohan Basu. Allahabad, India.

Kirtikar, K. R. & Basu, B. D. 1984. *Indian Medicinal Plants.* Vol. 1-4. Bishen Singh Mahendra Pal Singh, Dehradun.

Kothari, M. J. & Rao, K. M. 2000. Ethnobotanical studies of Thane district Maharashtra: 265-272. In Maheshwari, J. K. (*ed.*) *Ethnobotany and Medicinal Plants of Indian Subcontinent.* Scientific Publishers, Jodhpur (India).

Krishna, B. & Das, B. K. 1983. Fibre yielding plants of Sikkim. *J. Econ. Tax. Bot.* 4(1): 129-132.

Krishnamurti Naidu, G. 1917. *Commercial Guide to the Forest Economuc Products of Mysore.* Government Press, Bangalore.

Krishnamurthy, S. 1953. *Horticultural and Economic Plants of the Nilgiris.* Government Botanic Gardens Centenary Committee, Udhagamandalam.

Kshirsagar, S. R. 2003. Notes on miscellaneous uses of plants from tribal areas of Nashik district of Maharashtra: 894-895. In Singh, V. & Jain, A.P. (*eds.*) *Ethnobotany and Medicinal Plants of India and Orissa.* Vol. 2. Scientific Publishers, Jodhpur (India).

Kumar, S. 1999. *Economic Plants of North East India.* Scientific Publishers, Jodhpur (India).

Kumar, V. & Jain, S. K. 1998. A contribution to ethnobotany of Surguja district in Madhya Pradesh, India. *Ethnobotany 10* (1 & 2): 89-96.

Kumar, K. K., Sasidharan & Swarupanadan, K. 2000. Ethnobotanical studies on the hill tribes in the Shola forest of high ranges, Kerala, South India: 451-466. In Maheshwari, J. K. (*ed.*) *Ethnobotany and Medicinal Plants of Indian Subcontinent.* Scientific Publishers, Jodhpur (India).

Lata, S. 1998. *Grewia optiva* Drumm. (Bhimal): A boon for rural people in Kumaun hills, India: 471-472. In Jain, S. K. (*ed.*) *Ethnobotany in Human Welfare.* Deep Publications, New Delhi, India.

Lata, S. 1997. Indigenous knowledge about *Grewia optiva* in Indo- Nepal Himalaya. *Ethnobotany 9* (1 & 2): 112-116.

Lata, S. & Verma, K. R. 1995. Studies on fibres of *Grewia optiva* Drumm. growing in West Himalayan region. *J. Indian Bot. Soc. 74*: 45-47.

Maheshwari, J. K. 1987a. Interdisciplinary approaches in ethnobotany: 19-34. In Jain, S. K. (*ed.*) : *A Manual of Ethnobotany,* Jodhpur, India.

Maheshwari, J. K. 1996 (*ed.*). *Ethnobotany in South Asia.* Scientific Publishers, Jodhpur, India.

Maheshwari, J. K. 2000. *Ethnobotany and Medicinal Plants of Indian Subcontinent.* Scientific Publishers, Jodhpur (India).

Maheshwari, J. K. & Singh, U. 1965. *Dictionary of Economic Plants of India.* ICAR, New Delhi.

Maheshwari, J. K., Painuli, R. M. & Dwivedi, R. P. 1997. Notes on ethnobotany of the Oraon and Korwa tribes of Madhya Pradesh: 67-82. In Jain, S. K. (*ed.*) *Contribution to Indian Ethnobotany.* Scientific Publishers, Jodhpur (India).

Martin, G. J. 1994. *Ethnobotany: Method Manual.* Chapman & Hall, London.

Maya, S., Menon, V. S. & Nair, G. S. 2003. Economic importance of river vegetation of Kerala-A case study 796-803. In Singh, V. & Jain, A. P. (*eds.*) *Ethnobotany and Medicinal Plants of India and Nepal.* Vol. 2. Scientific Publishers, Jodhpur (India).

Maya, S., Pranmeela, K. S. & Menon, V. S. 2003. Ethnobotanical notes on the flora of sacred tanks of Kerala. *Ethnobotany 15* (1 & 2): 55-59.

Miller, D. F. 1958. *Composition of Cereal Grains and Forage*: National Academy of Sciences-National Research Council, Washington, D.C. Publication 585.

Mittre, V. 1981. Wild plants in Indian folk life- a historical perspective: 37-58. In Jain, S. K. (*ed.*) *Glimpses of Indian Ethnobotany.* Oxford & IBH Publishing Co., New Delhi.

Mittre, V. 1997. Wild plants in Indian folk life- a historical perspective: 35-53. In Jain, S. K. (*ed.*) *Contribution to Indian Ethnobotany.* Scientific Publishers, Jodhpur (India).

Momin, S. A. & Ray, S. C. 1943. Tree leaves as cattle fodder. 1. The seasonal variation in the composition of some edible tree leaves. *Indian J. Vet. Sci. 13*: 183-190.

Monandhar, N. P. 2000. Botany and ethnobotany of *Urtica diocia* L. (Stringing Nettle): 38-40. In: Maheshwari, J. K. (*ed.*) *Ethnobotany and Medicinal Plants of Indian Subcontinent.* Scientific Publishers, Jodhpur (India).

Nadkarni, K. M. 1927. *Indian Materia Medica*. Popular Book Depot, Bombay.

Nautiyal, S., Maikhuri, R. K., Rao, K. S. & Saxena, K. G. 2003. Ethnobotany of the Tolchha Bhotiya tribe of the Buffer Zone villages in Nanda Devi biosphere reserve, India: 119-142. In Singh, V. & Jain, A. P. (*eds.*) *Ethnobotany and Medicinal Plants of India and Nepal*. Vol. 1. Scientific Publishers, Jodhpur (India).

Nayar, M. P. Ramamurthy, K. & Agarwal, V. S. 1989. *Economic Plants of India*. Vol. I & II. Botanical Survey of India, Calcutta.

Negi, 1977. Fodder trees in Himachal Pradesh. *Indian Forest 103*(9): 616-622

Oommachan, M. 1977. *The Flora of Bhopal*. (*Angiosperm*). Anand Kumar Jain. J. K. Brothers. Bhopal (India).

Panda, S. 1996. Plant use- a recent perspectives from Eastern Himalayas. *J. Econ. Taxon. Bot. Add. Series* 12: 332-337. Vol. 2. Scientific Publishers, Jodhpur (India).

Parker, R. N. 1993. *A Forest Flora for the Punjab with Hazara and Delhi*. Bishen Singh Mahendra Pal Singh and Periodical Experts, Delhi.

Parmar, C. & Kaushal, M. K. 1982. *Wild Fruits of the Sub- Himalayan Region*. Kalyani Publishers. New Delhi.

Parrotta, J. A. 2001. *Healing Plants of Peninsular India*. CABI Publishing.

Patil, D. A. 1998. Vernicular plants names: their origin and utility in Dhule District (Maharashtra). *Ethnobotany 10* (1 & 2): 130-132.

Prajapati, N. D., Purohit, S. S., Sharma, A. K. & Kumar, T. 2003. *A Handbook of Medicinal Plants* (*A Complete Source Book*). Agrobios (India), Jodhpur.

Prakash & Hockins, D. 1986. Some favourite trees for fuel and fodder. Society for promotion of wastelands development, New Delhi, 117-124.

Prasad, V. K., Rajagopal, T. & Badarinath, K. V. S. 2002. Ethnobotany of Kondareddis of Rampa Agency, Eastern Ghats, Andhra Pradesh: 205-220. In Trivedi, P. C. (*ed.*) *Ethnobotany*. Aavishkar Publishers, Distributors, Jaipur (Raj.) India.

Prasad, V. K., Rajagopal, T. & Badarinath, K. V. S. 2003. Notes on the economic importance of wild plants of Rampa agency- East Godavari district, Andhra Pradesh, India: 603- 612. In Singh, V. & Jain, A. P. (*eds.*) *Ethnobotany and Medicinal Plants of India and Nepal*. Vol. 1. Scientific Publishers, Jodhpur (India).

Punjani, B. L. 2002. Plant used in contact therapy by tribals of Sabarkantha (Gujarat). *Ethnobotany 14*: 57-59.

Purohit, K. 1962a. Oaks that nourish Kumaun cattle in winter. *Indian Farming 11*(2): 20-21.

Purohit, K. 1962b. *Fodder Trees and Shrubs of Kumaon*, Deptt. of Animal Husbandry, U.P., Lucknow.

Purohit, K. & Samant, S. S. 1995. *Fodder Trees and Shrubs of Central Himalaya*. Gyanodaya Prakashan, Nainital.

Radhakrishnan, A. G., Pandurangan, A. G. & Pushpangadan, P. 2000. Tribal artifacts of Kerala. *Ethnobotany* 12: 67-71.

Raju, R. A. 1998. *Wild Plants of Indian Sub-Continent and Their Economic Use*. CBS Publishers & Distributors, New Delhi.

Ranjhan, S. K. 1977. *Animal nutrition and feeding Practices in India*. Vikas Publishing House Pvt., New Delhi, India.

Rao Vasudeva & Shanpru, R. 1981. Some plants in the life of the Garos of Meghalaya: 152-160. In Jain, S. K. (*ed.*) *Glimpses of Indian Ethnobotany*. Oxford, IBH Publishing Co., New Delhi.

Rao, N. R. & Henry, A. N. 1996. *The Ethnobotany of Eastern Ghats in Andhra Pradesh, India*. Botanical Survey of India, New Delhi.

Rao, V. M. K. & Shanpru, R. 1997. Some plants in the life of the Garos of Meghalaya: 179-185. In Jain. S. K. (*ed.*) *Contribution to Indian Ethnobotany*. Scientific Publishers, Jodhpur (India).

Rastogi, R. P. & Mehrotra, B. N. 1990. *Compendium of Indian Medicinal Plants*. Vol.1 (1960-1969). CDRI, Lucknow & Publications & Information Directorate, New Delhi. CSIR.

Rastogi, R. P. & Mehrotra, B. N. 1991. *Compendium of Indian Medicinal Plants*. Vol. 2 (1970-1979). CDRI, Lucknow & Publications & Information Directorate, New Delhi. CSIR.

Rastogi, R. P. & Mehrotra, B. N. 1993. *Compendium of Indian Medicinal Plants*. Vol. 3 (1980-1984). CDRI, Lucknow & Publications & Information Directorate, New Delhi. CSIR.

Rastogi, R. P. & Mehrotra, B. N. 1995a. *Compendium of Indian Medicinal Plants*. Vol. 4 (1985-1989). CDRI, Lucknow & Publications & Information Directorate, New Delhi. CSIR.

Rastogi, R. P. & Mehrotra, B. N. 1995b. *Compendium of Indian Medicinal plants*. Vol. 5 (1990-1994). CDRI, Lucknow & Publications & Information Directorate, New Delhi.CSIR.

Roy, B., Halder, A. C. & Pal, D. C. 1998. *Plants for Human Consumption in India*. Botanical Survey of India. Calcutta.

Sahni, K. C. 1998. *The Book of Indian Trees*. Bombay Natural History Society, Oxford University Press.

Saklani, A. & Jain, S. K. 1994. *Cross-Cultural Ethnobotany of Northeast India*. Deep Publication, New Delhi.

Samant, S. S. & Dhar, U. 1997. Diversity, endemism and economic potential of wild edible plants of Indian Himalaya. *Intern. J. Sustain. Dev. & World Ecology* 4: 179-191.

Samanta, A. K. & Das, D. C. 2003. Ethnobotanical studies on *Typha elephantina* Roxb. (Typhaceae) in the Southern parts of West Bengal, India: 576-579. In Singh, V. &

Jain, A. P. (*eds.*) *Ethnobotany and Medicinal Plants of India and Nepal*. Vol. 2. Scientific Publishers, Jodhpur (India).

Samati, H. 2004. Kitchen garden plants of Pnar tribe in Jaintia Hills district, Meghalaya. *Ethnobotany 16* (1 & 2): 125-130.

Sawhney, P. C. & Bedi, S. P. S. 1972. Trace element content of hill feeds and fodders. 1. Fodders of Kumaon range district Nanital, Uttar Pradesh. *Indian J. Anim. Sci. 42*(9): 659-662.

Sawhney, P. C. & Bedi, S. P. S. & Khan, S. A. 1973. A note on the content of trace elements in the feeds and fodders of Kumaon range, district Almora, Uttar Pradesh. *Indian J. Anim. Sci. 43*(1): 72-75.

Schultes, R. E. 1960. Tapping our heritage of ethnobotanical lore. *Econ. Bot. 14*: 257-262.

Schultes, R. E. 1962. The role of ethnobotanist in search for new medicinal plants. *Lloydia 35*: 257-266.

Sen, K. C. 1964. Nutritive values of Indian cattle feeds and feeding of animals. *Bull. Indian Coun. Agric. Res. 25*: 66-67; 114-117.

Shah, N. C. 1997. Ethnobotany of *Cannabis sativa* in Kumaon Himalaya region, India. *Ethnobotany 9*: 117-121.

Sharma, R. 2003. *Medicinal Plants of India. An Encyclopaedia.* Daya Publishing House, Delhi.

Sharma, S. K. 2004. *Ethnobotanical Studies on Trees, Shrubs and Climbers of Jogindar Nagar Mandi, (H.P.).* Ph.D. Thesis. Himachal Pradesh University, Shimla.

Sharma, P. P. & Singh, N. P. 1999. Less known ethnobotanical uses of plants in Dadra and Nagar Haveli (U.T.): Part 1 (A-K). *Ethnobotany 11* (1 & 2): 109-114.

Singh, G. S. 2000. Ethnobotanical study of useful plants of Kullu district in North Western Himalaya, India: 185-198. In Maheshwari, J. K. (*ed.*) *Ethnobotany and Medicinal Plants of Indian Sub- continent*. Scientific Publishers, Jodhpur (India).

Singh, G. S. 2002. Minor forest products of Sariska national park: An ethnobotanical profile: 221-238. In Trivedi, P. C. (*ed.*) *Ethnobotany*. Aavishkar Publishers, Distributors, Jaipur (Raj.) India.

Singh, G. S. 2003. Ethnobotanical inventory of Sariska National Park in the Aravalli Hills of Eastern Rajasthan, India: 181-195. In Singh, V. & Jain, A. P. (*eds.*) *Ethnobotany and Medicinal Plants of India and Nepal*. Vol. 1. Scientific Publishers, Jodhpur (India).

Singh, P. B. 1999. *Commercially Important Medicinal and Aromatic Plants of Himachal Pradesh.* Society of Herbal Medicine and Himalayan Biodiversity, Ram Nagar, Mandi, H.P.

Singh, R. V. 1982. *Fodder Trees of India*. Mohan Primlani for Oxford & IBH Publishing Co., New Delhi.

Singh, U. 1996. *Dictionary of Economic Plants in India.* Indian Council of Agricultural Research, Delhi.

Singh, K. K. & Prakash, A. 1996. Observation on ethnobotany of the Kol Tribes of Varanasi District, Uttar Pradesh, India. *J. Econ. Taxon. Bot. Add. Series 12*: 133-137. Scientific Publishers, Jodhpur (India).

Singh, B., Chaurasia, O. P. & Jadhav, K. L. 1996. An ethnobotanical study of Indus Valley. (Ladakh). J. *Econ. Taxon. Bot. Add. Series 12* (92-101). Scientific Publishers, Jodhpur (India).

Singh, J. P. & Shankar, V. 1996. Traditional forage resources of India: Exploitation Potential: 307-309. In Jain, S. K. *Ethnobotany in Human Welfare.* Deep Publication, New Delhi.

Singh, J. S. Singh, S. P. & Ram Jeet 1998. Fodder and fuelwood resources of Central Himalaya: problems and solutions. Report Submitted for Study Group on Fuel and Fodder, Planning Commission, Govt. of India, New Delhi.

Singh, K. K. & Kumar, K. 2000. *Ethnobotanical Wisdom of Gaddi Tribe in Western Himalaya.* Bishen Singh Mahendra Pal Singh, Dehradun.

Singh Tej, Handa, K. L. & Sobti, S. N. 1959. Chemical composition of some legumes of Jammu & Kashmir. *Indian J. Vet. Sci. 29*(2&3): 32-39.

Singh, U., Wadhwani, A. M. & John, B. M. 1983. *Dictionary of Economic Plants of India.* Indian Coun. Agric. Res., New Delhi.

Singh, V. & Pandey, R. P. 1998. *Ethnobotany of Rajasthan, India.* Scientific Publishers, Jodhpur (India).

Sood, S. K., Nath, R. & Kalia, D. C. 2001. *Ethnobotany of Cold Desert Tribes of Lahoul- Spiti (N.W. Himalaya).* Deep Publications, New Delhi.

Sood, S. K., Thakur, S. 2004. *Ethnobotany of Rewalsar Himalaya.* Deep Publications, New Delhi.

Sood, S. K., Parmar, S, & Lakhanpal, T. N. 2005. *Ethnic Plants of India Used in Cancer Cure-A Compendium.* Bishen Singh Mahendra Pal Singh, Dehradun (India).

Sundararaj, D. D. & Balasubramanyam, G. 1959. A Guide to the Economic Plants of South India. Amudha Vilayan (P) Ltd., Madras.

Sur, P. R. & Halder, A. C. 2002. Observation on the ethnobotany of Purulia district, West Bengal, India: 146-148. In Trivedi, P. C. (*ed.*) *Ethnobotany.* Aavishkar Publishers & Distributor, Jaipur (Raj.) India.

Thakur, R. S., Puri, H. S. & Husain, A. 1989. *Major Medicinal Plants of India.* C.I.M.A.P. Lucknow, India.

Thapliyal, A. B. 1962. *Composition of Kharsoo and Dudhiliya.* M. V. Sc. Thesis, Agra University, Agra.

Talbot, W. A. 1976a. *Forest Flora of The Bombay Presidency and Sindh.* Vol. 1. Today & Tomorrow's Printers and Publishers, New Delhi.

Talbot, W. A. 1976b. *Forest Flora of The Bombay Presidency and Sindh.* Vol. 2. Today & Tomorrow's Printers and Publishers, New Delhi.

Upadhye, A. S. & Kumbhojkar, M. S. 1998. *Schleichera oleosa* (Lour.) Oken. Tree: A boon for Tribals. *Ethnobotany 10* (1 & 2): 38-41.

Upadhyaya, R. & Chauhan, S.V.S. 2000. Ethnobotanical observations on koya tribe of Gundala Mandal of Khammam district, Andhra Pradesh. *Ethnobotany 12*: 93-99.

Uphof, J. C. T. 1986. *Dictionary of Economic Plants.* Second Edition. Revised and Enlarged. Stechert-Hafner Service Agency, Inc. New York.

Varghese, E. 1996. *Applied Ethnobotany. (A Case Study Among the Kharias of Central India.)* Deep Publications, New Delhi.

Varghese, E. & Hembrom, P. P. 2000. *Ethnobotanical Lore of the Paharias.* (Centre for Human Ecology : Mumbai, (India).

Watt, G. 1889-1896. *A Dictionary of Economic Products of India.* Vol. 2. Periodical Experts, Delhi (India).

Whyte, R. O. 1957. *The Grassland and Fodder Resources of India,* ICAR, New Delhi (Reprinted 1964).

Wickens, G. E. 1990. What is economic botany ? *Econ. Bot.* 44: 12-28.

Yoganarasimhan, S.N., Subramanyam, K. & Razi, B.A. (1981) *Flora of Chikmagular district, Karnataka, India.* International Book Distributors, Dehradun.

□□□

7

Plates

A. *Aesculus indica* Colebr.

B. *Alpina allughas* Rosc.

C. *Ammania baccifera* L

D. *Atriplex halimus* Linn.

Pl. 1 A-D. Fibre and Fodder Resources of India.

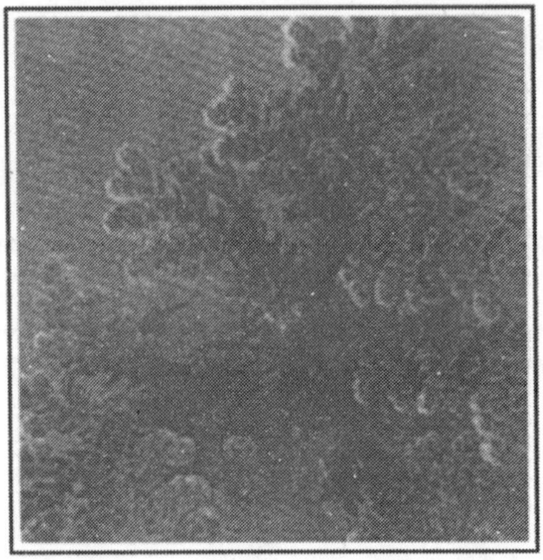

A. *Azolla pinnata* R. Br.

B. *Bambusa nutans* Wall. *ex* Munro

C. *Bauhinia racemosa* Linn.

D. *Betula utilis* Don.

Pl. 2A-D. Fibre and Fodder Resources of India.

A. *Calamus erectus* Roxb.

B. *Ceiba pentendra* (Linn.) Gaertn.

C. *Cyperus difformis* L.

D. *Cyperus iria* Linn.

Pl. 3A-D. Fibre and Fodder Resources of India.

A. *Derris trifoliata* Lour.

B. *Embelia ribes* Burm.

C. *Entada phaseoloides* Merrill.

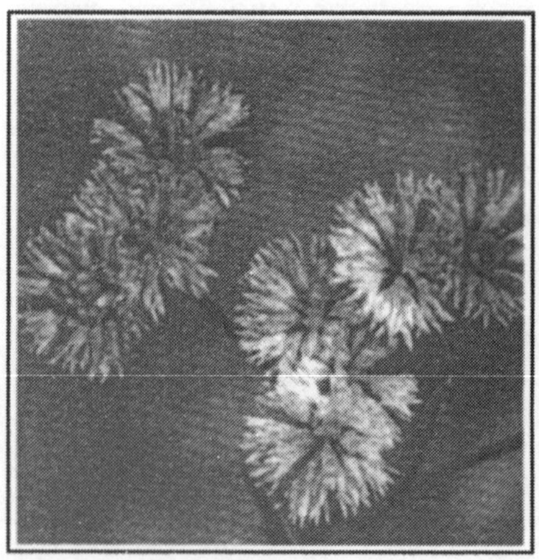

D. *Ethulia conyzoides* L.

Pl. 4A-D. Fibre and Fodder Resources of India.

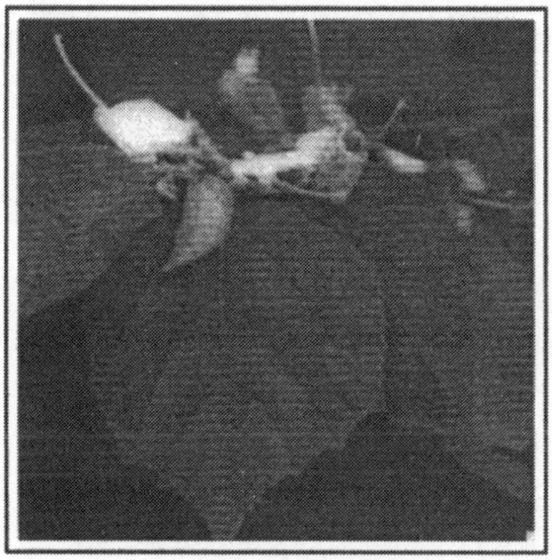

A. *Helicteres isora* L.

B. *Hiptage benghalensis* (Linn.) Kurz.

C. *Leucaena glauca* Benth.

D. *Lodoicea maldivica* (Poir.) Pers.

Pl. 5A-D. Fibre and Fodder Resources of India.

A. *Lolium perenne* Linn.

B. *Medicago sativa* Linn.

C. *Myrica esculenta* Buch.- Ham. *ex* Don

D. *Pandanus odoratimus* Linn. f.

Pl. 6A-D. Fibre and Fodder Resources of India.

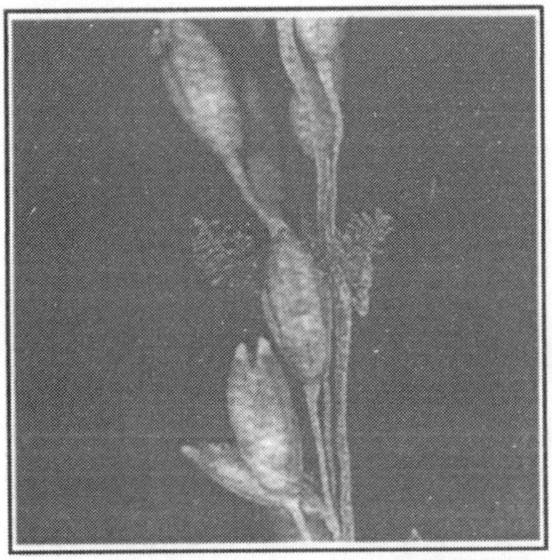

A. *Panicum maximum* Jacq.

B. *Phoebe lanceolata* Nees.

C. *Prosopis chilensis* Stuntz.

D. *Pterocarpus marsupium* Roxb.

Pl. 7A-D. Fibre and Fodder Resources of India.

A. *Sabal palmetto* Lodd. *ex* Roem. & Schult.

B. *Stemona tuberosa* Lour.

C. *Ulmus laevigata* Royle

D. *Vicia sativa* Linn.

Pl. 8A-D. Fibre and Fodder Resources of India.

A. *Vigna trilobata* (Linn.) Verdtcourt.

B. *Yucca filamentosa* Linn.

Pl. 9A-B. Fibre and Fodder Resources of India.

8

Appendices

Appendix I

Total Number of Genera and Species of Fibre and Fodder Resources of India under Various Divisions of Plant Kingdom

Division	No. of Family	Genus/ Species	Total No. of Species in a Genus
Gymnosperms	2	3/3	*Abies* (1), *Thuja* (1), *Taxus* (1).
Dicots	115	403/624	*Abelmoschus* (5), *Abroma* (1), *Abutilon* (8), *Acacia* (7), *Acer* (5), *Adansonia* (1), *Ademostemma* (1), *Adina* (1), *Aechmanthera* (1), *Aegle* (1), *Aeschynomene* (1), *Aesculus* (2), *Aganosoma* (1), *Agathis* (1), *Ageratum* (1), *Aidia* (1), *Ailanthus* (2), *Aisandra* (1), *Alangium* (2), *Albizia* (5), *Alhagi* (1), *Allmannia* (1), *Alnus* (2), *Alstonia* (1), *Alternanthera* (1), *Althaea* (1), *Alysicarpus* (1), *Amaranthus* (1), *Ammania* (1), *Ampelocissus* (1), *Anagallis* (1), *Ananas* (1), *Annona* (2), *Anodendron* (1), *Anogeissus* (2), *Anthocephalus* (1), *Antiaris* (1), *Aquilaria* (1), *Aetemisia* (1), *Artocarpus* (5), *Asclepias* (1), *Asparagus* (2), *Astilbe* (1), *Astragalus* (4), *Asystasia* (1), *Atriplex* (1), *Atylosia* (1), *Avicinnia* (1), *Azadirachta* (1), *Azana* (1), *Azolla* (1), *Bacopa* (1), *Barleria* (1), *Bauhinia* (4), *Beaumontia* (1), *Berberis* (5), *Berchemia* (1), *Berrya* (1), *Betula* (2), *Bidens* (1),

Contd...

Appendix I. Contd...

			Bischofia (1), *Bixa* (1), *Blepharis* (2), *Boehmeria* (5), *Bombax* (1), *Borassus* (1), *Borreria* (1), *Boswellia* (1), *Brassica* (2), *Bridelia* (2), *Bromelia* (1), *Broussonetia* (1), *Buchanania* (1), *Buddleia* (1), *Bupleurum* (1), *Butea* (2), *Caesalpinia* (1), *Cajanus* (1), *Calamus* (3), *Callicarpa* (2), *Calotropis* (2), *Cannabis* (1), *Capparis* (2), *Carallia* (1), *Carataeva* (1), *Cardaria* (1), *Carduus* (1), *Careya* (1), *Carica* (1), *Carissa* (2), *Caryota* (2), *Cassia* (2), *Castanopsis* (1), *Cedrela* (1), *Ceiba* (1), *Celastrus* (1), *Celtis* (4), *Chenopodium* (1), *Chonemorpha* (1), *Chorchorus* (2), *Cissampelos* (1), *Cissus* (3), *Clematis* (2), *Clinogyne* (1), *Coccinia* (1), *Coccothrinax* (1), *Cochlospermum* (1), *Colquhounia* (1), *Combretum* (1), *Cordia* (3), *Coriandrium* (1), *Coriaria* (1), *Cornus* (1), *Corypha* (2), *Cotoneaster* (1), *Cotula* (1), *Crotalaria* (2), *Cryptolepis* (1), *Cryptostegia* (1), *Cyamopsis* (1), *Cyathula* (1), *Dalbergia* (4), *Dalhousiea* (1), *Daphne* (2), *Debregeasia* (2), *Decalepis* (1), *Deeringia* (1), *Dendrocnide* (1), *Depatratium* (1), *Derris* (1), *Desmodium* (5), *Desmostachya* (1), *Dichrocephala* (1), *Dichrostachys* (1), *Diectomis* (1), *Dillenia* (1), *Diospyros* (1), *Diploknema* (1), *Dipsacus* (1), *Dolichandrone* (1), *Dolichos* (1), *Dracocephalum* (1), *Drega* (1), *Drypetes* (1), *Echinops* (1), *Edgeworthia* (1), *Ehertia* (2), *Embelia* (1), *Emblica* (1), *Entada* (1), *Equisetum* (2), *Erigeron* (1), *Erinocarpus* (1), *Eruca* (1), *Erythrina* (1), *Erythropsis* (2), *Ethulia* (1), *Eugenia* (1), *Euphorbia* (2), *Eurya* (1), *Fagopyrum* (1), *Feronia* (1), *Ficus* (16), *Flacourtia* (1), *Flemingia* (3), *Fragaria* (1), *Funtumia* (1), *Galinsoga* (1), *Garuga* (1),*Gaultheria* (1), *Gerbera* (1), *Girardina* (1), *Glinus* (1), *Gliricidia* (1), *Glochidion* (2), *Gnetum* (1), *Goniothalamus* (1), *Gonostegia* (1), *Gossypium* (3), *Grewia* (7), *Gyrinops* (1), *Guazuma* (1), *Haloptelea* (1), *Hedera* (1), *Helianthus* (1), *Helicteres* (1), *Hibiscus* (7), *Hiptage* (1), *Holboellia* (1), *Holostemma* (1),

Contd...

Appendix I. Contd...

			Hydrolea (1), *Hygrophilia* (1), *Hypserpa* (1), *Ichnocarpus* (1), *Ilex* (1), *Ilysanthus* (1), *Impatiens* (1), *Indigofera* (5), *Inula* (1), *Ipomoea* (3), *Ixora* (1), *Jasminum* (1), *Juglans* (1), *Kleinhovia* (1), *Kochia* (2), *Kydia* (1), *Lagerstroemia* (1), *Lannea* (1), *Laportea* (1), *Lasiosiphon* (1), *Lathyrus* (2), *Launaea* (1), *Leea* (1), *Lens* (1), *Lepidium* (2), *Leptadenia* (1), *Leucaena* (1), *Limnanthemum* (1), *Limnophilia* (2), *Lindera* (1), *Lindernia* (1), *Lithocarpus* (1), *Litsea* (4), *Livistonia* (2), *Lodoicea* (1), *Lonicera* (1), *Lotus* (1), *Luffa* (1), *Lycopersicon* (1), *Lyonia* (1), *Machilus* (1), *Madhuca* (1), *Maesia* (1), *Malachra* (1), *Mallotus* (2), *Malvestrum* (1) *Manilkara* (1), *Marsdenia* (2), *Mastersia* (1), *Mazus* (1), *Medicago* (2), *Melilotus* (2), *Meliosma* (1), *Melochia* (2), *Mikania* (1), *Miliusa* (1), *Millettia* (2), *Mirabilis* (1), *Mitragyana* (1), *Momordica* (1), *Moringa* (1), *Morus* (3), *Mucuna* (4), *Myrica* (1), *Myricareia* (1), *Naravelia* (1), *Nelumbo* (1), *Nepeta* (1), *Nypa* (1), *Ochna* (1), *Olea* (1), *Opomoea* (1), *Opuntia* (2), *Oreocnida* (2), *Oroxylum* (1), *Orthanthera* (1), *Ougeinia* (1), *Oxytropis* (1), *Pandanus* (1), *Parkinsonia* (1), *Parrotiopsis* (1), *Pavonia* (1), *Pegolettia* (1), *Persea* (3), *Persicaria* (1), *Phanera* (1), *Phaseolus* (2), *Philadelphus* (1), *Phoebe* (1), *Phoenix* (3), *Picrasma* (1), *Pilea* (1), *Pipturus* (1), *Pistacia* (1), *Pistia* (1), *Pithcellobium* (1), *Plumbago* (1), *Pogostemon* (1), *Polyalthia* (1), *Polygonum* (4), *Pongamia* (1), *Populus* (3), *Pouzolzia* (1), *Prangos* (1), *Premma* (1), *Prinsepia* (1), *Prosopis* (2), *Prunus* (3), *Pterocarpus* (2), *Pterocymbium* (1), *Pterospermum* (1), *Pueraria* (2), *Punica* (1), *Putranjiva* (1), *Pyrus* (2), *Quercus* (4), *Randia* (1), *Ranunculus* (1), *Raphia* (2), *Rhamnus* (4), *Rhizophora* (1), *Rhodiola* (1), *Rhus* (4), *Ribes* (2), *Robinia* (1), *Roccella* (1), *Rosa* (5), *Rotala* (2), *Rubia* (1), *Rubus* (3), *Rumex* (1), *Sabal* (1), *Sacosperma* (1), *Sageretia* (1), *Salix* (6), *Salmalia* (1), *Salsola* (1), *Salvadora* (1), *Samanea* (1), *Sapindus* (1), *Sapium* (1), *Saraca* (1),

Contd...

Appendix I. Contd...

			Sarcochlamys (1), *Sarcostemma* (1), *Saurauia* (2), *Saussurea* (1), *Scaphium* (1), *Schefflera* (1), *Schleichera* (1), *Scoparia* (1), *Sechium* (1), *Selinum* (1), *Sequoia* (1), *Sesbania* (4), *Shorea* (1), *Sida* (6), *Siegesbeckia* (1), *Smithia* (1), *Solanum* (1), *Soymida* (1), *Sparrmannia* (1), *Spartium* (1), *Spathodea* (1), *Spermadictylon* (1), *Sphaeranthus* (1), *Spiraea* (2), *Stemona* (1), *Stephania* (1), *Sterculia* (5), *Sterospermum* (1), *Streblus* (1), *Strobilanthus* (3), *Suaeda* (1), *Syagrus* (1), *Symphytum* (1), *Symplocos* (2), *Syzygium* (1), *Tacca* (1), *Tamarindus* (1), *Tamarix* (2), *Tanacetum* (1), *Tecomella* (1), *Terminalia* (4), *Thalictrum* (1), *Thespesia* (2), *Tilia* (1), *Tinospora* (1), *Trachelospermum* (1), *Trachycarpus* (1), *Trema* (2), *Trifolium* (5), *Trigonella* (1), *Triumfetta* (3), *Turpinia* (1), *Tylophora* (1), *Ulex* (1), *Ulmus* (1), *Urena* (1), *Urtica* (2), *Ventilago* (1), *Viburnum* (1), *Vicia* (2), *Vigna* (2), *Vitis* (1), *Wendlandia* (2), *Wikstroemia* (1), *Wissadula* (1), *Withania* (1), *Woodfordia* (1), *Wrightia* (1), *Xylia* (1), *Zanthoxylum* (1), *Zizyphus* (4).
Monocots	15	101/184	*Agave* (5), *Agropyron* (1), *Agrostis* (3), *Aloe* (1), *Alopecurus* (1), *Alpina* (1), *Apluda* (1), *Arundinaria* (2), *Arundinellia* (1), *Avena* (2), *Bambusa* (3), *Blyxa* (1), *Bothriochola* (2), *Brachiaria* (3), *Bromus* (2), *Canna* (1), *Carex* (2), *Cenchrus* (1), *Chimnobambusa* (1), *Chloris* (2), *Chrysopogon* (3), *Coix* (2), *Commelina* (1), *Cynodon* (1), *Cynotis* (1), *Cyperus* (9), *Dactylis* (1), *Dendrocalamus* (2), *Digitaria* (4), *Dinebra* (1), *Dracaena* (1), *Eichhornia* (1), *Eichinochola* (1), *Eleocharis* (1), *Ensete* (1), *Eragostiela* (1), *Eragrostis* (2), *Eremopogon* (1), *Erianthus* (2), *Eriochola* (1), *Euchlaena* (1), *Eulalia* (1), *Eulaliopsis* (1), *Floscopa* (1), *Furcarea* (3), *Glyceria* (1), *Hedychium* (1), *Hordeum* (1), *Hymenachne* (1), *Ichanthus* (1), *Imperata* (1), *Iris* (2), *Iseilema* (2), *Juncus* (1), *Kobresia* (1), *Koeleria* (1), *Kyllinga* (1), *Lasiurus* (1), *Leersia* (1), *Lolium* (2), *Mariscus* (1), *Microcos* (1), *Microstegium* (1),

Contd...

Appendix I. Contd...

			Muhelenbergia (1), *Muntinga* (1), *Musa* (2), *Nannorrhops* (1), *Ophiurus* (1), *Oryza* (3), *Panicum* (9), *Paspalidium* (1), *Paspalum* (3), *Pennisetum* (6), *Perotis* (1), *Phararis* (1), *Phormium* (1), *Phragmites* (2), *Plumeria* (2), *Poa* (4), *Pollinia* (2), *Raphidophora* (1), *Saccharum* (2), *Sacciolepis* (1), *Sansevieria* (3), *Schizachyrium* (1), *Scirpus* (2), *Sclerostachya* (1), *Sehima* (2), *Setaria* (2), *Smilax* (3), *Sorghum* (5), *Sporobolus* (4), *Thamnocalamus* (2), *Themeda* (2), *Thysanolaena* (1), *Trisetum* (2), *Typha* (2), *Urochola* (1), *Vossia* (1), *Yucca* (3), *Zea* (1).

Appendix II

Families of Fibre and Fodder Resources of India

Family Name	Genus/Species	Name of Genus/Species
Acanthaceae	6/9	*Aechmanthera (1), Asystasia* (1), *Barleria* (1), *Blepharis* (2), *Hygrophilia* (1), *Strobilanthus* (3).
Aceraceae	1/5	*Acer* (5)
Agavaceae	2/8	*Agave* (5), *Yucca* (3).
Alangiaceae	1/1	*Alangium* (1).
Amarantaceae	5/5	*Allmannia* (1), *Alternanthera* (1), *Amaranthus* (1), *Cyathula* (1), *Deeringia* (1).
Amaryllidaceae	1/3	*Furcarea* (3).
Anacardiaceae	4/7	*Buchanania* (1), *Lannea* (1), *Pistacia* (1), *Rhus* (4).
Annonaceae	4/5	*Annona* (2), *Goniothalamus* (1), *Miliusa* (1), *Polyalthia* (1).
Apiaceae	4/4	*Bupleurum* (1), *Coriandrium* (1), *Prangos* (1), *Selinum* (1).
Apocynaceae	10/11	*Aganosoma* (1), *Alstonia* (1), *Anodendron* (1), *Beaumontia* (1), *Carissa* (2), *Chonemorpha* (1), *Funtumia* (1), *Ichnocarpus* (1), *Trachelospermum* (1), *Wrightia* (1).
Aquifiliaceae	1/1	*Ilex* (1).
Araceae	2/2	*Pistia (1), Raphidophora* (1).
Araliaceae	2/2	*Hedera* (1), *Schefflera* (1).
Araucariaceae	1/1	*Agathis* (1).
Arecaceae	13/21	*Borassus* (1), *Calamus* (3), *Caryota* (2), *Coccothrinax* (1), *Corypha* (2), *Livistonia* (2), *Lodoicea* (1), *Nypa* (1), *Phoenix* (3), *Raphia* (2), *Sabal* (1), *Syagrus* (1), *Trachycarpus* (1).
Asclepiadaceae	11/13	*Asclepias* (1), *Calotropis* (2), *Cryptolepis* (1), *Decalepis* (1), *Drega* (1), *Holostemma* (1), *Leptadenia* (1), *Marsdenia* (2), *Orthanthera* (1), *Sarcostemma* (1), *Tylophora* (1).

Contd...

Appendix II. Contd...

Asparagaceae	1/2	*Asparagus* (2).
Asteraceae	21/21	*Ademostemma* (1), *Ageratum* (1), *Artemisia* (1), *Bidens* (1), *Carduus* (1), *Cotula* (1), *Dichrocephala* (1), *Echinops* (1), *Erigeron* (1), *Ethulia* (1), *Galinsoga* (1), *Gerbera* (1), *Helianthus* (1), *Inula* (1), *Launaea* (1), *Mikania* (1), *Pegolettia* (1), *Saussurea* (1), *Siegesbeckia* (1), *Sphaeranthus* (1), *Tanacetum* (1).
Azollaceae	1/1	*Azolla* (1).
Balsamiaceae	1/1	*Impatiens* (1).
Berberidaceae	1/5	*Berberis* (5).
Betulaceae	2/4	*Alnus* (2), *Betula* (2).
Bignoniaceae	5/5	*Spathodea* (1), *Stereospermum* (1), *Tecomella* (1), *Dolichandrone* (1), *Oroxylum* (1).
Bixaceae	1/1	*Bixa* (1).
Bombacaceae	4/4	*Adansonia* (1), *Bombax* (1), *Ceiba* (1), *Salmalia* (1).
Boraginaceae	3/6	*Symphytum* (1), *Cordia* (3), *Ehertia* (2).
Brassicaceae	4/6	*Brassica* (2), *Cardaria* (1), *Eruca* (1), *Lepidium* (2).
Bromeliaceae	2/2	*Ananas* (1), *Bromelia* (1).
Burseraceae	2/2	*Boswellia* (1), *Garuga* (1).
Cactaceae	1/2	*Opuntia* (2).
Cannabinaceae	1/1	*Cannabis* (1).
Cannaceae	1/1	*Canna* (1).
Capparidaceae	2/3	*Capparis* (2), *Crataeva* (1).
Caprifoliaceae	2/2	*Viburnum* (1), *Lonicera* (1).
Caricaceae	1/1	*Carica* (1).
Celastraceae	1/1	*Celastrus* (1).
Chenopodiaceae	5/6	*Atriplex* (1), *Chenopodium* (1), *Kochia* (2), *Salsola* (1), *Suaeda* (1).
Cochlospermaceae	1/1	*Cochlospermum* (1).

Contd...

Appendix II. Contd...

Combretaceae	3/7	*Anogeissus* (2), *Combretum* (1), *Terminalia* (4).
Commelinaceae	3/3	*Cynotis* (1), *Floscopa* (1), *Commelina* (1).
Convolvulaceae	2/4	*Ipomoea* (3), *Opomoea* (1).
Coriariaceae	1/1	*Coriaria* (1).
Cornaceae	1/1	*Cornus* (1).
Crassulaceae	1/1	*Rhodiola* (1).
Cucurbitaceae	4/4	*Coccinia* (1), *Luffa* (1), *Momordica* (1), *Sechium* (1)
Cyperaceae	7/17	*Carex* (2), *Cyperus* (9), *Eleocharis* (1), *Kobresia* (1), *Kyllinga* (1), *Mariscus* (1), *Scirpus* (2).
Dilleniaceae	1/1	*Dillenia* (1).
Dipsacaceae	1/1	*Dipsacus* (1).
Dipterocarpaceae	1/1	*Shorea* (1).
Ebnaceae	1/1	*Diospyros* (1).
Equisetaceae	1/2	*Equisetum* (2).
Ericaceae	2/2	*Gaultheria* (1), *Lyonia* (1).
Euphorbiaceae	9/12	*Bischofia* (1), *Bridelia* (1), *Drypetes* (1), *Emblica* (1), *Euphorbia* (2), *Glochidion* (2), *Mallotus* (2), *Putranjiva* (1), *Sapium* (1).
Fabaceae	58/110	*Acacia* (7), *Aeschynomene* (1), *Albizia* (5), *Alhagi* (1), *Alysicarpus* (1), *Astragalus* (4), *Atylosia* (1), *Bauhinia* (4), *Butea* (2), *Caesalpinia* (1), *Cajanus* (1), *Cassia* (2), *Crotalaria* (2), *Cyamopsis* (1), *Dalbergia*(4), *Dalhousiea* (1), *Derris* (1), *Desmodium* (5), *Dichrostachys* (1), *Dolichos* (1), *Entada* (1), *Erythrina* (1), *Flemingia* (3), *Gliricidia* (1), *Indigofera* (5), *Lathyrus* (2), *Lens* (1), *Leucaena* (1), *Lotus* (1), *Mastersia* (1), *Medicago* (2), *Melilotus* (2), *Millettia* (2), *Mucuna* (4), *Ougeinia* (1), *Oxytropis* (1), *Parkinsonia* (1), *Phanera* (1), *Phaseolus* (2), *Pithcellobium* (1), *Pongamia* (1), *Prosopis* (2), *Pterocarpus* (2), *Pueraria* (2), *Robinia* (1), *Samanea* (1), *Saraca* (1), *Sesbania* (4), *Smithia* (1), *Spartium* (1), *Tamarindus* (1), *Trifolium* (5), *Trigonella* (1), *Ulex* (1), *Ulmus* (1), *Vicia* (2), *Vigna* (2), *Xylia* (1).

Contd...

Appendix II. Contd...

Fagaceae	3/6	*Castanopsis* (1), *Lithocarpus* (1), *Quercus* (4).
Gentianaceae	1/1	*Limnanthemum* (1).
Gnetaceae	1/1	*Gnetum* (1).
Grossulariaceae	1/2	*Ribes* (2).
Hamamelidaceae	1/1	*Parrotiopsis* (1).
Hydrocharitaceae	1/1	*Blyxa* (1).
Hydrophyllaceae	1/1	*Hydrolea* (1).
Iridaceae	1/2	*Iris* (2).
Juncaceae	1/1	*Juncus* (1).
Junglandaceae	1/1	*Juglans* (1).
Lamiaceae	4/4	*Colquhounia* (1), *Dracocephalum* (1), *Nepeta* (1), *Pogostemon* (1).
Lardizabalaceae	1/1	*Holboellia* (1).
Lythraceae	4/5	*Ammania* (1), *Rotala* (2), *Lagerstroemia* (1), *Woodfordia* (1).
Lauraceae	5/10	*Lindera* (1), *Litsea* (4), *Machilus* (1), *Persea* (3), *Phoebe* (1).
Lecythidaceae	1/1	*Careya* (1).
Leeaceae	1/1	*Leea* (1).
Liliaceae	5/9	*Aloe* (1), *Dracaena* (1), *Phormium* (1), *Sansevieria* (3), *Smilax* (3).
Loganiaceae	1/1	*Buddleia* (1).
Malpighiaceae	1/1	*Hiptage* (1)
Malvaceae	14/39	*Abelmoschus* (5), *Abutilon* (8), *Althaea* (1), *Azana* (1), *Gossypium* (3), *Hibiscus* (7), *Kydia* (1), *Malachra* (1), *Malvestrum* (1), *Pavonia*(1), *Sida* (6), *Thespesia* (2), *Urena* (1), *Wissadula* (1).
Marantaceae	1/1	*Clinogyne* (1).
Meliaceae	3/3	*Azadirachta* (1), *Cedrela* (1), *Soymida* (1).
Menispermaceae	4/4	*Cissampelos* (1), *Hypserpa* (1), *Stephania* (1), *Tinospora* (1).

Contd...

Appendix II. Contd...

Moraceae	7/32	*Antiaris* (1), *Artocarpus* (5), *Broussonetia* (4), *Streblus* (1), *Ficus* (16), *Flacourtia* (1), *Morus* (4).
Moringaceae	1/1	*Moringa* (1).
Mulluginaceae	1/1	*Glinus* (1).
Musaceae	2/3	*Ensete* (1), *Musa* (2).
Myrcinaceae	1/1	*Embelia* (1).
Myricaceae	1/1	*Myrica* (1).
Myrsinaceae	1/1	*Maesia* (1).
Myrtaceae	2/2	*Syzygium* (1), *Eugenia* (1).
Nyctaginaceae	1/1	*Mirabilis* (1).
Nymphaeaceae	1/1	*Nelumbo* (1).
Ochnaceae	1/1	*Ochna* (1).
Oleaceae	2/2	*Jasminum* (1), *Olea* (1).
Pandanaceae	1/1	*Pandanus* (1).
Periplocaceae	1/1	*Cryptostegia* (1).
Pinaceae	2/2	*Abies* (1), *Thuja* (1).
Plumbaginaceae	1/1	*Plumbago* (1).
Poaceae	72/129	*Agropyron* (1), *Agrostis* (3), *Alopecurus* (1), *Apluda* (1), *Arundinaria* (2), *Arundinellia* (1), *Avena* (2), *Bambusa* (3), *Bothriochola* (2), *Brachiaria* (4), *Bromus* (2), *Cenchrus* (1), *Chimnobambusa* (1), *Chloris* (2), *Chrysopogon* (3), *Coix* (2), *Cynodon* (1), *Dactylis* (1), *Dendrocalamus* (2), *Desmostachya* (2), *Diecotomis* (1), *Digitaria* (4), *Dinebra* (1), *Eichinochola* (1), *Eragostiela* (1), *Eragrostis* (2), *Eremopogon* (1), *Erianthus* (2), *Eriochola* (1), *Euchlaena* (1), *Eulalia* (1), *Eulaliopsis* (1), *Glyceria* (1), *Hordeum* (1), *Hymenachne* (1), *Ichanthus* (1), *Imperata* (1), *Iseilema* (2), *Koeleria* (1), *Lasiurus* (1), *Leersia* (1), *Lolium* (2), *Microstegium* (1), *Muhelenbergia* (1), *Nannorrhops* (1), *Ophiuros* (1), *Oryza* (3), *Panicum* (9), *Paspalidium* (1), *Paspalum* (3), *Pennisetum* (6), *Perotis* (1), *Phararis* (1),

Contd...

Appendix II. Contd...

		Phragmites (2), *Plumeria* (2), *Poa* (4), *Pollinia* (2), *Sacciolepis* (1), *Saccharum* (2), *Schizachyrium* (1), *Sclerostachya* (1), *Sehima* (2), *Setaria* (2), *Sorghum* (5), *Sporobolus* (4), *Thamnocalamus* (2), *Themeda* (2), *Thysanolaena* (1), *Trisetum* (2), *Urochola* (1), *Vossia* (1), *Zea* (1).
Polygonaceae	4/7	*Fagopyrum* (1), *Persicaria* (1), *Polygonum* (4), *Rumex* (1).
Pontederiaceae	1/1	*Eichhornia* (1).
Primulaceae	1/1	*Anagallis* (1).
Punicaceae	1/1	*Punica* (1).
Ranunculaceae	4/5	*Clematis* (2), *Naravelia* (1), *Ranunculus* (1), *Thalictrum* (1).
Rhamnaceae	5/11	*Berchemia* (1), *Rhamnus* (4), *Sageretia* (1), *Ventilago* (1), *Zizyphus* (4).
Rhixophoraceae	2/2	*Carallia* (1), *Rhizophora* (1).
Roccellaceae	1/1	*Roccella* (1).
Rosaceae	8/18	*Cotoneaster* (1), *Fragaria* (1), *Prinsepia* (1), *Prunus* (3), *Pyrus* (2), *Rosa* (5), *Rubus* (3), *Spiraea* (2).
Rubiaceae	10/11	*Adina* (1), *Aidia* (1), *Anthocephalus* (1), *Borreria* (1), *Ixora* (1), *Mitragyana* (1), *Randia* (1), *Rubia* (1), *Spermadictylon* (1), *Wendlandia* (2).
Rutaceae	3/3	*Aegle* (1), *Feronia* (1), *Zanthoxylum* (1).
Sabiaceae	1/1	*Meliosma* (1)
Salicaceae	2/9	*Populus* (3), *Salix* (6).
Salvadoraceae	1/1	*Salvadora* (1).
Sapindaceae	3/4	*Aesculus* (2), *Sapindus* (1), *Schleichera* (1).
Sapotaceae	6/6	*Aisandra* (1), *Diploknema* (1), *Gonostegia* (1), *Madhuca* (1), *Manilkara* (1), *Sacosperma* (1).
Saurauiaceae	1/2	*Saurauia* (2).
Saxifragaceae	2/2	*Astilbe* (1), *Philadelphus* (1).
Scrophulariaceae	7/8	*Scoparia* (1), *Bacopa* (1), *Depatratium* (1), *Ilysanthus* (1), *Limnophilia* (2), *Lindernia* (1), *Mazus* (1).

Contd...

Appendix II. Contd...

Sirmarubaceae	2/2	*Ailanthus* (1), *Picrasma* (1).
Solanaceae	3/3	*Lycopersicon* (1), *Solanum* (1), *Withania* (1).
Stemonaceae	1/1	*Stemona* (1).
Sterculiaceae	10/16	*Abroma* (1), *Erythropsis* (2), *Guazuma* (1), *Helicteres* (1), *Kleinhovia* (1), *Melochia* (2), *Pterocymbium* (1), *Pterospermum* (1), *Scaphium* (1), *Sterculia* (5).
Styphyleaceae	1/1	*Turpinia* (1).
Symplocaceae	1/2	*Symplocos* (2).
Taccaceae	1/1	*Tacca* (1).
Tamaricaceae	2/3	*Myricaria* (1), *Tamarix* (2).
Taxaceae	1/1	*Taxus* (1).
Taxodiaceae	1/1	*Sequoia* (1).
Theaceae	1/1	*Eurya* (1).
Thymelaeaceae	6/7	*Aquilaria* (1), *Daphne* (2), *Edgeworthia* (1), *Grinops* (1), *Lasiosiphon* (1), *Wikstroemia* (1).
Tiliaceae	9/18	*Berrya* (1), *Chorchorus* (2), *Erinocarpus* (1), *Grewia* (7), *Microcos* (1), *Muntinga* (1), *Sparrmannia* (1), *Tilia* (1), *Triumfetta* (3).
Typhaceae	1/2	*Typha* (2).
Ulmaceae	3/7	*Celtis* (4), *Haloptelea* (1), *Trema* (2).
Urticaceae	11/18	*Boehmeria* (5), *Debregeasia* (2), *Dendrocnide* (1), *Girardina* (1), *Laportea* (1), *Oreocnida* (2), *Pilea* (1), *Pipturus* (1), *Pouzolzia* (1), *Sarcochlamys* (1), *Urtica* (2).
Verbenaceae	3/4	*Avicinnia* (1), *Callicarpa* (2), *Premma* (1).
Vitaceae	3/5	*Ampelocissus* (1), *Cissus* (3), *Itis* (1).
Zingiberaceae	2/2	*Alpina* (1), *Hedychium* (1).

❑❑❑

Appendix III

Fibre Resources of India

Botanical Name	
Abelmoschus crinitis	*Boehmeria nivea*
Abelmoschus esculentus	*Borassus flabellifer*
Abelmoschus ficulneus	*Bromelia penguin*
Abelmoschus manihot	*Broussonetia papyrifera*
Abelmoschus moschatus	*Butea parviflora*
Abies pindrow	*Caesalpinia crista*
Abroma angusta	*Calamus andamanicus*
Abutilon bedfordianum	*Calamus erectus*
Abutilon grandifolium	*Calotropis gigantean*
Abutilon hirtum	*Calotropis procera*
Abutilon indicum	*Canna orientalis*
Abutilon persicum	*Cannabis sativa*
Abutilon theophrastii	*Careya arborea*
Acacia mollissima	*Carica papaya*
Adansonia digitata	*Caryota mitis*
Aganosoma caryophyllata	*Caryota urens*
Agathis robusta	*Cassia auriculata*
Agave americana	*Ceiba pentandra*
Agave angustifolia	*Celastrus paniculatus*
Agave lurida	*Chonemorpha fragrans*
Agave sisalana	*Cissampelos pareira*
Agave veracruz	*Cissus adnata*
Alnus nitida	*Cissus pallida*
Aloe barbadensis	*Cissus repens*
Alstonia scholaris	*Clematis buchananiana*
Althaea officinalis	*Clinogyne dichotoma*
Ananas cosmosus	*Coccinia cordifolia*
Annona reticulata	*Coccothrinax argentea*
Annona squamosa	*Cochlospermum religiosum*
Anodendron paniculatum	*Corchorus capsularis*
Aquilaria agallocha	*Corchorus olitorius*
Artocarpus artilis	*Cordia rothii*
Arundinaria callosa	*Corypha umbraculifera*
Asclepias curassavica	*Corypha utan*
Azana lampas	*Crotalaria burhia*
Beaumontia grandiflora	*Crotalaria juncea*
Berrya ammonilla	*Crotalaria mucronata*
Bixa orellana	*Cryptolepsis buchanani*
Boehmeria glomerulifera	*Cryptostegia grandiflora*

Contd...

Cyperus articulatus
Dalhousiea papyracea
Daphne cannabina
Daphne mucronata
Decalepis hamiltonii
Dendrocalamus hamiltonii
Dendrocnide sinuta
Desmodium pannosus
Desmostachya bipinnata
Dipsacus fullonum
Dolichandrone spathacea
Dregea volubilis
Drypetes longifolia
Edgeworthia gardneri
Ensete vertricosum
Entada phaseoloides
Erinocarpus nimmonii
Erythrina suberosa
Erythropsis pallens
Eulaliopsis bipinata
Ficus tsiela
Funtumia elastica
Furcraea foetida
Furcraea hexapetala
Furcraea macrophylla
Gerbera lanuginosa
Girardinia heterophylla
Gnetum latifolium
Goniothalamus sequipedalis
Gonostegia pentendra
Gossypium arboreum
Gossypium barbadense
Gossypium herbaceum
Grewia asiatica
Grewia glabra
Grewia multiflora
Grewia rothii
Gyrinops walla
Hibiscus cannabinus
Hibiscus furcatus
Hibiscus macrophyllus
Hibiscus mutabilis
Hibiscus rosasinensis
Hibiscus sabdariffa
Hibiscus tilaceus
Hypserpa cuspidata
Ichnocarpus frutescens
Imperata cylindrical
Juncus effusus
Kleinhovia hospita
Lagerstoemia speciosa
Laportea crenulata
Lasiosiphon eriocephalus
Livistonia australis
Livistonia chinensis
Lodoicea maldivica
Luffa acutangula
Malachra capitata
Malvestrum coromandelianum
Maoutia puya
Mariscus albscens
Marsdenia roylei
Marsdenia tenacissima
Mastersia assamica
Melochia corchorifolia
Melochia umbellata
Microcos paniculata
Muntinga calabura
Musa textiles
Nannorrhops ritchieana
Naravelia zeylanica
Nelumbo nucifera
Nypa fructicans
Opuntia dillenii
Oreocnida frutiscens
Oreocnida integrifolia
Orthanthera viminea
Pandanus odoratimus
Panicum antidotale
Pavonia odorata
Phanera integrifolia
Philadelphus coronarius
Phoenix dactylifera
Phormium tenax
Phragmites karka
Pilea glaberrima
Pipturus incanus
Polyalthia suberosa
Pterocymbium tinctorium
Pterospermum xylocarpum

Contd...

Raphia farinifera
Raphia hookeri
Sabal palmetto
Saccharum officinarum
Saccharum spontaneum
Salix viminalis
Salmalia insignis
Sanseveria trifasciata
Sansevieria lyacinthoides
Sansevieria roxburghina
Sarcochlamys pulcherrima
Sarcostemma secamone
Scaphium scaphigerum
Sclerostachya fusca
Sechium edule
Sehima nervasum
Sequoia sempervirens
Sesbania cannabina
Sesbania grandiflora
Sesbania speciosa
Sida acuta
Sida codifolia
Sida cordata
Sida glutinosa
Sida ovata
Sida rhombifolia
Smilax aspera
Smilax ovalifolia
Soymida febrifuga
Sparrmannia africana
Spartium junceum
Spathodea campanulata
Stemona tuberosa
Stephania glabra
Sterculia guttata
Sterculia indica
Sterculia urens
Steruilia villosa
Syagrus coronata
Tacca leontoprtaloides
Thespesia lampas
Thuja plicata
Trachycarpus fortunei
Triumfetta pentendra
Triumfetta rhomboidea
Tylophora indica
Typha elephantina
Typha latifolia
Urena lobata
Urtica dioica
Ventilago denticulata
Wikstroemia indica
Wissadula lescheaultiana
Yucca aloiffolia
Yucca filamentosa

□□□

Appendix IV

Predominant Genera of Fibre Resources

Botanical Name	No. of Species	Botanical Name	No. of Species
Hibiscus	7	*Calamus*	2
Abutilon	6	*Calotropis*	2
Sida	6	*Caryota*	2
Abelmoschus	5	*Corypha*	2
Agave	5	*Daphne*	2
Grewia	4	*Livistonia*	2
Sterculia	4	*Marsdenia*	2
Cissus	3	*Melochia*	2
Crotolaria	3	*Oreocnida*	2
Furcarea	3	*Raphia*	2
Gossypium	3	*Saccharum*	2
Sansevieria	3	*Sesbania*	2
Yucca	3	*Smilax*	2
Annona	2	*Triumfetta*	2
Boehmeria	2	*Typha*	2

□□□

Appendix V

Fodder Resources of India

Botanical Name	
Abutilon fruticosum	*Anogeissus pendula*
Acacia caesia	*Anthocephalus chinensis*
Acacia jacquemontii	*Apluda mutica*
Acacia planifrons	*Artemisia sieversiana*
Acer acuminatum	*Artocarpus heterophyllus*
Acer caesium	*Artocarpus hirsuta*
Acer cappadocicum	*Arundinaria falcate*
Acer oblongum	*Arundinella nepalensis*
Acer villosum	*Asparagus curillus*
Ademostemma lavenia	*Asparagus racemosus*
Aegle marmelos	*Astilbe rivularis*
Aesculus igniarius	*Astragalus chorostachys*
Aesculus indica	*Astragalus grahamianus*
Ageratum conyzoides	*Astragalus marschallianus*
Agropyron cristatum	*Astragalus rhizanthus*
Agrostis canina	*Asystasia gangetica*
Agrostis gigantean	*Atriplex halimus*
Agrostis stolonifera	*Atylosia scarabaeoides*
Aidia tetrasperma	*Avena byzantina*
Ailanthus excelsa	*Avena sativa*
Aisandra butyracea	*Avicennia officinalis*
Alangium chinense	*Azadirachta indica*
Alangium salvifolium	*Azolla pinnata*
Albizia julibrissin	*Bacopa monniri*
Albizia lebbeck	*Bambusa nutans*
Albizia procera	*Barlaria cristata*
Albizia stipulata	*Berberis affinis*
Alhagi pseudalhagi	*Berberis asiatica*
Allmania nudiflora	*Berberis kumaonensis*
Alnus nepalensis	*Berberis lycium*
Alopecurus geniculatus	*Berberis petiolaris*
Alternanthera philoxeroides	*Berchemia floribunda*
Alysicarpus longifolius	*Betula alnoides*
Amaranthus viridis	*Bidens bipinnata*
Ammania baccifera	*Bischofia javanica*
Ammania multiflora	*Blepharis linariaefolia*
Ampelocissus rugosa	*Blepharis persica*
Anagallis arvensis	*Blyxa aubertii*
Anogeissus latifolia	*Boehmaria scbrella*

Contd...

Boehmeria rugolosa
Borreria articularis
Boswellia serrata
Bothriochloa pertusa
Bothriochola intermedia
Brachiaria distachya
Brachiaria paspaloides
Brachiaria ramosa
Brachiaria reptans
Brassica napus
Brassica oleracea
Bridelia montana
Bridelia retusa
Bromus catharticus
Bromus conifinis
Buchanania lanzan
Buddleia paniculata
Bupleurum candollei
Callicarpa arborea
Callicarpa macrophylla
Capparis sepiaria
Capparis zeylanica
Carallia brachiata
Cardaria draba
Carduus nutans
Carex cernud
Carex criciata
Carissa congesta
Carissa opaca
Cassia floribunda
Castanopsis tribuloides
Cedrela toona
Celtis australis
Celtis eriocarpa
Celtis setigerus
Celtis tertrandra
Cenchrus barbatus
Chenopodium album
Chloris incompleta
Chloris inflate
Chrysopogon gidarba
Chrysopogon lancearius
Chrysopogon orientalis
Clematis nepaulensis
Coix aquatica
Colquhounia coccinea
Combretum roxburghii
Commelina benghalensis
Cordia dichotoma
Cordia vestita
Coriandrium sativum
Coriaria nepalensis
Cornus capitata
Cotoneaster obtusa
Cotula hemispherica
Crataeva religiosa
Cyamopsis tetragonoloba
Cyanotis axillaris
Cyathula tomentosa
Cynodon dactylon
Cyperus compressus
Cyperus cuspidatus
Cyperus difformis
Cyperus distans
Cyperus haspans
Cyperus iria
Cyperus metzii
Cyperus niveus
Dactylis glomerata
Dalbergia latifolia
Dalbergia pinnata
Dalbergia sericea
Dalbergia sissoo
Deeringia amaranthoides
Depatratium junceum
Derris trifoliata
Desmodium gangeticum
Desmodium microphyllum
Desmodium motorium
Dichrocephala integrifolia
Diectomis fastigiata
Digitaria adscendens
Digitaria ciliaris
Digitaria jubata
Digitaria longifolia
Dinebra retroflexa
Diospyros melanoxylon
Diploknema buttracea
Dolichos lablab
Dracaena angustifolia

Contd...

Dracocephalum heterophyllum
Echinops echinatus
Ehretia acuminata
Ehretia laevis
Eichinochola stagnina
Eleocharis artopurpurea
Embelia ribes
Emblica officinalis
Equisetum arvense
Equisetum delibe
Eragrostiella bifaria
Eragrostis plumosa
Eragrostis tenella
Eremopogon foveolatus
Erianthus elephantinus
Erigeron alpinus
Eriochloa procera
Eruca sativa
Ethulia conyzoides
Euchlaena maxicana
Eugenia operculata
Eulalia cumingii
Euphorbia geniculata
Euphorbia royleana
Eurya acuminata
Fagopyrum esculentum
Feronia limonia
Ficus cunia
Ficus glomerata
Ficus hispida
Ficus nemoralis
Ficus palmata
Ficus retusa
Ficus sarmentosa
Ficus squamosa
Ficus subincisa
Ficus virens
Flacourtia indica
Flemingia fructiculosa
Flemingia macrophylla
Flemingia semialata
Floscopa scandens
Fragaria indica
Garuga pinnata
Glinus lotoides
Gliricidia sepium
Glochidion calocarpum
Glochidion velutinum
Glyceria tonglensis
Grewia flavescens
Grewia maderaspatana
Haloptelea integrifolia
Hedera helix
Helianthus annuus
Hiptage benghalensis
Holboellia angustifolia
Holostemma annularis
Hordeum vulgare
Hydrolea zeylanica
Hygrophila phlomoides
Hymenachne amplexicaulis
Ichnanthus vicinus
Ilex dipyrena
Ilysanthes parriflora
Impatiens tripetala
Indigofera artopurpurea
Indigofera cylindracea
Indigofera nummularifolia
Indigofera pulchella
Inula cappa
Ipomoea aquatica
Ipomoea digitata
Ipomoea eriocarpa
Iris lactea
Iseilema laxum
Iseilema prostratum
Ixora arborea
Jasminum multiflorum
Juglans regia
Kobresia duthiei
Kochia indica
Kochia scoparia
Koeleria cristata
Kyllinga brevifolia
Lannea coromandelica
Lasiurus hirsutus
Lathyrus aphaca
Lathyrus sativus
Launaea nudicaulis
Leea crispa

Contd...

Leersia hexandra
Lens culinaris
Lepidum apetalum
Lepidum sativum
Leucaena glauca
Limnanthemum nymphaeoides
Limnophila hirsuta
Limnophilia heterophylla
Lindernia crustacea
Lithocarpus dealbatus
Litsea glutinosa
Litsea monopetala
Litsea salicifolia
Litsea umbrosa
Lolium multiflorum
Lolium perenne
Lonicera quinquelocularis
Lotus corniculatus
Lycopersicon esculentum
Lyonia ovalifolia
Machilus gamblei
Madhuca indica
Maesia indica
Mallotus philippensis
Mallotus tetracoccus
Manilkara hexandra
Medicago hispida
Medicago sativa
Melilotus alba
Melilotus indica
Meliosma dilleniaefolia
Mikania cordata
Miliusa tomentosa
Millettia racemosa
Mirabilis himalaica
Mitragyana parvifolia
Momordica balsamina
Moringa oleifera
Morus australis
Morus laevigata
Morus serrata
Mucuna cochichinnensis
Mucuna deeringiana
Mucuna nigricans
Mucuna pruiens
Muhelenbergia huegelii
Myrica esculenta
Myricaria germanica
Nepeta floccosa
Ochna jabotapia
Olea ferruginea
Ophiuros extralatus
Opomoea aquatica
Opuntia ficusindica
Oroxylum indicum
Oryza officinalis
Oryza rupipogon
Oryza sativa
Oxytropis mollis
Panicum antidotale
Panicum atrosanguineum
Panicum colonum
Panicum incomtum
Panicum maximum
Panicum miliaceum
Panicum paludosum
Panicum sumatrense
Panicum virgatum
Parkinsonia aculeata
Parrotiopsis jackuemontiana
Paspalidium geminatum
Paspalum kora
Paspalum scrobiculatum
Paspalum vaginatum
Pegolettia senegalensis
Pennisetum cenchroides
Pennisetum divisum
Pennisetum holcoides
Pennisetum purpureum
Pennisetum typhoides
Pennisetum verticillatum
Perotis indica
Persea duthiei
Persea gammieana
Persea odoratissima
Persicaria capitata
Phararis minor
Phaseolus lathyroides
Phaseolus lunatus
Phoebe lanceolata

Contd...

Phragmites communis
Phragmites karka
Picrasma guassioides
Pistia stratiotes
Pithecellobium dulce
Plumbago auriculata
Plumeria annua
Plumeria pratensis
Poa annua
Poa khasiana
Poa sterilis
Poa trivialis
Pogostemon benghalenses
Pollinia argentea
Pollinia eriopoda
Polygonum alpium
Polygonum aviculare
Polygonum barbatum
Polygonum chinense
Pongamia pinnata
Populus ciliata
Populus euphratica
Populus laurifolia
Prangos pabularia
Premna latifolia
Prinsepia utillis
Prosopis chilensis
Prosopis cineraria
Prunus cerasoides
Prunus cornuta
Prunus persica
Pterocarpus marsupium
Pterocarpus santalinus
Pueraria tuberosa
Punica granatum
Putranjiva roxburghii
Pyrus foliolosa
Pyrus pashia
Quercus dilatata
Quercus glauca
Quercus incana
Quercus semecarpifolia
Randia spinosa
Ranunculus diffusus
Raphidophora glauca
Rhamnus pentapomica
Rhamnus purpurea
Rhamnus triqueter
Rhamnus virgatus
Rhizophora apiculata
Rhodiola imbricata
Rhus cotinus
Rhus mysurensis
Rhus parviflora
Rhus sinuta
Ribes glaciale
Ribes rubrum
Robinia pseudoacacia
Roccella montagnei
Rosa brunonii
Rosa glandulosa
Rosa macrophylla
Rosa sericea
Rosa webbiana
Rotala densiflora
Rotala rotundifolia
Rubia tinctorum
Rubus ellipticus
Rubus niveus
Rubus rosaefolius
Rumex hastatus
Sacciolepis interrupta
Sageretia oppositifolia
Salix acmophylla
Salix alba
Salix daphnoides
Salix tetrasperma
Salix wallichiana
Salsola baryosma
Salvadora olooides
Samanea saman
Sapindus mukorossi
Sapium insigne
Saraca asoca
Sarcosperma arboretum
Saurauria nepaulensis
Saurauria roxburghii
Saussurea costus
Schefflera impressa
Schizachyrium semiberbi

Contd...

Schleichera oleosa
Scirpus grossus
Scirpus tuberosus
Scoparia dulcis
Sehima nervasum
Sehima sulcatum
Selinum candollei
Setaria glauca
Setaria italica
Shorea robusta
Siegesbeckia orientalis
Smilax parvifolia
Smithia sensitiva
Solanum indica
Sorghum dochna
Sorghum milliforme
Sorghum purpureasericeum
Sorghum verticelliforum
Spermadictyon suaveolens
Sphaeranthus africanus
Spiraea conescens
Spiraea lindleyana
Sporobolus diander
Sporobolus maderaspatanus
Sporobolus pallidus
Sporobulus helvolus
Sterculia colorata
Stereospermum personatum
Strobilanthes artopurpureus
Strobilanthes wallichii
Strobilanthus glutinosis
Suaeda monoica
Symphytum asperum
Symplocos paniculata
Symplocos ramosissima
Syzygium cuminii
Tamarindus indica
Tanacetum longifolium
Taxus baccata
Tecomella undulata
Terminalia belerica
Terminalia chebula
Terminalia tomentosa
Thalictrum foliolosum
Thamnocalamus falconeri
Themeda cymbaria
Thysanolaena maxima
Tinospora cordifolia
Trachelospermum lucidum
Trifolium alexandrinum
Trifolium dubium
Trifolium pratense
Trifolium repens
Trifolium resupinatum
Trigonella polycerata
Trisetum aestivum
Trisetum flavescense
Turpinia nepalensis
Ulex europaeus
Urochola panicoides
Viburnum continifolium
Vicia hirsuta
Vicia sativa
Vigna acontifolia
Vigna trilobata
Vitis lanata
Vossia cuspidata
Wendlandia exserata
Wendlandia grandis
Wikstroemia indica
Withania sominifera
Woodfordia fruticosa
Xylia xylocarpa
Zanthoxylum acanthopodium
Zea mays
Zizyphus mauritiana
Zizyphus nummularia
Zizyphus rugosa
Zizyphus xylocarpus

Appendix VI

Predominant Genera of Fodder Resources

Botanical Names	No. of Species	Botanical Names	No. of Species
Ficus	11	*Strobolanthes*	3
Panicum	9	*Terminalia*	3
Cyperus	8	*Aesculus*	2
Pennisetum	6	*Alangium*	2
Acer	5	*Ammania*	2
Berberis	5	*Anogeissus*	2
Rosa	5	*Artocarpus*	2
Salix	5	*Asparagus*	2
Albizia	4	*Avena*	2
Astragalus	4	*Blepharis*	2
Brachiaria	4	*Boehmeria*	2
Celtis	4	*Brassica*	2
Dalbergia	4	*Bridelia*	2
Digitaria	4	*Bromus*	2
Indigofera	4	*Callicarpa*	2
Morus	4	*Capparis*	2
Litsea	4	*Carex*	2
Mucuna	4	*Carissa*	2
Poa	4	*Chloris*	2
Polygonum	4	*Cordia*	2
Quercus	4	*Dendrocalamus*	2
Rhus	4	*Ehertia*	2
Sorghum	4	*Equisetum*	2
Sporobolus	4	*Eragrostis*	2
Zizyphus	4	*Euphorbia*	2
Acacia	3	*Glochidion*	2
Agrostis	3	*Grewia*	2
Chrysopogon	3	*Iseilma*	2

Contd...

Appendix VI. Contd...

Desmodium	3	*Kochia*	2
Flemingia	3	*Lathyrus*	2
Ipomoea	3	*Lepidium*	2
Oryza	3	*Limnophila*	2
Paspalum	3	*Lolium*	2
Persea	3	*Mallotus*	2
Populus	3	*Medicago*	2
Prunus	3	*Melilotus*	2
Rubus	3	*Phragmites*	2
Phaseolus	2	*Scirpus*	2
Plumeria	2	*Sehima*	2
Pollinia	2	*Setaria*	2
Prosopis	2	*Spiraea*	2
Pterocarpus	2	*Symplocos*	2
Pyrus	2	*Trisetum*	2
Ribes	2	*Vicia*	2
Rotala	2	*Vigna*	2
Saurauria	2	*Wendlandia*	2

Appendix VII

Fibre and Fodder Resources of India

Botanical Name	
Abutilon glaucum	*Ficus auriculata*
Acacia leucophloea	*Ficus benghalensis*
Acacia nilotica	*Ficus rumphii*
Acacia senegal	*Ficus scandens*
Adina cordifolia	*Galinsoga parviflora*
Aechmanthera gossypina	*Grewia oppositifolia*
Aeschynomene indica	*Guazuma ulmifolia*
Ananas comosus	*Helicteres isora*
Antiaris toxicaria	*Indigofera gerardiana*
Artocarpus chama	*Kydia calycina*
Artocarpus lacucha	*Leptadenia pyrotechnica*
Bambusa arundinacea	*Millettia extensa*
Bambusa pallida	*Morus alba*
Bauhinia malabarica	*Musa paradisiaca*
Bauhina racemosa	*Ougeinia oogeinensis*
Bauhinia retusa	*Phoenix acaulis*
Bauhinia variegata	*Phoenix sylvestris*
Betula utilis	*Pistacia integerrima*
Boehmeria macrophylla	*Pueraria lobata*
Bombax ceiba	*Sesbania sesban*
Butea monosperma	*Sorghum bicolour*
Cajanus cajan	*Streblus asper*
Calamus tenius	*Tamarix dioica*
Chimnobambusa jaunsarensis	*Terminalia arjuna*
Coix lacrymajobi	*Thamnocalamus spathiflora*
Cordia dichotoma	*Themeda arundinacea*
Debregeasia hypoleuca	*Thespesia populnea*
Debregeasia longifolia	*Tilia europaea*
Dendrocalamus strictus	*Trema cannabina*
Desmodium elegans	*Trema politoria*
Dillenia pentagyna	*Triumfetta pilosa*
Eichhornia crassipes	*Ulmus laevigata*
Erianthus munja	*Urtica parviflora*
Erythropsis colorata	*Wrightia tomentosa*

□□□

Appendix VIII

Predominant Genera of Fibre and Fodder Resources

Botanical Name	No. of Species	Botanical Name	No. of Species
Bauhinia	4	*Bambusa*	2
Ficus	4	*Debregeasia*	2
Acacia	3	*Phoenix*	2
Artocarpus	2	*Trema*	2

□□□

Appendix IX

Index to Botanical Names

Botanical Name

Contd...

Contd...

Contd...

Contd...

Contd...

Contd...

Contd...

Contd...

Contd...

Appendix X

Index to English Names

Contd...

Contd...

Contd...

Contd...

Contd...

❑❑❑

Appendix XI

Index to Hindi Names

Hindi Name

Contd...

Contd...

Contd...

Contd...

Contd...

Contd...

Contd...

Appendix XII

Index to Sanskrit Names

Contd...

Contd...

Contd...

Contd...

Contd...

□□□